Lean Manufacturing and Industry 4.0

Lean Manufacturing and Industry 4.0

Editors

Luís Pinto Ferreira
Paulo Ávila
João Bastos
Francisco J. G. Silva
José Carlos Sá
Marlene Brito

MDPI • Basel • Beijing • Wuhan • Barcelona • Belgrade • Manchester • Tokyo • Cluj • Tianjin

Editors

Luís Pinto Ferreira
Department of Mechanical
Engineering
Polytechnic of Porto
Porto
Portugal

Paulo Ávila
Department of Mechanical
Engineering
Polytechnic of Porto
Porto
Portugal

João Bastos
Department of Mechanical
Engineering
Polytechnic of Porto
Porto
Portugal

Francisco J. G. Silva
Department of Mechanical
Engineering
Polytechnic of Porto
Porto
Portugal

José Carlos Sá
Department of Mechanical
Engineering
Polytechnic of Porto
Porto
Portugal

Marlene Brito
Department of Mechanical
Engineering
Polytechnic of Porto
Porto
Portugal

Editorial Office
MDPI
St. Alban-Anlage 66
4052 Basel, Switzerland

This is a reprint of articles from the Special Issue published online in the open access journal *Machines* (ISSN 2075-1702) (available at: www.mdpi.com/journal/machines/special_issues/lean_manufacturing_industry).

For citation purposes, cite each article independently as indicated on the article page online and as indicated below:

LastName, A.A.; LastName, B.B.; LastName, C.C. Article Title. *Journal Name* **Year**, *Volume Number*, Page Range.

ISBN 978-3-0365-7717-3 (Hbk)
ISBN 978-3-0365-7716-6 (PDF)

Contents

About the Editors

Luís Pinto Ferreira

Luís Pinto Ferreira received his PhD in Manufacturing Engineering from the University of Vigo, Spain; MSc in Industrial Engineering—Logistics and Distribution from the School of Engineering, University of Minho, Portugal; and bachelor's degree in Industrial Electronic Engineering from the School of Engineering, University of Minho, Portugal. He has been teaching in higher education institutions since 2000. He is currently the Deputy Director of the master's Program in Engineering and Industrial Management at the Polytechnic of Porto-School of Engineering (ISEP), Portugal, and he has supervised more than 100 MSc students. He has published more than 100 scientific papers (ISI/SCOPUS), book chapters, and 1 international book, and has presented at national and international events. In 2004, he was awarded the APDIO/IO prize by the Operation Research Portuguese Association.

Paulo Ávila

Paulo Avila is an Associate Professor in the Department of Mechanical Engineering at the Polytechnic of Porto-School of Engineering (ISEP), Portugal. He was previously the Director of the Department and is currently the Director of the bachelor's program in Industrial Engineering and Management. He received his PhD from the University of Minho, Portugal, in the field of Production and Systems. His scientific interests include resource selection; agile/virtual enterprises; production system optimization; computer-integrated manufacturing; sustainability; and production planning and control. He regularly publishes in international and national proceedings, journals, and books, and he is a co-chair and founding member of the Business Sustainability Conference. He is also a consultant for several enterprises and has participated in several national and international projects.

João Bastos

Joao Augusto Bastos is an Adjunct Professor in the Department of Mechanical Engineering at the Polytechnic of Porto-School of Engineering (ISEP), Portugal. He received his PhD in Industrial Engineering and Management, MSc in Industrial Informatics, and bachelor's degree in Mechanical Engineering from the Faculty of Engineering at the University of Oporto, Portugal. He regularly publishes in international and national conference proceedings, journals, and books. He is involved in consultancy projects for several industrial companies and has participated in multiple national and international projects, including EU-funded projects. His scientific interests include production planning and control; production system optimization; and machine learning and enterprise information systems.

Francisco J. G. Silva

Francisco J. G. Silva is an Associate Professor in Habilitation in the Department of Mechanical Engineering at the Polytechnic of Porto-School of Engineering (ISEP), Portugal. He was a member of the Scientific-Technical Council and the Pedagogical Council of ISEP. He was a Director of the master's program in Mechanical Engineering for eight years and of the bachelor's program for four years. He is a researcher at INEGI/LAETA. He has supervised more than 200 MSc theses and co-supervised 2 PhD theses in the field of Mechanical Engineering, and he currently has 9 PhD students under his supervision/co-supervision. He is a Founder and was previously an Editor-in-Chief of the journal *Coating Science Technology*. He is an author, co-author, or co-editor of seven books published by national and international publishers.

He is a co-author of more than 230 scientific papers published in international journals (H-index=36 in Scopus with more than 4000 citations). He is member of the Editorial Board of 10 indexed journals and a member of the Scientific Committee of several international conferences. He has been involved or is currently involved in eight national research projects, with his institution receiving a grant of EUR 1.5 million for the current project. He has extensive experience in manufacturing processes and industrial management in both industrial and scientific fields.

José Carlos Sá

Jose Carlos Sá received his MSc in Industrial Engineering and bachelor's degree in Production Engineering at the University of Minho, Portugal. In 2015, he was awarded the title of Specialist in Quality Management by the Polytechnic Institute of Viana do Castelo, Portugal. Currently, he is an Adjunct Professor and a Lean Six Sigma Post-Graduation Coordinator at the Polytechnic of Porto-School of Engineering (ISEP), Portugal. He has taught in higher education since 2005. He has over 20 years of industry experience, including performing functions related to production planning, process control, and quality management at a multinational company and several national SMEs. He is a member of the Editorial Advisory Board of *Technological Sustainability* (Emerald) and a member of the Editorial Board of the *Journal of Applied Research on Industrial Engineering*. He is also a member of the Review Board of the *International Journal of Environmental Research and Public Health* (MDPI). He has supervised more than 50 MSc students at ISEP-IPP and has published more than 80 papers (ISI/SCOPUS) on quality management, safety, and lean six sigma.

Marlene Brito

Marlene Brito has a PhD in Industrial Engineering and has over 12 years of experience leading operational change through training using lean and Kaizen philosophies in various manufacturing industries. She holds a post-graduate degree in Industrial Engineering and Management (IEM) from the University of Aveiro, Portugal, and an APICS Certification in Production and Inventory Management (CPIM), Lean Manufacturing System, and Lean Logistics (Pull + Levelling) from Robert Bosch GmbH, Stuttgart, Germany. She is currently a lecturer at the Polytechnic of Porto-School of Engineering (ISEP), Portugal. Her research interest lies in in the field of industrial engineering.

Preface to "Lean Manufacturing and Industry 4.0"

Existing competitiveness between companies in the global market requires a high degree of proactivity, with an aim of reducing costs, increasing quality, and providing the market with innovative products that are in line with the natural evolution of customer demands. The concepts of the Toyota Production System, which have led to the adoption of lean manufacturing, have helped industrial managers to implement tools and practices that eliminate or substantially reduce wastes that normally exist in conventional production lines. However, the need for successive increase in flexibility necessitates a more favorable evolution with technological advancements in the programming of equipment and devices, along with the use of robotics and the integration of information throughout the product manufacturing cycle using the IoT or powerful networks. This concept, usually referred to as Industry 4.0, is in line with the principles of lean manufacturing, although it can only be supported by a true technological revolution involving remote programming of machines, integration of manufacturing processes, and real-time control of the work performed through an entire production cycle. This Special Issue showcases important contributions in the area of lean manufacturing and Industry 4.0, with innovative concepts and applications that will certainly enthuse and strengthen readers' knowledge.

Luís Pinto Ferreira, Paulo Ávila, João Bastos, Francisco J. G. Silva, José Carlos Sá, and Marlene Brito

Editors

Review

Lean and Industry 4.0: A Review of the Relationship, Its Limitations, and the Path Ahead with Industry 5.0

André Moraes [1,*], André M. Carvalho [2,3,4] and Paulo Sampaio [1,5]

[1] Department of Production and Systems Engineering, School of Engineering, University of Minho, 4710-057 Braga, Portugal
[2] Department of Mechanical and Industrial Engineering, NOVA School of Science and Technology, NOVA University Lisbon, 2829-516 Caparica, Portugal; andremc@fct.unl.pt
[3] 2AI—School of Technology, IPCA, 4750-810 Barcelos, Portugal
[4] LASI—Associate Laboratory of Intelligent Systems, 4800-058 Guimarães, Portugal
[5] Centro Algoritmi, University of Minho, 4800-058 Guimarães, Portugal
* Correspondence: pg36490@alunos.uminho.pt

Abstract: This article aims to analyze the relationship between Lean and Industry 4.0, further exploring the opportunities for integration with the new concept of Industry 5.0. Departing from a literature review, it shows how the relationship between Industry 4.0 and Lean is—while unanimously positive—clearly orientated towards the more technological aspects. In this scenario, most studies on this relationship highlight the technological side of organizations, emphasizing the integration of Industry 4.0 technology to augment Lean methodologies and tools. As such, most of the apparent value of this relationship derives from the use of technology, and relatively limited inputs input are found on issues related to the human and social factors of organizations—such as leadership, people, integration, and training for new roles and new tasks. In the face of this reality, we evaluate the potential for integration between Lean and Industry 5.0, arguing how Lean may offer a proper perspective to support sustainability, resilience, and human orientation in Industrial contexts.

Keywords: lean manufacturing; quality management; Industry 4.0; Industry 5.0

Citation: Moraes, A.; Carvalho, A.M.; Sampaio, P. Lean and Industry 4.0: A Review of the Relationship, Its Limitations, and the Path Ahead with Industry 5.0. *Machines* **2023**, *11*, 443. https://doi.org/10.3390/machines11040443

Academic Editor: Dan Zhang

Received: 27 February 2023
Revised: 27 March 2023
Accepted: 29 March 2023
Published: 31 March 2023

1. Introduction

The move to Industry 4.0 led to a creation of context-specific variations for multiple philosophies, tools and approaches came to carry the same neologism "X 4.0" [1]. A few examples include Excellence 4.0 [2], Quality 4.0 [3], or Supply Chain Management 4.0 [4]). In most cases, these neologisms signify the application of technology to continuous improvement and operational management practices, and Lean was another common example of this trend [5,6]. As such, the relationship between Lean Management and Industry 4.0 has been frequently explored, to a point were several hundred articles and dozens of literature reviews are available [7]. However, a quick review of the most cited results quickly seems to indicate a reality where the relationship is mostly analyzed from a technological perspective [8–11].

Considering that a similar critique—the overshadowing of operational and social aspects of continuous improvement efforts in favor of technological adoption—has been made to, for example, Quality Management [12,13], we set out to identify if this trend is observable also in the literature on the relationship between Lean and Industry 4.0—and if so, to identify opportunities and risks for expanding this relation beyond its technological facet.

In order to validate this hypothesis, we set out to analyze the most recent reviews on this relationship. Still today, the literature shows a growing number of papers concerning "Lean and I4.0" [14], and in the first half of 2022 alone, eight literature reviews have been published on this relationship. In this paper, we analyze these reviews (and further literature), seeking to understand whether they are aligned and if their results support a common understanding, aiming to identify gaps and future research paths.

Based on this analysis, this article aims to comprehend whether this scientific production has converged to a single understanding, or if the debate is still active regarding the integration between Lean and Industry 4.0, its benefits, and its limitations. Furthermore, this work seeks to identify research gaps around this very same topic.

As a result, it is shown that most studies on this relationship highlight the technological side of organizations, emphasizing the integration of Industry 4.0 technology to augment Lean methodologies and tools. However, research on how Lean and Industry 4.0 contributes to social, environmental, and business continuity and sustainability are more limited. In the face of this finding, we then explore the opportunities for the integration of Lean and Industry 5.0 to overcome the existing limitations and support the acceleration of sustainability, resilience, and human-centered strategies.

This article aims to analyze the relationship between Lean and Industry 4.0, further exploring the opportunities for integration with the new concept of Industry 5.0. To this end, the article has five sections. To identify the necessary elements for the proper understanding of the potential additional value Lean Manufacturing can add to Industry 4.0, Section 2 offers a conceptual introduction to the topics of Lean and Industry 4.0. After this theoretical introduction, in Section 3, the relationship between Lean and Industry 4.0 is analyzed, departing from the eight most recent reviews at the date of the development of this article (September 2022). An analysis is conducted whereby the most frequent points of contact between Industry 4.0 and Lean are summarized, and the existing gaps in the literature about this relationship are identified. Next, Section 4 discusses the opportunities and challenges for Lean Manufacturing in Industry 5.0. A conclusion and perspectives for future work are presented in Section 5.

2. Conceptual Review

2.1. Industry 4.0

Industrial Revolutions are marked by a paradigm shift of the production process [15]. However, they further install themselves into society, to the point where their effects are clearly recognized outside industrial and business forums [16]. The 4th Industrial Revolution, also called Industry (I4.0), was declared as an opportunity for the transformation of the previous paradigm (born with the 3rd Industrial Revolution) at the Hannover Fair in 2011. This was followed by the launch of the Strategic Development Report for German Industry, published in April 2013 with the aim of allowing it to achieve a strategic global differential [17]. Based on the premise established in that work, the *Industrie 4.0* Program sought to give direction to companies not only on production process change but also at the level of investment, training, and technological development necessary for those industries to lead markets and supply products [17]. It is important to highlight that the concept "4.0" has taken dimensions for several business areas not limited to the industrial environment itself, and that it has become an icon ("4.0")—an identity sometimes greater than its original idea, even defining its evolution ("5.0") [1].

While it can be traced back to Germany [18], the concept quickly grew beyond the country, across the European Union, and beyond it. Each of the Union's members have a national agency working to improve the block's industrial competitiveness [19]. In Japan, the term Industry 4.0 has also gained attention, and inspired even broader societal approaches to a technologically-driven future, where the benefits of Industry 4.0 are exploited to serve society at large and create a "technology-based, human-centered society" [20]. In America and Asia, the terms "Smart Manufacturing" or "Smart Industry" are used frequently to signify a similar industrial and societal transition [21–23].

Smart Factories (SF) are considered as an essential component for the expression of I4.0, and, in them, the information transfer flows between people, machines, and resources are key. SF have internalized the data of origin and the date of manufacture; of the processing parameters to be applied and applied; how, where, when, and to whom they are to be delivered; and several others [17]. In this sense, interdisciplinary actions are strongly suggested, both in terms of technologies, processes, and the development of personal skills, and the creation of an Industry in the I4.0 paradigm requires multidimensional work [18] and not just a selection of technologies. However, Industry 4.0 has been studied predominantly from a technological point of view [24,25]. Earlier works in Industry 4.0 focused on understanding and framing the forces that influenced this paradigm shift, as well as its relevance [26]. Normally, they list a wide variety of technologies that will help improve performance and organization [27]. Nevertheless, strategies or initiatives deployed in the context of this transition tend to have a vast focus, often showing a disconnection between the advancements proposed by area experts and those derived from the technological push.

Another important aspect is the fact that Industry 4.0, from the perspective of a "revolution", undoubtedly represents the expected lasting and comprehensive economic impact; however, from the perspective of "innovation", Industry 4.0 is positioned more as a driving force of innovations than the result of innovations [28].

The transformation towards Industry 4.0 offers many opportunities in the way organizations integrate their processes and their production, and how their systems evolve [13]. In the face of this reality, several organizations are on the lookout for new and innovative approaches to help them navigate the transition towards Industry 4.0. For many, Lean is central for mastering such a transition [29].

2.2. Lean Manufacturing

Since the 1970s, the high performance of the Japanese automobile industry has created a standard of operations that industries should seek to maintain competitiveness and survival [30,31]. As a result, the term "Lean" gained increased attention [32], and the development of Lean models and tools for companies began to emerge to ensure competitiveness. Drivers were the elimination of waste from the value chain, stakeholder satisfaction, and continuous improvement [33–35]. The use of techniques and tools for process improvement were key for the achievement of the objectives of Lean, allowing performance to be maintained and improved in the long term; and a focus on the Culture and People was reinforced as a central aspect of industrial management [30]. Lean is thus supported by guiding principles and a strong culture aimed at innovation, permanent development, and respect for people and society [36].

The central goals of Lean are to eliminate any activities that represent waste, to ensure a fluid operation, to achieve minimum set-up times and eliminate downtime due to breakage and defects, to level stocks and production quantities, or to maintain and develop standardized operations [33,37]. From these core perspectives, there are several methods and tools—Jidoka, Kaisen, pull system, Value Stream Mapping (VSM), Just-in-Time (JIT), Andon, Gemba, Kanban, Hoshin Kanri, Kata, etc.—that support it, reinforced by a culture where trust, respect, and empowerment of employees and partners is constant [38,39].

Lean Manufacturing was an adaptation to the mass production management model advocated in the 3rd Industrial Revolution. Differently, Industry 4.0 is a set of technological principles and tools that guides production to automation, integration, and connectivity to their highest levels. Therefore, Lean and I4.0 are different things, but complementary to each other—in spite of the imminent conflict between them, since in I4.0, production seems to promote a reduction of the dependence on employees, while Lean considers the human factor as a competitive differential and a valuable asset [30,33,37,40,41].

Because of these factors, for years major global consultancies [18,42,43] and academic contributions [44–49] have been trying to explore the relationship between Lean and Industry 4.0, and to understand it at both the theoretical and empirical levels—most often, with the hope of demonstrating the capacity that Lean can add competencies to I4.0. While there seems to be a wide agreement in relation to some aspects of this relationship, such as the benefit of technology for improved performance, improved data collection and analysis, and reduced human error [50], practical results are most often tied to increased automation [8]. Integration between Lean and Industry 4.0 concerning the social and environmental aspects remains mostly at the theoretical level [11,51].

3. Literature Review

3.1. Methodology

There are many articles on the numerous aspects of the relationship between Lean Manufacturing and Industry 4.0. A search using the Scopus database produced 764 results on the query "Lean AND Industry 4.0", all between 2014 and 2022. A quick analysis of the results showed a variety of themes, with a total of 104 different keywords occurring 10 or more times.

The purpose of this work, however, is not to dive into specific topics within the relationships between Industry 4.0 and Lean, but rather to gain an overall perspective of the state of the art. Our goal is to understand if, after almost a decade of research on this topic, there is a clear, common perspective on how Lean and Industry 4.0 interact. As such, a review of reviews methodology was followed [52,53]. 42 reviews relating to the topics of Lean and Industry 4.0 were identified. However, after excluding those unrelated to our research, only 12 articles remained—all ranging between 2021 and 2022. Amongst the exclusion criteria were a focus in narrow or specific trade or industry, research agendas, or narrow perspectives on the relationship between Lean and Industry 4.0. Exclusion criteria removed neighboring topics such as Lean Six Sigma, or particular aspects such as "Lean-Green" or "Sustainability".

In order to use the most recent works as our base, we selected the literature reviews written in 2022. From there, we moved on to older works. Departing from the most recent literature reviews, we explored the main trends, opportunities, and limitations in this relationship.

The following literature reviews on the subject of "Lean" and "Industry 4.0", published this year (2022), were identified (Table 1).

Table 1. Articles published in 2022—"Lean" and "Industry 4.0" literature review.

Authors	Title
Tailise, M.M.; Mergulhão, R.C.; Mano, A.P.; Silva, A.A.A.	The integration of technologies Industry 4.0 technology and Lean Manufacturing: A systematic literature review.
Rajaba, S.; Afy-Shararaha, M.; Salonitisa, K.	Using Industry 4.0 Capabilities for Identifying and Eliminating Lean Wastes.
Lucantoni, L.; Antomarioni, S.; Ciarapica, F.E.; Bevilacqua, M.	Implementation of Industry 4.0 Techniques in Lean Production Technology: A Literature Review.
Terra, J.D.R.; de Melo, C.C.; Berssaneti, F.T.	Are Lean, World Class Manufacturing and Industry 4.0 are related?
Nedjwa, E.; Bertrand, R.; Boudemagh, S.S.	Impacts of Industry 4.0 technologies on Lean management tools: a bibliometric analysis.
Yürekli, S.; Schulz, C.	Compatibility, opportunities and challenges in the combination of Industry 4.0 and Lean Production
Komkowski, T.; Antony, J.; Garza-Reyes, J.A.; Tortorella, G.L.; Pongboonchai-Empl, T.	The integration of Industry 4.0 and Lean Management: a systematic review and constituting elements perspective
Rossi, A. H. G.; Marcondes, G. B.; Pontes, J.; Leitão, P.; Treinta, F. T.; de Resende, L. M. M.; Mosconi, E.; Yoshino, R.T.	Lean Tools in the Context of Industry 4.0: Literature Review, Implementation and Trends

In most articles, the results focused on the contribution of different technologies of Industry 4.0 to Lean. Since different names were used for similar technologies, and in order to avoid confusion, we have grouped the different technologies into seven broader technological areas: (1) Cybernetics, (2) Connectivity and Integration, (3) Big Data, (4) Industrial Automation, (5) Administrative Process Automation, (6) Simulation and Augmented Reality, and (7) Additive Manufacturing. The correlation table of the technologies presented in the articles in Table 1 and the "Seven broader technological areas" are indicated in Appendix A.

3.2. The Relationship between Lean Manufacturing and Industry 4.0: Review of Reviews

From the above listed articles, it was possible to validate the connection between Lean and I4.0 in distinct dimensions. In "The integration of Industry 4.0 technologies and Lean Manufacturing: A systematic literature review" [54], we find an analysis of nine articles published between 2017 and 2021, seven of which demonstrate empirical examples of the joint use of I4.0 technologies and Lean tools. This result is presented in Table 2, adapting the original dimensions to the dimensions of Cybernetics, Connectivity, and Integration, Big Data, Industrial Automation, Administrative Process, Automation, Simulation and Augmented Reality, Additive Manufacturing. The intensive use of Connectivity and Integration, Big Data, Simulation, and Virtual and Augmented Reality are the most adopted dimensions in the empirical cases observed in conjunction with Lean. In addition to the integration between I4.0 technologies and Lean practices and tools, this article highlights other important topics within its review results. Other topics are (1) the extent to which I4.0 design principles are supporting Lean Manufacturing tools and (2) environmental factors in the integration of Industry 4.0 and Lean manufacturing.

Table 2. Analysis of the article "The integration of technologies Industry 4.0 technology and Lean Manufacturing: A systematic literature review" [54].

Industry 4.0 Technology	Cybernetics	Connectivity and Integration	Big Data	Industrial Automation	Administrative Process Automation	Simulation and Augmented Reality	Additive Manufacturing
Article							
Continuous Improvement Programs and Industry 4.0: Descriptive Bibliometric Analysis		X	X	X	X	X	
How Industry 4.0 Can Enhance Lean Practices	X	X	X	X	X	X	X
Impact of Industry 4.0 Concept on the Levers of Lean Manufacturing Approach in Manufacturing Industries	X	X	X	X	X	X	X
Impacts of Industry 4.0 technologies on Lean principle	X	X	X			X	X
Implementation of Industry 4.0 and lean production in Brazilian manufacturing companies		X	X	X		X	X
Industry 4.0 and Lean Manufacturing: A systematic literature review and future research directions	X	X	X		X	X	X
The Relationship between Lean and Industry 4.0: Literature Review		X	X		X	X	X

The next review, "Using Industry 4.0 Capabilities for Identifying and Eliminating Lean Wastes" [55], presents several examples of the integration between I4.0 technologies and the elimination of inherent waste from production processes, based on the 7 + 1 classical waste perspective. Based on the literature review between 2008 and 2021, 53 articles were analyzed, with the results showing a greater contribution to the elimination of classic Lean waste in the Big Data dimension, followed by Administrative Process Automation and Simulation and Augmented Reality (Table 3).

Table 3. Analysis of the article "Using Industry 4.0 Capabilities for Identifying and Eliminating Lean Wastes" [55].

Technology	Cybernetics	Connectivity and Integration	Big Data	Industrial Automation	Administrative Process Automation	Simulation and Augmented Reality	Additive Manufacturing
Lean wastes and Industry 4.0							
Defects				X	X		
Overproduction					X		
Waiting			X				
Transportation	X		X				X
Over-processing			X				
Inventory		X					
Underutilized Skills						X	
Motion						X	X

The review "Implementation of Industry 4.0 Techniques in Lean Production Technology: A Literature Review" [56] was conducted on articles published between 2018 and 2021. This review sought to identify combinations of Lean practices and I4.0 technologies. The result is presented in Table 4, and shows the intensive use of Lean tools in conjunction with I4.0 technologies. In this work, the strong contribution of the I4.0 technologies to the analyzed Lean tools is observed—including the areas of Industrial Automation and Additive Manufacturing, which have a smaller but no less relevant contribution.

Table 4. Analysis of the article "Implementation of Industry 4.0 Techniques in Lean Production Technology: A Literature Review" [56].

Technology	Cybernetics	Connectivity and Integration	Big Data	Industrial Automation	Administrative Process Automation	Simulation and Augmented Reality	Additive Manufacturing
Lean Tools							
Value Stream Mapping	X	X	X	X	X	X	X
Cellular Manufacturing	X	X	X		X	X	X
Kanban	X	X	X		X	X	
Jidoca	X	X	X		X	X	
5S		X	X		X	X	
Total Productive Maintenance	X	X	X	X			X
Just In Time	X	X		X	X		
Poka Yoke	X		X		X	X	

The review "Are Lean, World Class Manufacturing and Industry 4.0 related?" [14] looked at 165 articles published between 1984 and 2021. By analyzing its results, it was verified that the relationship between Lean and Industry 4.0 is yet limited if looking at the overall publishing scenario on these topics. The results (shown in Table 5) demonstrate, among other things, high academic interest in the topics themselves, but still some limitation when it comes to the relationship of Lean and I4.0 (15.5% of the articles reviewed show a focus on the two concepts) (Table 5). Even focusing on a shorter time range, from 2016 to 2021, it is observed that the articles in this subset that have a focus on I4.0 and Lean still represent less than 20% of the total. In addition to these findings, the article provided little contribution regarding how different Industry 4.0 areas interact with Lean Manufacturing (Table 6).

Table 5. Analysis of the article "Are Lean, World Class Manufacturing and Industry 4.0 related?" [14].

Main Themes Covered in Sample (1984–2021)	N	%	Ac.%
Industry 4.0	34	20.61%	20.61%
Lean and Industry 4.0	25	15.15%	35.76%
Lean	24	14.55%	50.30%
World Class Manufacturing	22	13.33%	63.64%
Structure Process	17	10.30%	73.94%
Lean and World Class Manufacturing	8	4.85%	78.79%
Manufacturing	7	4.24%	83.03%
Business Process Mapping	6	3.64%	86.67%
Lean and Manufacturing	6	3.64%	90.30%
World Class Manufacturing and Industry 4.0	5	3.03%	93.33%
Semi-structure Process	4	2.42%	95.76%
Multi-Criteria Decision Analysis	2	1.21%	96.97%
Lean and Agile	1	0.61%	97.58%
Lean and Information Technology	1	0.61%	98.18%
Lean and Six sigma	1	0.61%	98.79%
Lean and Value Stream Mapping	1	0.61%	99.39%
Process Structure	1	0.61%	100.00%

Table 6. Analysis of the article "Are Lean, World Class Manufacturing and Industry 4.0 related?" [14]—subset 2016–2021.

Main Themes Covered in Sample (2016–2021)	N	%	Ac.%
Industry 4.0	34	28.10%	28.10%
Lean and Industry 4.0	23	19.01%	47.11%
Lean	21	17.36%	64.46%
Structure Process	8	6.61%	71.07%
World Class Manufacturing	8	6.61%	77.69%
Lean and World Class Manufacturing	7	5.79%	83.47%
Lean and Manufacturing	6	4.96%	88.43%
World Class Manufacturing and Industry 4.0	5	4.13%	92.56%
Business Process Mapping	2	1.65%	94.21%
Manufacturing	2	1.65%	95.87%
Lean and Agile	1	0.83%	96.69%
Lean and Information Technology	1	0.83%	97.52%
Lean and Six sigma	1	0.83%	98.35%
Lean and Value Stream Mapping	1	0.83%	99.17%
Process Structure	1	0.83%	100.00%
Multi-Criteria Decision Analysis	0	0.00%	100.00%
Semi-structure Process	0	0.00%	100.00%

In "Impacts of Industry 4.0 technologies on Lean management tools: a bibliometric analysis" [57], a review is made considering articles from 2011 to 2020. As a result, the relationship between Lean methodologies and techniques and I4.0 technologies are presented in Table 7. The article is different from the results presented before, since it expands the scope of the analysis to include other Lean tools and methodologies, and presents similar results. As in the previous work, the dimensions that promote the greatest impact on Lean are Big Data and Connectivity and Integration at the top, and Industrial Automation as the one that, although impactful, has the least contribution in relation to the other dimensions. The other tools, despite not being at the top, have a relevant contribution, being basically equivalent to each other. In this last group, Additive Manufacturing has a slightly lower impact than the others.

Table 7. Analysis of the article "Impacts of Industry 4.0 technologies on Lean management tools: a bibliometric analysis" [57].

Technology	Cybernetics	Connectivity and Integration	Big Data	Industrial Automation	Administrative Process Automation	Simulation and Aug. Reality	Additive Manufacturing
Methodology & Tools							
Continuous improvement	X	X	X	X	X	X	X
Heijunka	X	X	X		X	X	
TPM (Total Productive Maintenance)	X	X	X		X	X	X
Communication and Information sharing	X	X	X	X	X		
Jidoka	X	X	X	X	X	X	
JIT	X	X	X			X	X
5S		X	X			X	X
Andon	X	X	X		X	X	
Kanban	X	X	X			X	X
Poka-yoke	X	X	X			X	X
Pull flow	X	X	X	X		X	X
Standardization work	X	X	X	X			
VSM		X	X		X	X	
Waste reduction		X	X		X	X	
CIM (Computer Integrated Manufacturing)		X	X		X		X
Decreased operation and waiting times	X	X	X	X			X
Decreased stocks and inventory management	X	X	X			X	X
Problems solving	X	X	X		X	X	
Quality control		X	X		X	X	
Standardization		X	X		X	X	
Increased flexibility	X		X	X		X	
KPI	X		X			X	
Statistical control process	X	X	X		X		
Cellular manufacturing		X	X				X
Empowerment and involvement of workers		X	X		X		
Improved human resources			X			X	X
Set up reduction used (SMED)		X				X	X
Supermarket	X	X	X				
WIP reduction		X					X
Automation					X		
Machine and human separation							X
Production Smooth					X		
Supplier development					X		
Takt time						X	

The article "Compatibility, opportunities and challenges in the combination of Industry 4.0 and Lean Production" [58] analyzed 15 articles published between 2015–2020, and among the results, it presented a correlation between I4.0 principles with some Lean tools. The compatibility is presented on a Likert scale from 0 to 3, with "0" being no compatibility and "3" being high compatibility, and the result is presented in Table 8. Strong overall compatibility is perceived, with Kanban and Total Productive Maintenance (TPM) being present in all I4.0 principles.

Table 8. Analysis of the article "Compatibility, opportunities and challenges in the combination of Industry 4.0 and Lean Production" [58].

Lean Tools vs. Industry 4.0 Principles	Real-Time Capability	Interoperability	Decentralization	Virtualization	Modularity
Kanban	3	3	3	2	3
TPM	3	3	3	2	1
Just-in-time-production	3	0	2	3	3
One-piece-flow	3	3	2	0	3
SMED	2	3	2	1	3
Andon	3	3	2	3	0
Kaizen	3	3	1	2	0
Poka Yoke	1	1	3	3	0

Two more literature reviews were analyzed. In line with the broader goals of this work, their focus was less on the relationship between Industry 4.0 technologies and Lean Manufacturing practices and methods, and more focused on the integration between the two within an organization. These reviews provided different results. "The integration of Industry 4.0 and Lean Management: a systematic review and constituting elements perspective" [59] sought, through analysis of 111 articles published between 2015 and 2021, to identify the key components for the integration process between Lean and I4.0. This article draws a link between 'what' elements of Lean and I4.0 organizations should integrate and 'how' they may do it. The findings indicate that integrations cover the essential constituting elements of LM. However, serious gaps were identified concerning the operational level. The authors identify as major constraints the lack of enabling processes, routines, and implementation pathways. It further identifies Change Management as the key component of the integration process, given the transformational nature that both Lean and I4.0 represent for the organization.

Finally, in "Lean Tools in the Context of Industry 4.0: Literature Review, Implementation and Trends" [51], a different analysis was identified. While still approaching the relationship between I4.0 and Lean, this article is focused on the implementation of a lean philosophy for the digital environment, where it may be used to tackle problems related to inefficient digitalization within organizations. Through the analysis of 53 articles, a proposed framework for the implementation of Lean to the digital environment is elaborated. While the work presents the phases and objectives to be achieved, a strong technological perspective is present, emphasizing automation, including at the administrative level. In this proposition, not only does Lean become Lean 4.0, but its tools and methodologies are also adapted to this digital environment.

3.3. Lean Manufacturing and Industry 4.0: Review, Analysis, and Discussion

The articles reviewed presented visions of Lean and I4.0 from different dimensions and perspectives, but always demonstrating added value in the relationship between them. Most articles are focused on the technological side of the relationship between Lean and

Industry 4.0. Nevertheless, some connections to other factors—especially at the human and social level—were identified.

To better understand the relationship, in this section we first dive into the contributions of the different I4.0 technologies to Lean. Looking at these reviews, we must highlight the relationship between specific Industry 4.0 technologies and Lean methods and tools. In fact, these were the most frequent associations between the literature reviews analyzed, with six of the reviews presenting this perspective. Amongst the technological areas most prone to support Lean Manufacturing, the most frequently indicated in the literature are: Big Data (27), Connectivity and Integration (26), Administrative Process Automation (21), Cybernetics (18), and Augmented Reality (17). Regarding these technologies, in Appendix B we indicate the main contributions. Although the presentation is by I4.0 technology, the contribution is considered systemically—the more tools applied, the more iterations, the more value.

These results show the central aspect of technology in the relationship between Lean and Industry 4.0. They demonstrate that technology offers performance improvement opportunities when integrated with traditional industrial control and management practices [60], often in such a way that new approaches and tools are created [61,62]. Digital Lean Manufacturing (DLM) is one example of such "new" analytics applications where Connectivity and Integration are central. A DLM system relies on new data acquisition, integration, processing, and visualization capabilities to detect, fix, predict, and prevent ambiguous parameters and avoid quality issues inside defined tolerance ranges. These capabilities lead to fostering substantial feedback loops for quality assurance and quality management digitalization [63]—allowing the reliable functioning of several of the technologies presented above. Highly reliable, secure, and clean data ensure quality is guaranteed in the use of machine learning, high-performance computing, predictive modeling, correlations, and pattern recognition, neural networks, and others [64–66]. Similarly, Closed-loop Manufacturing (CLM) also allows the use of Big data, gathered during manufacturing in the production machine, to be shared across the different systems along the product lifecycle [67]. This allows increased connectivity, with immediate information sharing with product development activities, reducing variability and the risk of defects in the process [68].

The performance and stability of processes may also be improved using increased Administrative Process Automation—such as Robotic Process Automation (RPA). RPA allows the elimination of operational risk and brings companies the opportunity to better manage their resources, attaining savings in time and cost [69]. RPA will deliver a higher quality by standardizing operations and processes and reducing human errors by diminishing or eliminating the possibility of a process being performed in the wrong way or by an operator without proper knowledge [70,71].

Taking advantage of Cybernetics, organizations may create interconnected systems, where technologies are integrated in a collaborative way. They allow information to be closely monitored and synchronized between the physical factory floor and the cyber computational space, allowing for enhanced equipment efficiency, reliability, and quality [72].

Another area that gains interest from a systems perspective is that of extended digital support systems [73,74]. Technologies supporting such systems include collaborative robots (COBOTs), Augmented Reality (AR), and Smart Human Interfaces (SHI), and a number of smart technologies such as screens, 3D glasses, or exoskeletons. Such systems allow companies to achieve standardization, attain superior performance, and avoid human errors [75,76].

Despite this strong technological partiality, the relationship between Lean and Industry 4.0 is not encapsulated by the adaptation of technology to Lean, or vice versa. While results are more limited in terms of absolute numbers, there are important takes in the Literature about the importance of other aspects in this relationship—such as social, environmental, and business sustainability and continuity. In fact, some authors argue that even technology should not be regarded only to the extent of its technical applications, being instead used as catalysts of people, products, processes' efficiency, performance, and innovation [77]. In this sense, technology, processes, and people must be integrated in the I4.0 transition [78]. Looking back at our review of reviews, very little input is perceived in issues such as leadership, people, and cultural management, integration (downstream and upstream) in the value chain, innovation and design, and problem solving. At most, we found important points that indicate that issues of leadership and people management and culture are keys to the success of a Lean to I4.0 integration model. Along these lines, we selected a few excerpts from "The integration of Industry 4.0 and Lean Management: a systematic review and constituting elements perspective" and from "Lean Tools in the Context of Industry 4.0: Literature Review, Implementation and Trends":

- Lean Management "maturity introduces success factors" supporting the integration of Lean Management with I4.0: "a learning culture, senior management leadership, cross-functional team development, change governance frameworks, and training activities" [59];
- "Concerning elements of change" management, a series of factors were "identified [as] essential issues to be covered in integrating I4.0: transformation strategy, design, delivery, governance, and leadership" [59].
- "Processes only focused on digital technologies tend to be less efficient than processes based on Lean and that are rethought by people. In this way, workers are the center of innovation in a sustainable way, it is from them that comes the ability to improve processes and develop specific improvements for organizations, and it is up to them to focus on improving their skills and training them to deal with digital technologies" [51].
- "As in the past, without thinking people there is no Lean, and without Lean, there is no waste control, whether digital or not. This has the potential to impact the transition from the fourth industrial revolution to a new bias of Industry 5.0 or Society 5.0, in which, in addition to machines, employees are also a fundamental part of the industrial process, and Lean 4.0 converges along this same line of thought. Thus, there is a need for further research to understand what impacts Lean 4.0 has on the new way of thinking about Society 5.0 and if they can go together and what their intersection points are (...)" [51].

Moving on to environmental sustainability, a similar scenario is observed. The relationship between Lean and Sustainability has been explored from multiple perspectives [79]—including the use of technology to improve green production [80–82]. A significant part of these studies is focused on the supply chain [83,84]. The search for "Corporate Sustainability" has led multiple frameworks and instrumental studies on the causality between sustainability and performance, and on managerial practices and on tools to manage them [85,86]. However, these efforts have seen mixed results. While theoretical studies were able to pinpoint both the synergies and trade-offs between environment and performance, instrumental studies failed to provide convincing evidence, either because of inaccurate measurement systems or because they were often representative of a single company or sector, and they failed to provide a cross-industry perspective [85,87].

However, the impact of the relationship between Lean and Industry 4.0 has been more limited, and better integration between Industry 4.0 business models and performance management frameworks is still necessary [88]. Examples of limited approaches abound, and include the listing of paperless organizations as a benefit of the sustainability-productivity link, or claiming sustainability as a result of predictive analytics in production and operations management, resulting in less waste. However, these perspectives create a false sense integration, a feeble solution instead of the larger promotion of a structural change in the management practices and methods of organizations.

Business continuity is another topic with limited mentions across the literature, with the topic being absent even from the most recent reviews. Nevertheless, some important notes exist on how Lean supports long-term success and the transition to Industry 4.0. It has been identified that organizations that have been implementing lean productions extensively are more likely to concurrently adopt Industry 4.0 technologies [89], demonstrating that Lean may be critical for the resilience of organizations, including in helping to navigate large industrial transitions [13]. However, the contribution of the relationship between Lean and Industry 4.0 towards resilience has only had limited mention in the literature [90,91].

4. The Path Ahead with Industry 5.0

4.1. A New Strategic Era: Industry 5.0

While Industry 4.0 is a recent technological and societal transformation—affecting not only our industries but society at large—other more ambitious initiatives have been proposed since its inception. In 2015, the "2030 Agenda for Sustainable Development", adopted by UN member countries, defined 17 Sustainable Development Goals (SDGs) to improve health, education, and inequality that stimulate economic development, without losing focus on combating climate change and preserving oceans and forests [92]. In the same year, Michael Rada [93] posted his reflection on the lack of a man-machine perspective in a synergistic way of Industry 4.0 and coined this evolutionary vision as "Industry 5.0". In 2016, Japan, in the "5th Science and Technology Basic Plan", presented the concept of Society 5.0 (S 5.0), which started from the hunter/gatherer evolving to the partisan/farmer, and then to the industrial and later to the information ages [94,95]. This proposition of S5.0 that has as its foundation the SDGs and represents a society that keeps humanity as the central actor, promoting happiness and a sense of value to them, through the use of technology, which is the instrument for the promotion of these actions [94]. From there, the proposed definition of Industry 5.0 as it is known today begins [96,97].

In the unfolding of the S5.0 concept, there is the proposal of Industry 5.0 (I5.0). Proposed by the European Union [98], the definition of I5.0 [99] presents industry as an active element in the societal change process towards a more sustainable, resilient, and human-centered perspective [99]. In this new concept, industries take an active role in the achievement of targets beyond employment and growth, coming to consider also the productive limits of the planet, supporting research and innovation at the service of sustainability, and promoting a human-centered digitalization and a resilient society [100,101]. The key activators for I.5.0 and their relation with the SDGs are presented in Table 9 [101].

Since I5.0 is conceptually oriented towards an innovative, resilient, and human-centric industry, new points of attention regarding problems that are reflected in the technological, social, and governance spheres tend to arise. This is one of the areas where understanding the relationship between I4.0 and I5.0 becomes critical, as we need to identify the key technologies that facilitate the transition. Examples are [101]:

- Human-centric solutions and human-machine interaction technologies that connect and combine the strengths of humans and machines.
- Bio-inspired technologies and smart materials using recyclable materials with embedded sensors and enhanced features.
- Simulation and real-time digital twins for modeling systems.
- Cyber-secure data analysis, transmission, and storage technologies that can manage the interoperability of systems and data.
- Artificial intelligence, such as the ability to find causal relationships in complicated dynamic systems and produce useful information.

Reliable, energy-efficient autonomy technologies are also needed because key technology enablers will use a lot of energy.

A point of attention is the fact that despite the fact that the Industry 5.0 theme is incipient and, as a result, there is little literature on the subject, the growth of research in this regard has grown significantly. This requires converging understanding on the subject in order to avoid neologisms and derivations that tend to create confusion and conceptual failure on the subject [102].

Table 9. Sustainable Development Goals link with Industry 5.0 & Society 5.0 Key Enables.

Technologies vs. Sustainable Development Goals	1 No Poverty	2 Zero Hunger	3 Good Health & Well-Being	4 Quality Education	5 Gender Equality	6 Clean Water & Sanitation	7 Affordable & Clean Energy	8 Decent Work & Economic Growth	9 Industry, Innovation & Infrastructure	10 Reduce Inequalities	11 Sustainable Cities & Communities	12 Responsible Consumption & Production	13 Climate Action	14 Life below Water	15 Life Land	16 Peace, Justice & Strong Institutions	17 Partnerships for the Goals
Advanced Materials	X			X	X	X	X								X		
Artificial Intelligence		X	X		X				X		X			X	X	X	X
Augmented Reality				X							X				X		
Big Data & Analytics			X								X		X			X	
Blockchain	X						X			X						X	
Cloud											X						
Cybersecurity				X				X									X
Data Democratization				X												X	X
Digital Twin			X									X					
Drones		X												X	X		
E-Governance							X										X
Electric Vehicles													X	X			
Innovation strategies				X			X								X		
Internet of Things		X								X			X		X		X
IoT enabled supply chain process optimization & automation	X						X	X	X	X							X
Maintenance		X				X	X				X	X		X	X		
Robotics		X				X		X	X					X			
Smart Sensors						X											
Virtual Reality				X							X						
Wearables			X														

The three dimensions of I5.0 [103]—Sustainable, Resilience, and Human-centric—need further detailed description in order to identify the impact of their implementation. In this regard, Table 10 [104] presents a proposed framework for I5.0. These are exactly the characteristics that make Industry 5.0 a different type from previous Industrial Revolutions, to the point that it prevents the consideration of it as a natural "chronological evolution" resulting from the previous Revolution—as happened from Industry 1.0 to Industrial 4.0. When dealing, for example, with the question of the productive process centered on the human being, where man and machine "co-labor" and "co-exist", it is something far out of the context of the previous Industrial Revolutions, and there is a great opportunity for studies and projects in the construction of this path [103,105].

Table 10. Industry 5.0 Framework.

Industry 5.0 Dimensions	Society Level	Network Level	Plant Level	Organization—Level and Dimensions
Resilience	Viability of intertwined supply networks	Supply chain, resilience Reconfigurable supply chain	Resilience of manufacturing and logistics facilities Reconfigurable plants.	Resilient Value Creation and Usage
Sustainability	Sustainable usage of resources and energy on the earth	Supply chain sustainability Life cycle assessment of value-adding chains	Reduction of CO_2 emissions Energy-efficient manufacturing and logistics.	Sustainable Manufacturing and Society
Human-Centricity	Viability of human-centric ecosystems	Cyber-physical supply chains Digital supply chains	Human-machine collaboration Health protection standard and layouts	Human Well-Being

Table 11 presents the enablers that support the transition between I4.0 and I5.0 [101].

Although I5.0 seems to still be at a theoretical level—since its ability to deliver its results is yet to be proved—studies have shown that it has the ability to support the development of the proposed values to promote sustainable development [100]. An unplanned "trial run" of the benefits that I5.0 can bring to society as a whole was the critical pandemic scenario that plagued society between 2020 and 2021 [106]. The contingency scenario that nations and companies required accelerated the virtualization of work activities in a way not previously imaginable—in the form of remote work and meetings and in virtual rooms, intensive use of corporate networks, etc. These activities, which hitherto depended on travel, were now performed virtually. The reduction in the use of cars in general in the cities, the use of airplanes, etc. changed the appearance of the big centers. The reason for this action was to ensure sanitary conditions and preserve the health of employees, customers, and partners—a vision focused on man as the center of society. This generated learning and developed knowledge and skills previously unused or unknown—that is, resilience. Because of the low consumption/production promoted by the general activity level of society, it promoted the reduction of pollution in general (air and water) [107]. Some fruits of this transformation have altered how we work beyond the pandemic scenario.

Table 11. Key Enabling Technologies—transition from Industry 4.0 to the concept of Industry 5.0.

Authors	Impact
Edge Computing	Low latency Ensure cybersecurity Expanded interoperability Reduce storage cost
Artificial Intelligence	Intelligent automation Greater efficiency Quality control Quick decision making Increased productivity
Cobots	Robustness Enhanced dexterity More consistent and accurate

Table 11. *Cont.*

Authors	Impact
5G & Boyond	Knowledge discovery
	Smart resource management
	Low latency
	Ultra-high reliability
Digital Twins	Reduce cost
	Predicting future errors
	Design customization
	Predictive maintenance
Blockchain	Decentralized management
	Operational transparency
	Create digital identities
	Compartmentalized approach
Internet of Everything	Asset Productivity
	Cost reduction
	Supply-chain and logistics
	Reflect intelligence in network
Big Data Analytics	Customization
	Faster, better decision making
	Foster competitive pricing
	Real-time forecasting

4.2. Opportunities and Challenges for Lean Manufacturing in Industry 5.0

Lean, as a set of methodologies and tools with a strong cultural basis, is a production management model oriented to the elimination of waste, focusing on value for the customer, honoring and respecting its employees, partners, and society [30,36]. This is different from the I4.0 the I5.0 models, which are not a production management model and not even fully implemented—given the high degree of "economy of scale" required for full adoption. As such, they end up representing, at most, a vision of the future supported by actions or processes more or less fragmented in relation to the integrated, fully-tested perspective of Lean. Despite these differences, Lean may serve as a bridge between the I5.0 vision and the reality of the productive system management of a company.

The number of works that integrate I5.0 to Lean is appreciably low. However, in these few articles it is possible to see the correlation potential between both. Compared to the scenario of I4.0, in which the focus of technology overshadows Lean and tends to integrate it mostly in the use of methodologies and tools, Lean may find its real value I5.0, as in the case of the analysis, which demonstrates that Lean, due to its people/partner-oriented perspective, is a facilitator in the implementation of I5.0 [108]. In this context, the Lean principles orbit among the "core" elements of I5.0 by cause-effect relationships.

In "Industry 4.0 and Industry 5.0 from the Lean Perspective", of the 13 articles analyzed regarding I5.0 and Lean, 12 of them indicate the "Human-centric" perspective as that most related to Lean. This is in line with the principles of Lean [30,32,39,40] where people—not only employees, but customers, partners, suppliers, and society in general—are relevant to the success of an organization and should be strongly considered.

Regarding "Sustainability", "Lean Green" (and its variations, such as "Sustainability & Lean", "Eco-Efficiency and Lean", "Sustainability & Eco-Efficiency & Lean"), the link between Industry 5.0 and Lean may help tackle the limitations still standing in the integration of Lean and a more ambitious sustainability policy. The literature demonstrates the benefits related to environmental performance in several dimensions, highlighting emissions (air), energy use, solid disposal, reduction in water pollution, reduction of toxic chemicals, and improvement in water and material use, among others. These results are strong opportunities for setting impactful Lean Green practices [109].

As for "Resilience", research conducted in 2022 regarding the impact of the COVID-19 pandemic in three different companies in the construction industry showed that the companies that were better qualified in relation to Lean principles were also better rated in relation to the dimensions of "Resilience", and, moreover, they had less impact on business and sometimes performed better during the critical period [110]. This result corroborates previous findings (from 2016), which demonstrated the positive synergy between Lean and resilience, and concluded that a company with a high level in implementing the dimensions of resilience and a high breadth in Lean implementation will be more likely to perform better after disruption or a contingency situation [110,111].

5. Conclusions

5.1. Lean, Industry 4.0 and Industry 5.0

In this article, we set out to analyze the relationship between Lean Management and Industry 4.0, exploring how Lean is framed within the so-called 4th Industrial Revolution, and uncovering any remaining opportunities and limitations. As a departure point, we analyzed the most recent reviews on the relationship between Lean and Industry 4.0. The eight literature reviews analyzed here unanimously highlighted the positive aspects of the relationship between Industry 4.0 and Lean. However, a clear orientation towards the integration of Industry 4.0 technologies with Lean methods and tools was shown. A common trend was uncovered in the different articles reviewed: the relationship between Lean and Industry 4.0 gains most of its apparent value based on the use of technology, and the increase oin such value is considered systemically—the more tools and technology are integrated, the more iterations, and consequently the most value. In opposition, very little or no input is perceived on issues related to the human and social factors of organizations—such as leadership, people, integration, and training for new roles and new tasks [112]. Similar gaps were observed regarding sustainability—whether in its social, financial/operational, or environmental aspects. Concerning business sustainability and resilience—paying special attention to the long-term success of the business—no reference was identified. As for environmental sustainability, the single reference to the topic across the most recent reviews is precisely the identification of the topic as a limitation of the literature on the relationship between Lean and Industry 4.0. The social focus, as mentioned above, is most often lacking.

These gaps became more obvious in the current *zeitgeist*. In fact, they have all pointed to Industry 4.0 itself, to the point of giving rise to a new concept, Industry 5.0., which aims to attack the limitations and missed integration opportunities in Industry 4.0 regarding human orientation, resilience, and environmental sustainability. Unlike the three previous Industrial Revolutions, I4.0 and I5.0 are not only the result of a disruptive reality in the production management model but also the product of technical and/or political decisions with the intention of adopting, in a targeted manner, larger societal strategies [103,113]. In the case of Industry 4.0, the clear intention was that of raising the level of the automation of industry—virtually to the limit of having a "lights out" production system, where decisions and production stages would be integrated and autonomous, and with very limited need for people. Naturally, this model has some particular restrictions in its integration with Lean, as people are central in this philosophy. Furthermore, the fact that I.4.0 is not a production management model but a technological vision creates limitations to its integration with Lean. The so-called I4.0 implementations are also often limited to a few areas or productive segments, a step away from the value chain perspective that is promoted in Lean.

Industry 5.0, unlike the former, has a broader perspective. It supports the vision of a more sustainable and balanced society, where environmental, societal, and governance challenges are further shared by all societal agents, including industrial companies. Lean Manufacturing, when implemented in a broad and sustainable way in an organization, yields the basic components of I5.0: Human-orientation, Resilience, and Sustainability [98]. By applying the Lean Management model, this broader vision enables the promotion of a more sustained digital transition—one where the leadership sees people, the environment, and society as elements that must be considered in the long term as a condition for the business itself to prosper. Continuous improvement will promote that people and machines, without conflicting roles, can operate in a complementary way. These synergies will improve the business process, developing the competence of people and creating a strong culture to support this management model permanently [114].

Attempts to implement a I4.0 without considering Lean were a way to speed up and automate industries [115]. However, such improvements are achieved in a way that better integrates the social and sustainability aspects of continuous improvement. As shown in this article, most references in the relationship between Lean and Industry 4.0 verge towards technology implementation and integration—an issue that has already been discussed by researchers in related areas, such as Quality Management and Operational Excellence [12,13]. As a result, the perspectives on this relationship are heavily skewed towards the technical aspects, diminishing the comprehensiveness of Lean Management. It is in the face of this reality that we argue for the strong potential of aligning Lean Management with Industry 5.0. By taking the ongoing shift in attention from I4.0 to I5.0, there is a clear opportunity to renew the role of Lean Management in today's industries, promoting, at the same time, an acceleration of sustainability, resilience, and human-centered strategies.

5.2. Considerations and Future Work

This work is one of the first efforts to explore the relationship between Industry 5.0 and Lean as a way to address the limitations of the integration of Lean and Industry 4.0. In our effort to summarize the relationship between Industry 4.0 and Lean, we have opted to center our work on the most recent reviews around this relationship. While we understand that this may be seen as a limitation, we would like to highlight that it is not the purpose of this article to offer another review—even of it is a review of reviews—on this relationship, describing in detail its many perspectives. As argued in this article, there have been many such works, including those reviewed in this document. Instead, our goal is to efficiently digest the different perspectives on this relationship, identifying and explaining them, uncovering their limitations, and setting the path ahead.

We believe we were able to successfully meet our goals. We have dissected the main trend in the relationship between Lean and Industry 4.0—that related to technological integration—by pointing which technologies are the most cited and how they relate to different Lean tools, methods, and practices. Furthermore, we have highlighted the uncovered topics—such as social, environmental, and business sustainability—that are the most lacking in terms of past research. Finally, this article points towards Industry 5.0 as a way of further integrating Technology and Lean, but by assuming a broader perspective where people are the central focus of the transition, and where the single focus on productivity is replaced by a joint perspective of sustainability, resilience, and productivity.

In terms of future work, the next steps are to deepen the study on the integration between Lean and Industry 5.0. Our goal is to create a framework where the triple perspective sustainability-resilience-productivity may be easily followed by organizations, helping them set goals and anticipate tradeoffs as they seek further (sustained) development.

Author Contributions: Conceptualization, A.M., A.M.C. and P.S.; methodology, A.M., A.M.C. and P.S.; formal analysis, A.M., A.M.C. and P.S.; investigation A.M., A.M.C. and P.S.; writing—original draft preparation, A.M.; writing—review and editing, A.M.C. and P.S. All authors have read and agreed to the published version of the manuscript.

Funding: This paper was funded by national funds, through the FCT—Fundação para a Ciência e a Tecnologia and FCT/MCTES under the scope of the project LASI-LA/P/0104/2020. This paper was funded by national funds (PIDDAC), through the FCT—Fundação para a Ciência e a Tecnologia and FCT/MCTES under the scope of the projects UIDB/05549/2020 and UIDP/05549/2020.

Data Availability Statement: Any research-related data may be provided by the authors upon reasonable request.

Conflicts of Interest: The authors declare no conflict of interest.

Appendix A

Technologies (Articles Table 1)/Broader Technological Areas	Cybernetics	Connectivity and Integration	Big Data	Industrial Automation	Administrative Process Automation	Simulation and Aug. Reality	Additive Manufacturing
The integration of technologies Industry 4.0 technology and Lean Manufacturing: A systematic literature review	"Cyber Physical System"	"Cloud" "Vertical/Horizontal Integration" Machine-to-machine communications" "Radio Frequency"	"Big Data"	"Autonomous Robotics"	"Internet of Things"	"Simulation" "Virtual and augmented reality"	"3D Printing"
Using Industry 4.0 Capabilities for Identifying and Eliminating Lean Wastes	"Cyber Physical System"	"Cloud"	"Big Data"	"Autonomous Robotics"	"Internet of Things"	"Simulation" "Virtual and augmented reality"	"3D Printing"
Implementation of Industry 4.0 Techniques in Lean Production Technology: A Literature Review	"Cyber Physical System"	"Cloud"	"Big Data"	"Autonomous Robotics"	"Internet of Things"	"Virtual Simulation" "Virtual and augmented reality"	"Additive manufacturing"
Impacts of Industry 4.0 technologies on Lean management tools: a bibliometric analysis	"Cyber Physical System"	"Cloud" "Sensors"	"Big Data"	"Robotics"	"Internet of Things" "Internet of Services"	"Simulation" "Augmented reality and Virtual reality"	"3D Printing"

Appendix B. Contribution of Different Technologies to Lean Manufacturing—Summary of the Articles Listed in Table 1

Technology	Contributions	Articles
Cybernetics	• Help Lean correct failures through interconnected systems and employ self-optimizing manufacturing systems aiming at zero defects, ensuring process security.	[54]
	• System integration refers to the increased connectivity and linking of cyber-physical systems in the manufacturing operation.	[55]
	• Cyber Physical System for a flexible Kanban production scheduling.	[56]
	• Lean tools most related to 4.0 technologies in the industry are Heijunka, Continuous Improvement, Just in Time and Jidoka, Communication Information Sharing and Waste Reduction. Among these categories, the technologies most used to improve Lean Management tools are: Big Data, Cyber Physical System, Sensors, Augmented Reality, Data Analytics and 3D Printing with, Cloud, Internet of Things. • Lean Manufacturing Automation (Jidoka) using Cyber Physical Systems is considered a cost- effective and efficient approach to improve system flexibility in increasingly challenging global economic conditions. This is the reason for intelligent Lean manufacturing automation powered by Cyber Physical Systems technologies, which is also based on Jidoka's analysis and the intelligence capability of Cyber Physical Systems technologies. • Able to collect real-time data on service needs and automatically send signals to the service (e.g., automatic notifications) of the machines when errors or defects are detected. • Can be used in combination with CPS networks providing real-time data to establish a virtual VSM. Intended to improve work processes and conditions, as well as to enable faster decision making. VR	[57]
Connectivity and Integration	• Help Lean with integrating and sharing with different sectors of the plant, perform remote maintenance management of complex equipment, reduce the space for data storage, stores, and share data between different companies (data warehouse), share company's data and information with the internal and external public, and establish communication between ERP systems of different companies.	[54]
	• Guarantees better estimates for product and predicted inventory amounts compared to traditional software downloads on personalized computers. • Increases the cost-effectiveness of inventory handling by eliminating the need for servers, as well as human-related services rendered in inventory management on-site. • Inventory amounts could be accessed anytime, anywhere, and from any device with a high degree of accuracy. • All processes relevant to inventory organization, arrangement, and ordering could be handled in one location, the Cloud, where every stakeholder has access to the needed information.	[55]
	• Bring more benefits in product design and development, and how condition monitoring and Cloud computing contribute to enhancing TPM in electric drives production.	[56]
	• Lean tools most related to 4.0 technologies in the industry are Heijunka, Continuous Improvement, Just in Time and Jidoka, Communication Information Sharing, and Waste Reduction. Among these categories, the technologies most used to improve Lean Management tools are: Big Data, Cyber Physical System, Sensors, Augmented Reality, Data Analytics and 3D Printing with, Cloud, Internet of Things. • Allows the collection of data on resources, ensuring the configuration, implementation, and flexibility of the Jidoka performance system. Enables an autonomous global supply chain with improved process, efficiency, and zero defects.	[57]

Technology	Contributions	Articles
	• Helps to manage data to optimize the maintenance of complex equipment; reduces the time to make decisions based on history; helps in the creation of new products based on customer relationship management (CRM) and their preferences, in addition to performing market analysis and monitoring different degrees of customer satisfaction in relation to horizontal/vertical integration; helps to integrate different sectors of the company such as engineering and production; integrates information technology systems horizontally and vertically to obtain productivity, cost and quality gains and shares data within the entire value stream.	[54]
Big Data	• Decrease lead time in the manufacturing process. • Manufacturing time at each phase of the batch input manufacturing decreased. • Consumers may channel their preferences along the manufacturing process allowed by intelligent data systems that integrate consumers' desires in the manufacturing process of products. • Big data analytics decreased waiting time in tracking products, their locations, and features. • Positive relationship between the implementation of big data and sustainable supply chain practice—big data applications improve real-time knowledge about problematic routes, equipment, vehicles, personnel, and suppliers making manufacturers correct their choices to guarantee smooth transportation quickly. • Big data tools could be linked with machines to autonomously correct any deviations from standardized processes using the information generated and recommended by the intelligent data-based systems.	[55]
	• Provides support to Lean, increasing productivity. • Big Data and 5S integration with only a few practical applications. • Few researchers have developed robust and sustainable Cellular Manufacturing.	[56]
	• Lean tools most related to 4.0 technologies in the industry are Heijunka, Continuous Improvement, Just in Time and Jidoka, Communication Information Sharing and Waste Reduction. • Cited in some publications as facilitating problem solving and promoting accountability to improve services, reducing waste of human activities and material resources, while improving the quality of the patient experience and reducing costs. • Can be attributed to waste disposal.	[57]
	• Helps in solving workstation problems through the use of devices such as tablets, smart glasses, or smartphones; supports carrying out maintenance remotely through knowledge sharing and technical guidance, and simulation; facilitates the construction of prototypes and samples; and also simulates projects and processes in different production and programming scenarios.	[54]
Simulation and Virtual and augmented Reality	• Additive manufacturing refers to the creation of high-dimensional objects by the disposition of materials such as 3D printing. Eliminate wastes in the production planning phase by the evaluation of alternative planning strategies to determine the optimal course of action. • Allows manufacturers to integrate siloed operations within the general system by eliminating excess resources, people, or equipment additive manufacturing saves large amounts of resources and minimizes the waste of operator motion, resources, costs, and energy used in production. Virtual and augmented reality application in Jidoka principle. • Few case studies have been implemented, focusing mainly on JIT, TPM, and VSM	[55]
	• Bring more benefits in product design and development.	[56]
	• Lean tools most related to 4.0 technologies in the industry are Heijunka, Continuous Improvement, Just in Time and Jidoka, Communication Information Sharing and Waste Reduction. Among these categories, the technologies most used to improve Lean Management tools are: Big Data, Cyber Physical System, Sensors, Augmented Reality, Data Analytics and 3D Printing with, Cloud, Internet of Things.	[57]

Technology	Contributions	Articles
	• Strong impact on real-time tracking of customer demand and Work in progress and finished product inventory. Solution difficulties, such as inadequate management and poorly organized manufacturing systems, and, in addition to collaborating with data storage correctly, it reduces the time between failure notification and failure occurrence, makes systematic management of the supply chain more efficient, and supplies and improves the supply chain. • Help Lean to perform inventory control and material traceability and data sharing with the network to optimize preventive maintenance, and provides real-time information to support management decision making.	[54]
Administrative Process Automation	• Reduces information defects and increases product traceability. Tested the use of a blockchain platform that allows producers, logistics services providers, and consumers to participate in information certification on improving products' information integrity and traceability. • Advanced sensors that read many variables and act autonomously can reduce risks of gas leaks or unexpected overflows. Increases the real-time data collection of customer needs, making the manufacturing of only desired products realized. Enhancement of machine-to-machine communication increases optimal production generating only needed amounts requested by customers. Enhances the implementation of Just-in-Time manufacturing, and thus avoids overproduction. Reduces excessive communication among varying departments in the enterprise. • Lean tools most related to 4.0 technologies in the industry are Heijunka, Continuous Improvement, Just in Time and Jidoka, Communication Information Sharing, and Waste Reduction.	[55]
	• Bring more benefits in product design and development. Management system is considering Agile–Kanban.	[56]
	• Enables an autonomous global supply chain with improved processes, efficiency, and zero defects. • Supported by networked communication between production and supply by providing real-time data on operations and machines. Using the available information, we are able to optimize processes, reduce costs, and minimize resource consumption. • Offers enormous possibilities for providing real-time data for analysis, eliminating waste and the need for human intervention.	[57]

References

1. Bongomin, O.; Yemane, A.; Kembabazi, B.; Malanda, C.; Mwape, M.C.; Mpofu, N.S.; Tigalana, D. Industry 4.0 Disruption and Its Neologisms in Major Industrial Sectors: A State of the Art. *J. Eng.* **2020**, *2020*, 8090521. [CrossRef]
2. Carvalho, A.M.; Sampaio, P.; Rebentisch, E.; Saraiva, P. 35 years of excellence, and perspectives ahead for excellence 4.0. *Total Qual. Manag. Bus. Excell.* **2019**, *32*, 1215–1248. [CrossRef]
3. Dias, A.M.; Carvalho, A.M.; Sampaio, P. Quality 4.0: Literature review analysis, definition and impacts of the digital transformation process on quality. *Int. J. Qual. Reliab. Manag.* 2021, *ahead-of-print*. [CrossRef]
4. Frazzon, E.M.; Rodriguez, C.M.T.; Pereira, M.M.; Pires, M.C.; Uhlmann, I. Towards Supply Chain Management 4.0. *Braz. J. Oper. Prod. Manag.* **2019**, *16*, 180–191. [CrossRef]
5. Javaid, M.; Haleem, A.; Singh, R.P.; Rab, S.; Suman, R.; Khan, S. Exploring relationships between Lean 4.0 and manufacturing industry. *Ind. Robot* **2022**, *49*, 402–414. [CrossRef]
6. Mayr, A.; Weigelt, M.; Kühl, A.; Grimm, S.; Erll, A.; Potzel, M.; Franke, J. Lean 4.0-A conceptual conjunction of lean management and Industry 4.0. *Procedia CIRP* **2018**, *72*, 622–628. [CrossRef]
7. Vahid, T.; Beauregard, Y. The relationship between lean and industry 4.0: Literature review. In Proceedings of the 5th North American Conference on Industrial Engineering and Operations Management, Detroit, MI, USA, 10–14 August 2020.
8. Rossini, M.; Costa, F.; Tortorella, G.L.; Valvo, A.; Portioli-Staudacher, A. Lean Production and Industry 4.0 integration: How Lean Automation is emerging in manufacturing industry. *Int. J. Prod. Res.* **2021**, *60*, 6430–6450. [CrossRef]
9. Kaio, J.; Dossou, P.-E.; Chang, J., Jr. Using lean manufacturing and machine learning for improving medicines procurement and dispatching in a hospital. *Procedia Manuf.* **2019**, *38*, 1034–1041.
10. Bitsanis, I.A.; Ponis, S.T. The Determinants of Digital Transformation in Lean Production Systems: A Survey. *Eur. J. Bus. Manag. Res.* **2022**, *7*, 227–234. [CrossRef]
11. Ciano, M.P.; Dallasega, P.; Orzes, G.; Rossi, T. One-to-one relationships between Industry 4.0 technologies and Lean Production techniques: A multiple case study. *Int. J. Prod. Res.* **2021**, *59*, 1386–1410. [CrossRef]
12. Gunasekaran, A.; Subramanian, N.; Ngai, W.T.E. Quality management in the 21st century enterprises: Research pathway towards Industry 4.0. *Int. J. Prod. Econ.* **2019**, *207*, 125–129. [CrossRef]
13. Carvalho, A.M.; Sampaio, P.; Rebentisch, E.; Oehmen, J. Technology and quality management: A review of concepts and opportunities in the digital transformation. In Proceedings of the International Conference on Quality Engineering and Management, Braga, Portugal, 15–17 July 2020.
14. Terra, J.D.R.; de Melo, C.C.; Berssaneti, F.T. Are Lean, World Class Manufacturing and Industry 4.0 are related? In Proceedings of the 5th International Conference on Quality Engineering and Management, Braga, Portugal, 14–15 July 2022; pp. 760–778.
15. Rudrapati, R. Industry 4.0: Prospects and challenges leading to smart manufacturing. *Int. J. Ind. Syst. Eng.* **2022**, *42*, 230–244. [CrossRef]
16. Lasi, H.; Fettke, P.; Kemper, H.-G.; Feld, T.; Hoffmann, M. Industry 4.0. *Bus. Inf. Syst. Eng.* **2014**, *6*, 239–242. [CrossRef]
17. Kagermann, D.; Wahlster, W.; Helbig, J. *Securing the Future of German Manufacturing Industry-Recommendations for Implementing the Strategic Initiative INDUSTRIE 4.0*; Final Report of the Industrie 4.0 Working Group; National Academy of Science and Engineering: Frankfurt/Main, Germany, 2013.
18. Gates, D.; Bremicker, M. Beyond the Hype-Separating Ambition from Reality in Industry 4.0. 2017 KPMG LLP, a UK Limited. 2017. Available online: https://assets.kpmg.com/content/dam/kpmg/xx/pdf/2017/05/beyond-the-hype-separating-ambition-from-reality-in-i4.0.pdf (accessed on 23 October 2022).
19. Davies, R. *Industry 4.0 Digitalisation for Productivity and Growth*; EPRS—European Parliamentary Research Service: Belgium, Brussels, 2015.
20. Cabinet Office-Government of Japan. *Society 5.0*; Cabinet Office: Tokyo, Japan, 2018.
21. Tantawi, K.H.; Fidan, I.; Tantawy, A. Status of smart manufacturing in the United States. In Proceedings of the 2019 IEEE 9th Annual Computing and Communication Workshop and Conference, CCWC 2019, Las Vegas, NV, USA, 7–9 January 2019. [CrossRef]
22. Singapore Smart Industry Readiness EDB. *The Sinapore Smart Industry Readiness Index: Catalysing the Transformation of Manufacturing*; EDB: Singapore, 2017; p. 46.
23. Singapore Smart Industry Readiness EDB. *The Smart Industry Readiness*; EDB: Singapore, 2018; p. 46.
24. Liao, Y.; Deschamps, F.; Loures, E.d.F.R.; Ramos, L.F.P. Past, present and future of Industry 4.0-a systematic literature review and research agenda proposal. *Int. J. Prod. Res.* **2017**, *55*, 3609–3629. [CrossRef]
25. Schroeder, A.; Bigdeli, A.Z.; Zarco, C.G.; Baines, T. Capturing the benefits of industry 4.0: A business network perspective. *Prod. Plan. Control* **2019**, *30*, 1305–1321. [CrossRef]
26. Zhou, K.; Liu, T.; Zhou, L. Industry 4.0: Towards future industrial opportunities and challenges. In Proceedings of the 2015 12th International Conference on Fuzzy Systems and Knowledge Discovery (FSKD), Zhangjiajie, China, 15–17 August 2015; pp. 2147–2152.
27. Nicole, R. *Connected, Intelligent, Automated: The Definitive Guide to Digital Transformation and Quality 4.0.*; Quality Press: Welshpool, WA, USA, 2020.
28. Reischauer, G. Industry 4.0 as policy-driven discourse to institutionalize innovation systems in manufacturing. *Technol. Soc. Chang.* **2018**, *132*, 26–33. [CrossRef]

29. Ghobakhloo, M.; Fathi, M. Corporate survival in Industry 4.0 era: The enabling role of lean-digitized manufacturing. *J. Manuf. Technol. Manag.* **2020**, *31*, 1–30. [CrossRef]

30. Liker, J. *O Modelo Toyota: 14 princípios de Gestão do Maior Fabricante do Mundo*, 1st ed.; Bookman: Porto Alegre, Brasil, 2005.

31. Womack, J.; Jones, D.; Roos, D. *The Machine That Changed the World*, 2007th ed.; Free Press: New York, NY, USA, 1991. [CrossRef]

32. Liker, J.; Ross, K. *The Toyota Way to Service Excellence*, 1st ed.; McGraw Hill: New York, NY, USA, 2017.

33. Ohno, T. *Toyota Production System: Beyond Large Scale Production*; Taylor & Francis Group: Philadelphia, PA, USA, 1988.

34. Salaiz, C. Lean operations at Delphi. *Manuf. Eng.* **2003**, *131*, 97–104.

35. Uhrin, A.; Bruque-Camara, S.; Moyano-Fuentes, J. Lean production, workforce development and operational performance. *Manag. Decis.* **2017**, *55*, 103–118. [CrossRef]

36. Corporation, T.M. Toyota Motor Corpotation. 2020. Available online: https://global.toyota/en/company/vision-and-philosophy/ (accessed on 22 May 2020).

37. Shingo, S.; Dillon, A. *A Study of the Toyota Production System from an Industrial Engineering Viewpoint*; Productivity Press: New York, NY, USA, 1989.

38. Liker, J.K.; Morgan, J.M. The Toyota Way in Services: The Case of Lean Product Development. *Acad. Manag. Perspect.* **2006**, *20*, 5–20. [CrossRef]

39. Liker, J.K.; Hoseus, M. Human Resource development in Toyota culture. *Int. J. Hum. Resour. Dev. Manag.* **2010**, *10*, 34–50. [CrossRef]

40. Industries, T. Toyota Industries Corporation. 2020. Available online: https://www.toyota-industries.com/ (accessed on 22 May 2020).

41. Sá, G. *O Valor das Empresas*, 1st ed.; Rio de Janeiro, Brazil, 2001.

42. Küpper, D.; Heidemann, A.; Ströhle, J.; Knizek, C.; Spindelndreier, D. When Lean Meets Industry 4.0 Next Level Operational Excellence. Boston Consulting Group 2022. 2017. Available online: https://www.bcg.com/publications/2017/lean-meets-industry-4.0 (accessed on 9 October 2022).

43. Rüßmann, M.; Lorenz, M.; Gerbert, P.; Waldner, M.; Justus, J.; Engel, P.; Harnisch, M.; Justus, J. Industry 4.0: The Future of Productivity and Growth in Manufacturing Industries. *Boston Consult. Group* **2015**, *9*, 54–89. Available online: https://www.bcg.com/publications/2015/engineered_products_project_business_industry_4_future_productivity_growth_manufacturing_industries (accessed on 9 October 2022).

44. Rossini, M.; Costa, F.; Tortorella, G.; Portioli-Staudacher, A. The interrelation between Industry 4.0 and lean production: An empirical study on European manufacturers. *Int. J. Adv. Manuf. Technol.* **2019**, *102*, 3963–3976. [CrossRef]

45. Shahin, M.; Chen, F.F.; Bouzary, H.; Krishnaiyer, K. Integration of Lean practices and Industry 4.0 technologies: Smart manufacturing for next-generation enterprises. *Int. J. Adv. Manuf. Technol.* **2020**, *107*, 2927–2936. [CrossRef]

46. Vita, R. *Intregation of Industry 4.0 and Lean Manufacturing and the Impact on Organizational Performance*; Universidade do Porto: Porto, Portuga, 2018.

47. De Souza, R.O.; Ferenhof, H.A.; Forcellini, F.A. Industry 4.0 and Industry 5.0 from the Lean Perspective. *Int. J. Manag. Knowl. Learn.* **2022**, *11*. [CrossRef]

48. Mesquita, L.L.; Lizarelli, F.L.; Duarte, S.; Oprime, P.C. Exploring relationships for integrating lean, environmental sustainability and industry 4.0. *Int. J. Lean Six Sigma* **2022**, *13*, 863–896. [CrossRef]

49. Huang, Z.; Jowers, C.; Kent, D.; Dehghan-Manshadi, A.; Dargusch, M.S. The implementation of Industry 4.0 in manufacturing: From lean manufacturing to product design. *Int. J. Adv. Manuf. Technol.* **2022**, *121*, 3351–3367. [CrossRef]

50. Adam, S.; Elangeswaran, C.; Wulfsberg, J.P. Industry 4.0 implies lean manufacturing: Research activities in industry 4.0 function as enablers for lean manufacturing. *J. Ind. Eng. Manag. JIEM* **2016**, *9*, 811–833.

51. Rossi, A.H.G.; Marcondes, G.B.; Pontes, J.; Leitão, P.; Treinta, F.T.; De Resende, L.M.M.; Mosconi, E.; Yoshino, R.T. Lean Tools in the Context of Industry 4.0: Literature Review, Implementation and Trends. *Sustainability* **2022**, *14*, 12295. [CrossRef]

52. Schäfer, M.; Löwer, M. Ecodesign—A Review of Reviews. *Sustainability* **2021**, *13*, 315. [CrossRef]

53. Deblois, S.; Lepanto, L. Lean and Six Sigma in acute care: A systematic review of reviews. *Int. J. Health Care Qual. Assur.* **2016**, *29*, 192–208. [CrossRef]

54. Tailise, M.M.; Mergulhão, R.C.; Mano, A.P.; Silva, A.A.A. The integration of technologies Industry 4.0 technology and Lean Manufacturing: A systematic literature review. In Proceedings of the 5th International Conference on Quality Engineering and Management, Braga, Portugal, 14–15 July 2022; pp. 647–661.

55. Rajaba, S.; Afy-Shararaha, M.; Salonitisa, K. *Using Industry 4.0 Capabilities for Identifying and Eliminating Lean Wastes*; Elsevier: Amsterdam, The Netherlands, 2022. [CrossRef]

56. Lucantoni, L.; Antomarioni, S.; Ciarapica, F.E.; Bevilacqua, M. Implementation of Industry 4.0 Techniques in Lean Production Technology: A Literature Review. *Manag. Prod. Eng. Rev.* **2022**, *13*, 83–93. [CrossRef]

57. Nedjwa, E.; Bertrand, R.; Souad, S.B. Impacts of Industry 4.0 technologies on Lean management tools: A bibliometric analysis. *Int. J. Interact. Des. Manuf. IJIDeM* **2008**, *16*, 135–150. [CrossRef]

58. Yürekli, S.; Schulz, C. Compatibility, opportunities and challenges in the combination of Industry 4.0 and Lean Production. *Logist. Res.* **2022**, *15*, 14. [CrossRef]

59. Komkowski, T.; Antony, J.; Garza-Reyes, J.A.; Tortorella, G.L.; Pongboonchai-Empl, T. The Integration of Industry 4.0 and Lean Management: A systematic Review and Constituting Elements Perspective. *Total Qual. Manag. Bus. Excell.* **2022**. [CrossRef]

60. Yadav, N.; Shankar, R.; Singh, S.P. Hierarchy of Critical Success Factors (CSF) for Lean Six Sigma (LSS) in Quality 4.0. *Int. J. Glob. Bus. Compet.* **2021**, *16*, 1–14. [CrossRef]

61. Shin, W.S.; Dahlgaard, J.J.; Dahlgaard-Park, S.M.; Kim, M.G. A Quality Scorecard for the era of Industry 4.0. *Total Qual. Manag. Bus. Excell.* **2018**, *29*, 959–976. [CrossRef]

62. Tortorella, G.L.; da Silva, E.F.; Vargas, D.B. An empirical analysis of Total Quality Management and Total Productive Maintenance in Industry 4.0. In Proceedings of the International Conference on Industrial Engineering and Operations Management, Pretoria/Johannesburg, South Africa, 29 October–1 November 2018; Volume 2018, pp. 742–753.

63. Romero, D.; Gaiardelli, P.; Powell, D.; Wuest, T.; Thürer, M. Total Quality Management and Quality Circles in the Digital Lean Manufacturing World. *IFIP Adv. Inf. Commun. Technol.* **2019**, *566*, 3–11. [CrossRef]

64. Miloslavskaya, N.; Tolstoy, A. Big Data, Fast Data and Data Lake Concepts. *Procedia Comput. Sci.* **2016**, *88*, 300–305. [CrossRef]

65. Kibria, M.G.; Nguyen, K.; Villardi, G.P.; Zhao, O.; Ishizu, K.; Kojima, F. Big Data Analytics, Machine Learning, and Artificial Intelligence in Next-Generation Wireless Networks. *IEEE Access* **2018**, *6*, 32328–32338. [CrossRef]

66. Armani, C.G.; de Oliveira, K.F.; Munhoz, I.P.; Akkari, A.C.S. Proposal and application of a framework to measure the degree of maturity in Quality 4.0: A multiple case study. In *Advances in Mathematics for Industry 4.0*; Elsevier: Amsterdam, The Netherlands, 2021; pp. 131–163. [CrossRef]

67. Danjou, C.; Le Duigou, J.; Eynard, B. Closed-loop manufacturing process based on STEP-NC. *Int. J. Interact. Des. Manuf.* **2017**, *11*, 233–245. [CrossRef]

68. Saif, Y.; Yusof, Y. Integration models for closed loop inspection based on step-nc standard. *J. Phys. Conf. Ser.* **2019**, *1150*, 012014. [CrossRef]

69. Jovanović, S.Z.; Đurić, J.S.; Šibalija, T.V. Robotic Process Automation: Overview and Opportunities. *Int. J. Adv. Qual.* **2018**, *46*, 34–39.

70. Mendling, J.; Decker, G.; Reijers, H.A.; Hull, R.; Weber, I. How do machine learning, robotic process automation, and blockchains affect the human factor in business process management? *Commun. Assoc. Inf. Syst.* **2018**, *43*, 297–320. [CrossRef]

71. Teja, Y.R. The RPA and AI automation. *Int. J. Creat. Res. Thoughts IJCRT* **2018**, *6*, 2320–2882.

72. Lee, J.; Bagheri, B.; Kao, H.-A. Recent Advances and Trends of Cyber-Physical Systems and Big Data Analytics in Industrial Informatics. In Proceedings of the Int. Conference on Industrial Informatics (INDIN) 2014, Porto Alegre, Brazil, 27–30 July 2014. [CrossRef]

73. Gorecky, D.; Schmitt, M.; Loskyll, M.; Zühlke, D. Human-Computer-Interaction in the Industry 4.0 Era. In Proceedings of the 12th IEEE International Conference on Industrial Informatics (INDIN), Porto Alegre, RS, Brazil, 27–30 July 2014; pp. 289–294.

74. Kadir, B.A.; Broberg, O.; Da Conceição, C.S. Designing human-robot collaborations in industry 4.0: Explorative case studies. *Proc. Int. Des. Conf. Des.* **2018**, *2*, 601–610. [CrossRef]

75. Djuric, A.M.; Rickli, J.L.; Urbanic, R.J. A Framework for Collaborative Robot (CoBot) Integration in Advanced Manufacturing Systems. *SAE Int. J. Mater. Manuf.* **2016**, *9*, 457–464. [CrossRef]

76. Dalenogare, L.S.; Benitez, G.B.; Ayala, N.F.; Frank, A.G. The expected contribution of Industry 4.0 technologies for industrial performance. *Int. J. Prod. Econ.* **2018**, *204*, 383–394. [CrossRef]

77. Radziwill, N. Let's Get Digital: The many ways the fourth industrial revolution is reshaping the way we think about quality. *arXiv* **2018**, arXiv:1810.07829. [CrossRef]

78. de Souza, F.F.; Corsi, A.; Pagani, R.N. Total quality management 4.0: Adapting quality management to Industry 4.0. *TQM J.* **2021**, *34*, 749–769. [CrossRef]

79. Amani, P.; Lindbom, I.; Sundström, B.; Östergren, K. Green-Lean Synergy-Root-Cause Analysis in Food Waste Prevention. *Int. J. Food Syst. Dyn.* **2015**, *6*, 99–109. [CrossRef]

80. Boye, J.I.; Arcand, Y. Current Trends in Green Technologies in Food Production and Processing. *Food Eng. Rev.* **2013**, *5*, 1–17. [CrossRef]

81. van der Goot, A.J.; Pelgrom, P.J.; Berghout, J.A.; Geerts, M.E.; Jankowiak, L.; Hardt, N.A.; Keijer, J.; Schutyser, M.A.; Nikiforidis, C.V.; Boom, R.M. Concepts for further sustainable production of foods. *J. Food Eng.* **2016**, *168*, 42–51. [CrossRef]

82. Jagtap, S.; Bhatt, C.; Thik, J.; Rahimifard, S. Monitoring potato waste in food manufacturing using image processing and internet of things approach. *Sustainability* **2019**, *11*, 3173. [CrossRef]

83. Pejić, V.; Lerher, T.; Jereb, B.; Lisec, A. Lean and Green Paradigms in Logistics: Review of Published Research. *Promet Traffic –Traffico.* **2016**, *28*, 593–603. [CrossRef]

84. Rodrigues, V.S.; Kumar, M. The Management of Operations Synergies and misalignments in lean and green practices: A logistics industry. *Prod. Plan. Control.* **2019**, *30*, 369–384. [CrossRef]

85. Salzmann, O.; Ionescu-Somers, A.M.; Steger, U. The business case for corporate sustainability: Literature review and research options. *Eur. Manag. J.* **2005**, *23*, 27–36. [CrossRef]

86. Abdel-Basset, M.; Mohamed, R.; Zaied, A.E.N.H.; Smarandache, F. A hybrid plithogenic decision-making approach with quality function deployment for selecting supply chain sustainability metrics. *Symmetry* **2019**, *11*, 903. [CrossRef]

87. Jody, G.; Serafeim, G. Research on corporate sustainability: Review and directions for future research. *Found. Trends®Account.* **2020**, *14*, 7–127.

88. Abson, D.J.; Fischer, J.; Leventon, J.; Newig, J.; Schomerus, T.; Vilsmaier, U.; von Wehrden, H.; Abernethy, P.; Ives, C.D.; Jager, N.W.; et al. Leverage points for sustainability transformation. *Ambio* **2017**, *46*, 30–39. [CrossRef]

89. Tortorella, G.L.; Fettermann, D. Implementation of Industry 4.0 and lean production in Brazilian manufacturing companies. *Int. J. Prod. Res.* **2018**, *56*, 2975–2987. [CrossRef]
90. Reyes, J.; Mula, J.; Díaz-Madroñero, M.; Iaz-Madro, M.D. Development of a conceptual model for lean supply chain planning in industry 4.0: Multidimensional analysis for operations management. *Prod. Plan. Control.* **2021**. [CrossRef]
91. Metaxas, I.N.; Chatzoglou, P.D.; Koulouriotis, D.E. Proposing a new modus operandi for sustainable business excellence: The case of Greek hospitality industry. *Total Qual. Manag. Bus. Excell.* **2019**, *30*, 499–524. [CrossRef]
92. United Nations. The 17 Goals. Department of Economic and Social Affairs-Sustainable Development Report. 2015. Available online: https://sdgs.un.org/goals (accessed on 2 December 2022).
93. Rada, M. Industry 5.0-from Virtual to Physical. 2015. Available online: https://www.linkedin.com/pulse/industry-50-from-virtual-physical-michael-rada/ (accessed on 16 March 2023).
94. Shiroishi, Y.; Uchiyama, K.; Suzuki, N. Society 5.0: For Human Security and Well-Being. *Computer* **2018**, *51*, 91–95. [CrossRef]
95. Pereira, A.G.; Lima, T.M.; Santos, F.C. *Society 5.0 As a Result of the Technological Evolution: Historical Approach*; Springer: Berlin/Heidelberg, Germany, 2020; Volume 1018. [CrossRef]
96. Rada, M. Industry 5.0 Definition. 2017. Available online: https://www.linkedin.com/pulse/industrial-upcycling-definition-michael-rada/ (accessed on 16 March 2023).
97. Vollmer, M. What Is Industry 5.0? 2018. Available online: https://www.linkedin.com/pulse/what-industry-50-dr-marcell-vollmer/ (accessed on 16 March 2023).
98. Breque, M.; De Nul, L.; Petridis, A. *Industry 5.0-Towards a Sustainable, Human-Centric and Resilient European Industry*; Publications Office of the European Union: Brussels, Belgium, 2021. [CrossRef]
99. European Commission. Industry 5.0-Human-Centric, Sustainable and Resilient. 2021. Available online: https://research-and-innovation.ec.europa.eu/research-area/industrial-research-and-innovation/industry-50_en#what-is-industry-50 (accessed on 2 December 2022).
100. Ghobakhloo, M.; Iranmanesh, M.; Mubarak, M.F.; Mubarik, M.; Rejeb, A.; Nilashi, M. Identifying industry 5.0 contributions to sustainable development: A strategy roadmap for delivering sustainability values. *Sustain. Prod. Consum.* **2022**, *33*, 716–737. [CrossRef]
101. Mourtzis, D.; Angelopoulos, J.; Panopoulos, N. A Literature Review of the Challenges and Opportunities of the Transition from Industry 4.0 to Society 5.0. *Energies* **2022**, *15*, 6276. [CrossRef]
102. Madsen, D.Ø.; Berg, T. An Exploratory Bibliometric Analysis of the Birth and Emergence of Industry 5.0. *Appl. Syst. Innov.* **2021**, *4*, 87. [CrossRef]
103. Xu, X.; Lu, Y.; Vogel-Heuser, B.; Wang, L. Industry 4.0 and Industry 5.0-Inception, conception and perception. *J. Manuf. Syst.* **2021**, *61*, 530–535. [CrossRef]
104. Ivanov, D. The Industry 5.0 framework: Viability-based integration of the resilience, sustainability, and human-centricity perspectives. *Int. J. Prod. Res.* **2022**, *61*, 1683–1695. [CrossRef]
105. Lu, Y.; Zheng, H.; Chand, S.; Xia, W.; Liu, Z.; Xu, X.; Wang, L.; Qin, Z.; Bao, J. Outlook on human-centric manufacturing towards Industry 5.0. *J. Manuf. Syst.* **2022**, *62*, 612–627. [CrossRef]
106. Mehendale, R.; Radin, J. Welcome to the virtual age: Industrial 5.0 is changing the future of work. *Delloite*, 11 June 2020.
107. Raza, T.; Shehzad, M.; Abbas, M.; Eash, N.S.; Jatav, H.S.; Sillanpaa, M.; Flynn, T. Impact assessment of COVID-19 global pandemic on water, environment, and humans. *Environ. Adv.* **2023**, *11*, 15. [CrossRef]
108. Mladineo, M.; Ćubić, M.; Gjeldum, N.; Žižić, M.C. Human-centric approach of the Lean management as an enabler of Industry 5.0 in SMEs. *Mech. Technol. Struct. Mater.* **2021**, *2021*, 111–117.
109. Dieste, M.; Panizzolo, R.; Garza-Reyes, J.A.; Anosike, A. The relationship between lean and environmental performance: Practices and measures. *J. Clean. Prod.* **2019**, *224*, 120–131. [CrossRef]
110. Simeão, I.; Ferreira, K.A. Lean construction and resilience while coping with the COVID-19 pandemic: An analysis of construction companies in Brazil. *Int. J. Lean Six Sigma* **2022**, 2040–4166. [CrossRef]
111. Birkie, S.E. Operational resilience and lean: In search of synergies and trade offs. *J. Manuf. Technol. Manag.* **2016**, *27*, 185–207. [CrossRef]
112. Domingues, J.P.T.; Correia, F.D.; Uzdurum, I.; Sampaio, P. The Profile of Forthcoming Quality Leaders: An Exploratory Factor Analysis. In Proceedings of the IEEE International Conference on Industrial Engineering and Engineering Management, Macao, China, 15–18 December 2019; pp. 1606–1610. [CrossRef]
113. Pedro, C.; Bessa, C.; Landeck, J.; Silva, C. Industry 5.0: The Arising of a Concept. *Procedia Comput. Sci.* **2023**, *217*, 1137–1144.
114. Tortorella, G.L.; Narayanamurthy, G.; Thurer, M. Identifying pathways to a high-performing lean automation implementation: An empirical study in the manufacturing industry. *Int. J. Prod. Econ.* **2021**, *231*, 107918. [CrossRef]
115. Antoniou, F.; Lopes, J.P.; Marinelli, M. Human–Robot Collaboration and Lean Waste Elimination: Conceptual Analogies and Practical Synergies in Industrialized Construction. *Buildings* **2022**, *12*, 2057. [CrossRef]

Article

Lean Manufacturing in Industry 4.0: A Smart and Sustainable Manufacturing System

Benedictus Rahardjo [1], **Fu-Kwun Wang** [1,*], **Ruey-Huei Yeh** [1] and **Yu-Ping Chen** [2]

[1] Department of Industrial Management, National Taiwan University of Science and Technology, Taipei City 10607, Taiwan
[2] Arit Machinery Co., Ltd., New Taipei City 24143, Taiwan
* Correspondence: fukwun@mail.ntust.edu.tw

Abstract: Background: Exploring the impact of combining Industry 4.0 technologies and Lean Manufacturing tools on organizational performance has been a popular topic in recent years. Design/Methodology/Approach: We propose a novel Smart and Sustainable Manufacturing System (SSMS) to provide management insights related to social impact, economic performance, and environmental impact. Some tools called Dynamic Lean 4.0 tools, such as Sustainable Value Steam Mapping (VSM), Extended Single Minute Exchange of Die (SMED), and Digital Poka-Yoke, are presented as outputs of synergistic relationships that optimize production processes. Originality/Research gap: There are few studies on the application of SSMS. This work presents a case study, aiming to fill this gap. A case study of vacuum degassing equipment fabrication is presented to demonstrate the improvement of utilizing the Define-Measure-Analyze-Improve-Control (DMAIC) method with Digital Poka-Yoke. Key statistical results: The implementation of this project increased the process capability index, Cpk, from 1.278 to 2. Practical Implications: It was concluded that the company successfully implemented a smart and sustainable manufacturing system, and created a safer working environment and new job opportunities, while increasing production yield from 99.44% to 100%, improving worker utilization, and directly saving NT$68,000. Limitations of the investigation: This paper is the use of a single case study. More applications of Dynamic Lean 4.0 tools in SSMS should be explored.

Keywords: industry 4.0; smart and sustainable manufacturing system; dynamic lean 4.0 tools

Citation: Rahardjo, B.; Wang, F.-K.; Yeh, R.-H.; Chen, Y.-P. Lean Manufacturing in Industry 4.0: A Smart and Sustainable Manufacturing System. *Machines* **2023**, *11*, 72. https://doi.org/10.3390/machines11010072

Academic Editors: Luís Pinto Ferreira, Paulo Ávila, João Bastos, Francisco J. G. Silva, José Carlos Sá, Marlene Brito and Mosè Gallo

Received: 22 November 2022
Revised: 18 December 2022
Accepted: 3 January 2023
Published: 6 January 2023

1. Introduction

Lean Manufacturing (LM) refers to a set of management techniques and principles designed to eliminate waste and simplify activities that add value to products from a customer perspective. By optimizing process steps and eliminating waste, only useful value is added at each production stage. LM principles are now applied across industries, and these tools are recognized worldwide as a successful operational framework for reducing waste, increasing productivity, and continuously improving organizations [1]. Due to LM, waste and costs are minimized, thereby increasing productivity and profit [2]. From an economic point of view, LM reduces waste, which in turn increases market share and profits [3]. In terms of social impact, LM improves occupational health and safety, thereby improving the quality of life in society [4]. The LM concept has been shown to be critical to achieving sustainability in any organization [5]. On the environmental front, LM can lead to lower levels of pollutants and slower resource consumption due to improved products and reduced storage materials. Lean culture has shown a significant impact on the sustainability performance of different industries, including plastics manufacturing, and lean culture has a moderating effect [6]. Several industries have benefited from the implementation of LM and resulted in better organizational performance [7–9].

Industry 4.0 offers organizations the opportunity to improve their current manufacturing operations and practices to a more advanced level by leveraging emerging technologies.

As the global market continues to evolve, manufacturing systems become smarter, more flexible, digital, agile, and are able to keep pace with market volatility [10]. Industry 4.0 drives smart manufacturing, enabling manufactures to maximize output from existing production capacities and develop next generation production capabilities necessary to compete in the digital economy. Manufacturers around the world are experiencing dramatic changes by leveraging Industry 4.0 technologies, such as cloud computing, big data analytics, robotics, and the Internet of Things (IoT). Industry 4.0 technologies have a significant impact on the sustainability performance of the manufacturing industry as they produce better machines, enhanced communications, improved working conditions, and product quality [11]. IoT, sensors, and big data can enhance the environmental, social, and economic aspects of the Brazilian plastics industry [12]. Through digitalization, Industry 4.0 technologies can reduce production and transportation costs and lead times, thereby increasing customer satisfaction and organization's profits [13]. With respect to the environmental aspect, data sharing among supply chain stakeholders, and the availability of real-time data facilitate the efficient allocation of raw materials, water, energy, and labor time, thereby reducing resource consumption and waste generation [14]. As a social concept, Industry 4.0 offers people with new technologies to improve motivation and morale by providing safe working conditions [15].

The importance of combining LM and Industry 4.0 has become as a hot topic [16]. Integrating LM into Industry 4.0 requires more research, although it is considered an enabler of Industry 4.0 or a pre-requisite for its introduction [17]. A systematic review of the literature found that they have many symbioses and synergies [18]. Organizations are expected to benefit greatly from Industry 4.0 technologies and LM together. Their combination can also reduce waste and cost in areas where LM alone is not feasible [19]. Additionally, Industry 4.0 technologies are more expensive to implement without LM principles, so further integration of them is expected to reduce implementation costs. Although these two factors are important in different manufacturing industries, few studies have examined their relationship [20]. Furthermore, there are few real case studies on LM and Industry 4.0 applications and their impact on organizational performance [21]. Organizations can benefit from new innovative and automated manufacturing techniques, namely lean digital transformation [22].

The work involved in LM and Industry 4.0 can be divided into two categories: (i) conceptual and theoretical discussions [11]; and (ii) application-oriented use case study [23]. This paper covers both categories. First, there is a lack of research on the combined impact of Industry 4.0 and LM principles on corporate performance. In this regard, the novelty of this study lies in the in-depth exploration of possible relationships between Industry 4.0 technologies and useful LM tools. We propose a Smart and Sustainable Manufacturing System (SSMS) with Dynamic Lean 4.0 tools as the outputs of synergistic relationships for optimized production processes. Second, as a validation of the proposed framework and integrated tools, a case study of vacuum degassing equipment fabrication is presented. The need to improve process capability and production yield became the background for the selection of the case study. The rest of the paper is organized as follows: the related work on lean tools, followed by Industry 4.0 technologies. In Section 3, we describe a SSMS framework and some Dynamic Lean 4.0 tools. Section 4 presents the implementation of the proposed method for a real case study. The final section draws the conclusion.

2. Related Work

2.1. Lean Manufacturing Tools

LM aims at producing products and services at the lowest cost and as fast as required by the customer. Various lean tools effectively eliminate the organization's waste. In the context of Industry 4.0, Value Stream Mapping (VSM) 4.0 is recently developed as collaborative value stream tool for lean management. By utilizing VSM 4.0, companies can map flow components and process boundaries, and plan full-scale implementation digitally [24]. Single Minute Exchange of Die (SMED) describes the activities necessary to prepare a production line for manufacturing a product, while setup time refers to the time spent between the end of a previous process and the beginning of the next process [25]. Poka-Yoke devices prevent errors from occurring or make them apparent [26]. The Poka-Yoke system can be used to identify any errors, prevent them from moving to the next process, and manage the identification of the causes of any errors occurring [27].

2.2. Industry 4.0 Technologies

Industry 4.0 refers to a new stage of industrialization in which companies can achieve greater industrial performance by integrating vertical and horizontal manufacturing processes. A key feature of Industry 4.0 is to enable regular machines to become self-aware and self-learning so they can perform better and monitor their maintenance more efficiently. Industry 4.0 mainly focuses on real-time monitoring of data, tracking the status and location of products, and controlling production processes [28]. A number of digital technologies have become enablers of Industry 4.0. The use of autonomous robots allows for more precise execution of autonomous production methods and operations where the work of human workers is limited [29]. In the context of Industry 4.0, simulation reaches the next frontier, known as the digital twin, which is a virtual representation of a physical object using digital data. To create higher quality products, digital simulation tools can enable faster, more flexible and efficient processes by integrating with production systems. Using sensory data in digital simulation can improve the efficiency of production planning and execution by increasing the credibility of production system [30]. IoT includes the dynamic management of complex systems through the real-time interaction of people, machines, objects, and information and communication technology systems. The growth of cyber-physical systems within Industry 4.0 means that a cybersecurity market is emerging [31]. A key component of Industry 4.0 is additive manufacturing, as it reduces the complexity of manufacturing and saves time and money. This allows rapid prototyping and highly decentralized production processes [32]. In a digitalized and easily comprehensible manner, Augmented Reality (AR) provides remote maintenance support through numerous applications for technical knowledge dissemination [33]. An important aspect of Industry 4.0 is the application of big data, which can examine enormous quantities of information to discover hidden patterns and correlations [34].

3. Smart and Sustainable Manufacturing System

This section presents a novel SSMS framework and provides managerial insights about organizational performance with social impact, economic performance, and environmental impact. The SSMS framework adopted the concept of the Define-Measure-Analyze-Improve-Control (DMAIC) methodology and some tools called Dynamic Lean 4.0 tools as the outputs of synergistic relationships for optimized production processes.

3.1. SSMS Framework

In the Industry 4.0 digital environment, smart factories play an important role in machines and devices can automate and optimize processes. Real-time data and connections between machines support automated and analytical manufacturing. By combining powerful, in-depth and accurate analytics, manufacturers will be able to unlock new possibilities and business advantages by leveraging the autonomous smart manufacturing capabilities of Industry 4.0.

From an economic point of view, smart factories have a direct impact on production, services, and final products. The concept of the smart factory involves the use of smart devices, such as smart machines, robots and workpieces, that communicate in a continuous manner during the production process. Integrated production optimizes operations through self-organization and adaptation, resulting in smart products that can be easily tracked using Radio-Frequency Identification (RFID) tags. In addition, smart production ensures customer satisfaction by producing high-quality products, custom designs, and short production lead times and lead times.

From a societal perspective, smart factories affect a wide range of issues, including the job market, worker safety, and labor laws. Using the smart factory model, the company is able to automate and optimize its operations. With the development of smart devices, certain tasks and functions are being replaced by smart devices, thereby changing the job market and reducing the number of low-skilled workers. Additionally, as smart devices become more commonplace, more skilled workers are required to operate them. When factories become smart factories, workers will face another job hazard. Therefore, it is important to improve worker safety at work and ensure compliance with human rights while reducing workplace risks.

Sustainable manufacturing is the responsibility of the manufacturer, both in terms of operations and product design. As a result of industrial operations, it is clear that industry contributes to the overuse of the earth's resources. In addition, pollution, excessive waste, and improper disposal can lead to further environmental damage. When the industry is under pressure to reduce its environmental footprint, ignoring environmental concerns represents a huge risk. As climate change and environmental concerns increase, the need for sustainable manufacturing will increase. Manufacturers must make meaningful changes to meet local and government needs. As a result of these changes, people are turning to smart factories, and digital transformation is becoming more common.

Creating value in manufacturing is enabled by smart and sustainable operations that reduce waste and improve the environment. Enabling lean techniques in Industry 4.0 helps organizations achieve more smart and sustainable operations with data-driven decision making and optimizing manufacturing and operations, as depicted in Figure 1.

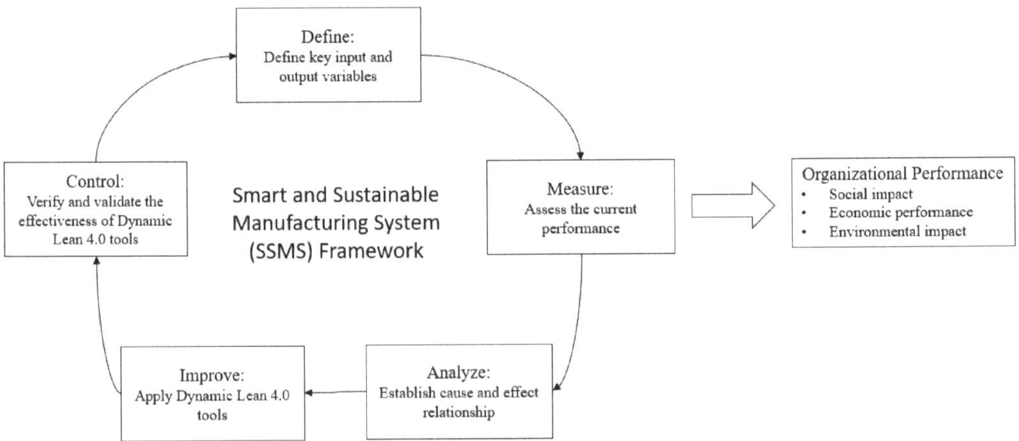

Figure 1. Smart and sustainable manufacturing system framework.

3.2. DMAIC Methodology on SSMS Framework

A standard improvement model, such as DMAIC, is very helpful for deploying improvements in a company because it provides an improvement roadmap. Project definition leads to the identification of key process characteristics and the benchmarking of these characteristics during the measure and analyze phase. In the improve phase, a process is transformed for better performance, while in the control phase, gains are monitored and maintained. Therefore, we propose a DMAIC model based on SSMS framework and Dynamic Lean 4.0 tools to improve organizational performance. The DMAIC model consists of the following basic elements and key tools.

During the define phase, the goal is to identify opportunities for the project and verify or confirm their viability. The project must have a significant impact on both the customer and the business. Potential uses of the project require consent from stakeholders and downstream customers. The define phase can also be aided by graphical aids, such as process maps and SIPOC diagrams. A flowchart describes an overview of a company's approach to meeting customer requirements. The SIPOC diagram consists of Supplier, Input, Process, Output, and Customer, and is used to refine the project scope and boundaries. It can help visualize and understand the essential elements of a process: Supplier—an entity that provides any information, material, or item processes in a process. Input—information or material provided. Process—the steps required to transform an input into an output. Output—the product or service that is sent to the customer, considering the critical quality attributes. Customer—the next step or end customer of the business.

The measure phase evaluates and understands the current process state. It involves gathering information on quality, cost, and timeliness from various sources. In order to be able to analyze and understand current process performance in relation to key metrics, sufficient data is required. Observational studies are often necessary to collect current data. Process data can be collected over a period of time using continuous data collection or sampling methods. The collected data is used to determine the baseline performance of the current process. A run chart can be used to display observed data in chronological order. In addition, the capability of the process should be assessed. This can be done using process capability analysis, such as Process Capability Ratio (PCR), to measure the actual capability of the process.

Data collected during the measure phase is used in the analyze phase to begin to identify causal relationships in the process and understand the different sources of variability. Specifically, the analyze phase aims to identify potential causes of defects, quality issues, customer issues, cycle time, and throughput issues, waste, and inefficiencies leading to the project. Various tools are available during the analyze phase, such as Five Whys and Quality Function Development (QFD). Five Whys is a problem-solving approach that explores the underlying cause-and-effect relationships of a particular problem. QFD is a structured method for defining customer requirements and translating design specifications or product control characteristics in the form of a planning matrix. The House of Quality (HOQ) matrix involves collecting and analyzing the voice of customer to define the relationship between customer needs and product or company capabilities.

The improve phase should identify specific opportunities and root causes for improvement. As part of the improve phase, improvement strategies are developed and tested in practice. At this phase, think creatively about what can be changed and how the performance of the process can be improved to achieve the desired effect. Digital transformation is essential for manufacturers to evolve into smart and sustainable businesses. We introduce some novel tools such as an integration of lean tools and Industry 4.0 technologies, called Dynamic Lean 4.0, including Sustainable VSM, Extended SMED, and Digital Poka-Yoke. The three steps to apply Dynamic Lean 4.0 tools are as follows:

Step 1: Technology roadmaps describe related technologies to help organizations plan and implement technology development. In addition, it can be used as a forecasting tool for technical trends. By analyzing the current state of technology, customer needs, and expected market entry strategies, technology roadmaps can be used to identify alternative technologies, competitors, and market entry opportunities. Additionally, it can help understand how organizational goals, organizational technology resources, and rapidly changing market conditions relate to each other.

Step 2: Technology classification is conducted by evaluating the technology's main functions. Due to advances in operations and information technology, a well-established sequence of layers has been established. There are four levels, including sensors, manufacturing data acquisition, monitoring and control, and operations management.

Step 3: As shown in Table 1, the application of Dynamic Lean 4.0 tools is expected to help manufacturers solve certain problems. By separating the elements of production that are actively involved in the manufacturing process from those that are passive, Lean 4.0 tools can evolve into an intelligent and adaptable Industry 4.0 production system. An active asset system can be viewed as a dynamic, manageable resource system for digital control, real-time tracking, and processes transparency. Manufacturing cells are transformed into Lean-Industry 4.0 by updating specific active physical components, corresponding digital components and activating Lean 4.0 tools. It allows Lean 4.0 tools to interact with manufacturing cloud databases, customize according to manufacturing scenarios, integrate with applications, or implement different algorithms. The transition from lean to smart, adaptable Industry 4.0 production systems makes Lean 4.0 tools dynamic, flexible and intelligent. Ejsmont et al. [35] pointed out several Industry 4.0 technologies such as digital twin, big data, IoT and how they interact with lean tools. However, not all authors agree on which are the main Industry 4.0 technologies affecting lean [36]. Langlotz et al. [37] emphasized that these digital technologies, when properly integrated into a lean environment, can improve processes of a pulled nature to increase efficiency. Thus, Dynamic Lean 4.0 tools perform the interaction of Industry 4.0 technologies with lean tools. Sustainable VSM enables the management and control of the daily operations of the production line facilitate the integration of the entire value stream in the organization chain and its related dependents. Digital Poka-Yoke uses digital data from digital technologies to facilitate error-proofing processes by integrating employees and Industry 4.0. Combining advanced technology and lean tools, smart and autonomous production systems are possible, and are in line with the dynamic demands of the global economy. Companies will benefit from lower production costs, improved regulatory compliance, and long-term resilience, while better connections to target customers will yield better growth opportunities.

Table 1. Dynamic Lean 4.0 tools and its function.

Industry 4.0 Technologies	Dynamic Lean 4.0 Tools		
	Sustainable VSM	Extended SMED	Digital Poka-Yoke
Digital twin	Provide greater visibility for all stakeholders in a digital twin project by generating a comparison of reliable results for future scenarios.	Analysis of collected data and comparison of improved process.	• Identify faults more accurately and improve manufacturing intelligence. • Enhance the production quality by effectively monitoring equipment in workshops.
Augmented reality	-	• Make simple the complicated elements of the manual changeover. • Improve workers understanding of each step of the changeover process.	Design the parts that suggest to the operator how to assemble the product.
Additive manufacturing	Provide the overview for derivation of improvement measures.	• Produce varying workpieces with minimum setup time. • Omit times for selection, search tools, and work-pieces adjustment.	Get the visually impaired into work by embedding blind-friendly fixtures for assemblies.
System integration	• Strategy development for sustainability of production systems or supply chains. • Broader knowledge concerning the manufacturing line performance indicators.	-	-
Autonomous robots	-	-	Ensure safety can create superior conditions for elimination of human errors through advanced automation.
Big data analytics	Use of operational intelligence to see possible process improvement and performance metrics through data collection and analysis.	Use RFID to recognize each die and know their storage address.	Produce enriched data sets to optimize efficiencies in an automated process, increase productivity and minimize errors by connecting to the traceability system.
Cyber security	-	-	• Detect breach before it occurs. • Explain the data and IP will be locked once a breach is detected.
Internet of things	Smart real-time monitoring IT solution with intelligent aspects concerning lean targets to build an action plan with stakeholders.	Speed up the process become more efficient, reduce human errors, and organize complex system.	Enable immediate control of error-proofing devices.

The goal of the control phase is to complete all remaining work on the project, ensuring that the benefits of the project contribute to the tracking of the process and subsequent improvements. Before and after data on key process indicators should be provided and a validation check after project completion is recommended. Data needs to be collected to compare and demonstrate that improvements achieve better results. Preliminary results must remain stable to continue to have a positive financial impact. Keeping a good list of items is essential to keep the process improving.

4. Case Study

The case company, called Company-T, is located in Taiwan. The company manufactures vacuum degassing equipment (see Figure 2), where its key function is to support in several systems, such as vacuum, ice water, hydraulic, air pressure, and electrical system. This company faces a problem that the storage volume in this equipment does not reach the target, therefore causing an anomaly status. The tube with fault label judged by the machine will be removed, and the storage volume value will be abnormal. A clear roadmap can be achieved through the use of the DMAIC process, which provides a useful framework for running selected projects. An explanation of the DMAIC method for problem solving and the importance of achieving SSMS is provided in this case study. The following explains how DMAIC works in this case study.

Figure 2. Vacuum degassing equipment.

4.1. Define

During the define phase, product design and development process issues are identified. Subsequently, the results of product construction and trial operation were reviewed. As part of this process improvement project, the project team used SIPOC-related activities (see Table 2) to identify all relevant elements. It helps define a complex project that may not have a good scope and provides additional detail. The operation process of heat pipe water injection is shown in Figure 3.

Table 2. SIPOC diagram.

Supplier	Input	Process	Output	Customer
• Customer discussion • Processor • Off-the-shelf manufacturer • Design surface • Power distribution drawing • Design program • Program test	• Designer • Purchasing and processing manufacturer • Purchasing off-the-shelf manufacturer • Assembler • Distribution staff • Programmer • Testers	• Design • Outsourcing manufacturer • Procurement • Assemble • Power distribution • Program • Test	• Heat pipe filling and vacuuming machine	• Vacuum degasser (yield test and improvement)

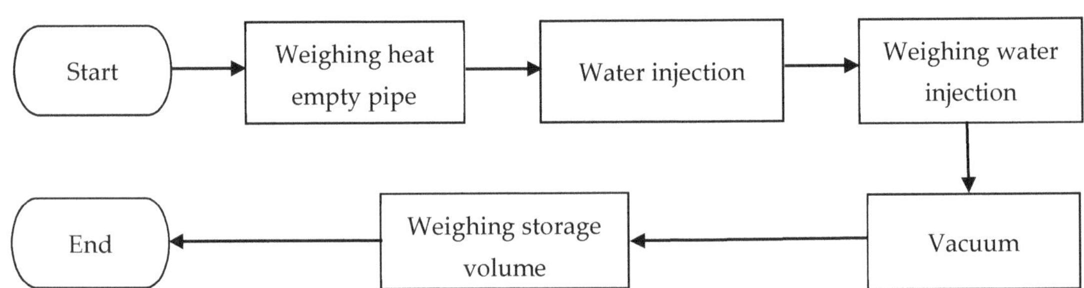

Figure 3. Heat pipe water injection process.

4.2. Measure

At this phase, customer specifications are confirmed and current performance is assessed. A total of 1810 samples were collected, and the production yield was 99.44%. Table 3 provides the initial state of the machine. It showed that the short-term capability PCR-Cpk of MA whole line was 1.278. Therefore, the current process capability is not that good compared to the guideline for minimum value, which is 1.33. Note: Cpk = min{(Process mean − LSL)/3s, (USL-process mean)/3s}, LSL = lower specification limit, and USL = upper specification limit. The storage data for each machine is shown in Figure 4. It shows that the amount of storage is significantly from the target value of 0.82. Therefore, it is impossible to meet the customer's specification requirements for production.

Table 3. The performance of initial state of the machine.

Storage Volume	Ma Whole Line	MA-1	MA-2	MA-3	MA-4
Specification limits	0.82 ± 0.03	0.82 ± 0.03	0.82 ± 0.03	0.82 ± 0.03	0.82 ± 0.03
USL	0.85	0.85	0.85	0.85	0.85
LSL	0.79	0.79	0.79	0.79	0.79
Process means	0.827	0.826	0.824	0.827	0.83
Standard deviation	0.006	0.007	0.007	0.006	0.005
Cpk	1.278	1.143	1.238	1.278	1.333

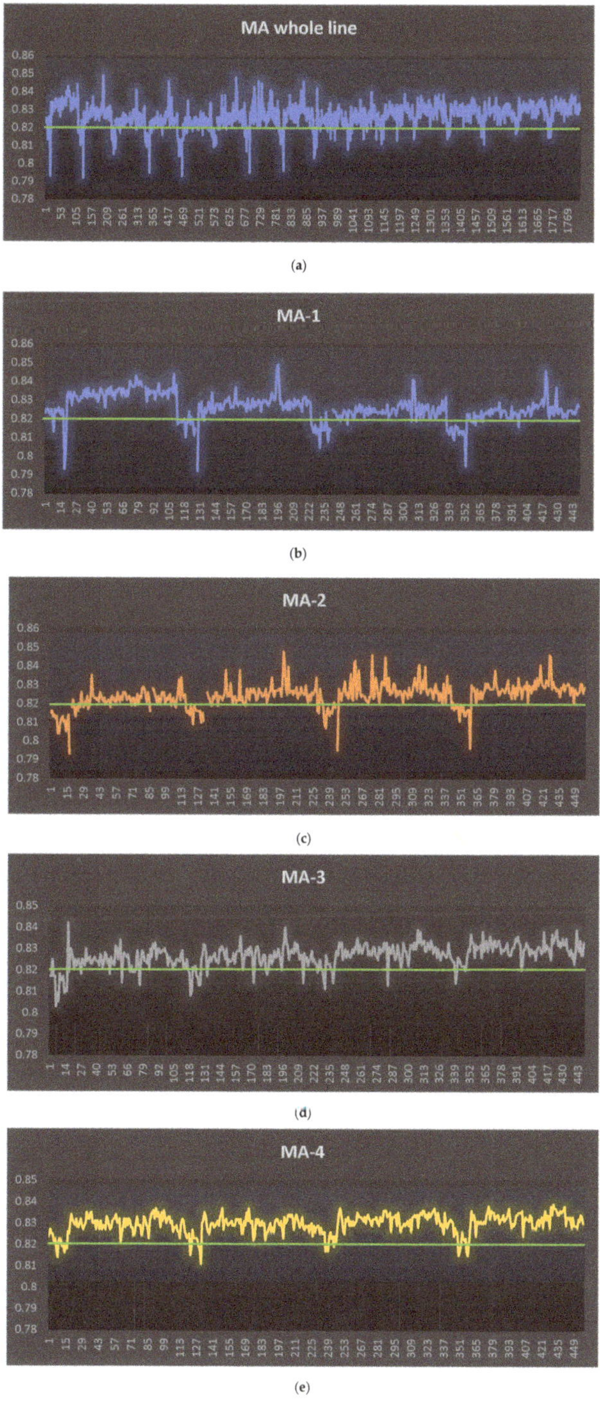

Figure 4. Initial state-run chart of storage volume on machines (**a**) MA whole line, (**b**) MA-1, (**c**) MA-2, (**d**) MA-3, and (**e**) MA-4.

4.3. Analyze

At this phase, the specifications and product characteristics are analyzed. The two steps involved in the phase are as follows:

Step 1: Identify root cause—The root cause of the vacuum system abnormal problem is determined by using the Five Whys tool, as shown in Figure 5.

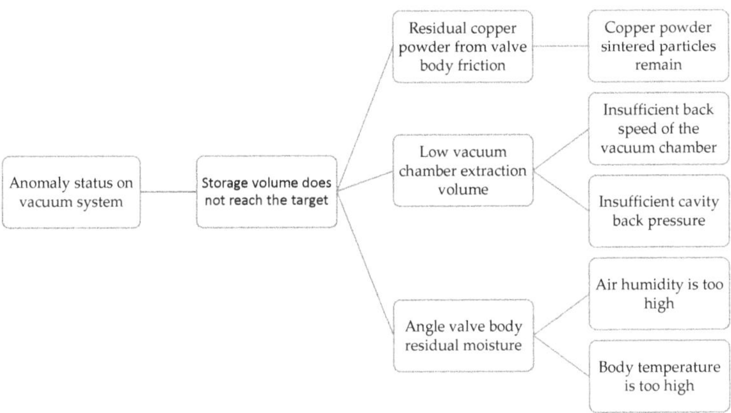

Figure 5. The root cause identification using the Five Whys tool.

Step 2: Planning process—QFD is used to define and translate customer requirements into specific plans to produce a product that meets the voice of the customer. As shown in Figure 6, a summary of customer requirements is described in the house of quality matrix. Among the design factors, vacuum has the largest weight at 35.2%.

Quality factor	Shape size	Weight	Durability	Vacuum	Operability	Design	Safety	Importance	Company-T	Company-A	Company-B	Company-C	Planning quality	Level improvement rate	Sales focus	Absolute weight	% Req. quality weight
1. Air humidity	1	0	0	5	0	0	0	5	4	5	3	4	4	1	1.2	6	21
2. Body temperature	3	0	0	3	0	1	3	5	3	4	3	3	4	1.30	1	6.67	24
3. Sintered copper powder	0	1	5	5	0	3	1	4	4	3	3	3	5	1.30	1.5	7.5	27
4. Back speed	1	1	1	5	3	3	1	3	4	3	3	3	4	1.00	1	3	11
5. Back pressure	0	3	1	3	0	3	5	4	4	4	2	3	4	1.00	1.2	4.8	17
Importance	104	89	162	418	32.2	188	195	1188							Total	28	100
% Weights	8.73	7.5	13.6	35.2	2.71	15.8	16.4										
Important items				1		3	2										
Company-T	4	3	4	5	3	3	5										
Company-A	3	4	4	3	4	3	3										
Company-B	3	3	3	3	2	3	2										
Company-C	3	3	2	3	3	4	3										
Design quality	13	13	13	14	12	13	13										

Independent configuration

Related	USA	Japan	other
Strong	9	4	5
Middle	3	2	3
Weak	1	1	1

Figure 6. House of quality matrix.

4.4. Improve

During the improve phase, we applied a Dynamic Lean 4.0 tools called Digital Poka-Yoke to improve the process performance. The three steps involved in the phase are as follows:

Step 1: Technology roadmap, which is based on the root causes and customer needs identified during the analysis phase, we identified five Industry 4.0 technologies that are a foolproof match. The five technologies are autonomous robotics, digital twins, IoT, big data analytics, and additive manufacturing to reduce process variation.

Step 2: Technology classification, which is based on technical relevance to Poka-Yoke. The technology architecture is divided into four layers: sensor, integration, intelligence, and response (see Figure 7). At the sensor layer, physical and environmental data are monitored locally. Once the sensor layer collects data, it is passed to the integration layer for processing. The intelligence layer uses data analytics and algorithms to make predictions based on aggregated information. The response layer then develops various applications and services on top of the other three layers.

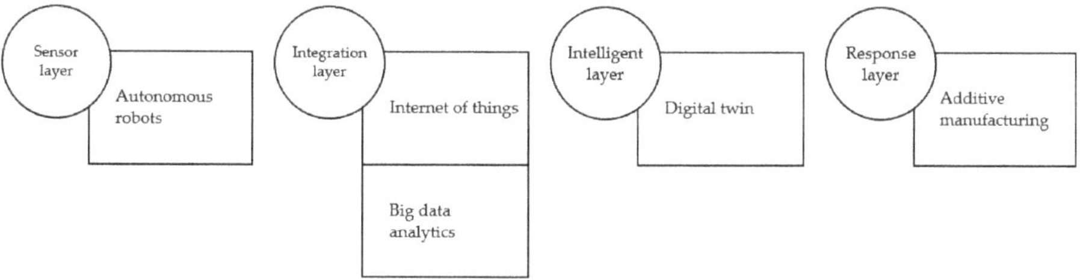

Figure 7. Technology architecture and classification.

Step 3: Digital Poka-Yoke application—Autonomous robots make a significant contribution to increasing productivity and streamlining processes by integrating automated intelligent systems into transportation and handling operations. They facilitate continuous flow and improve operator performance through standardized processes. The role of worker just input the material to the machine, then the robots do all the process operations until vacuum degassing equipment can run well. Using high-volume, interconnected, or even integrated robots with sensors to ensure job safety can eliminate human error through advanced automation that eliminates the need for human intervention. This means the worker has no longer exist on the production floor.

In the context of current applications of Digital Poka-Yoke, this company realizes a manufacturing production system based on digital twins. An innovative concept of digital twin is introduced into equipment fault diagnosis and trend prediction as a means of improving fault identification accuracy and increasing manufacturing intelligence. Production quality can be effectively improved and production efficiency can be increased through the application of the digital twin to error proofing management. An integrated manufacturing workshop management and control system with digital twin technology and IT is designed to optimize manufacturing equipment monitoring so that errors do not occur in production. In addition to the digital twin, data transmission and workshop data collection are used to predict equipment faults. Real-time machine status data is captured by a factory-level digital twin model that enables virtual and real-time interaction feedback. In spite of the fact that error-proofing sensors have been existed, IoT advancements have made them easier to install and maintain. With the advent of IoT, error-proofing relies less on manual mechanisms and more on IoT-enabled sensors interconnected across all devices in the shop floor (see Figure 8). In Figure 8, it indicates that the current process is out of control because the metric is greater than the specification limit. An IoT-enabled sensor can send this false alarm information automatically. Initially, the machine runs under the operator's control. By applying Digital Poka-Yoke, the machine no longer needs to be manually controlled by the operator. The operator does not need to wait and pay attention to the machine status. On the other hand, the operator can do other added-value work to improve utilization. If the measured value is greater than the nominal value, the machine will automatically stop and an alarm will be issued to inform the operator. After that, the operator will check and eliminate the anomaly. Using IoT, error-proofing devices can be controlled immediately. With such immediacy, error-proofing devices are less likely to be bypassed. As the company next improvement in the future, sensors will be able to be adjusted remotely from any location at any time. Through the integration of sensors directly into processes, error detection has been greatly improved. Flexible manufacturing involves Digital Poka-Yoke devices based on specific product versions. Sensors provide true data to the control system or offer remote programming capabilities. Traceability is also integrated into the system, so they know exactly what version of the product is being manufactured. By using digital and wireless industrial communication network, sensors can capture and process enriched data. The data then is used to increase productivity, optimize efficiency and minimize errors in automated processes.

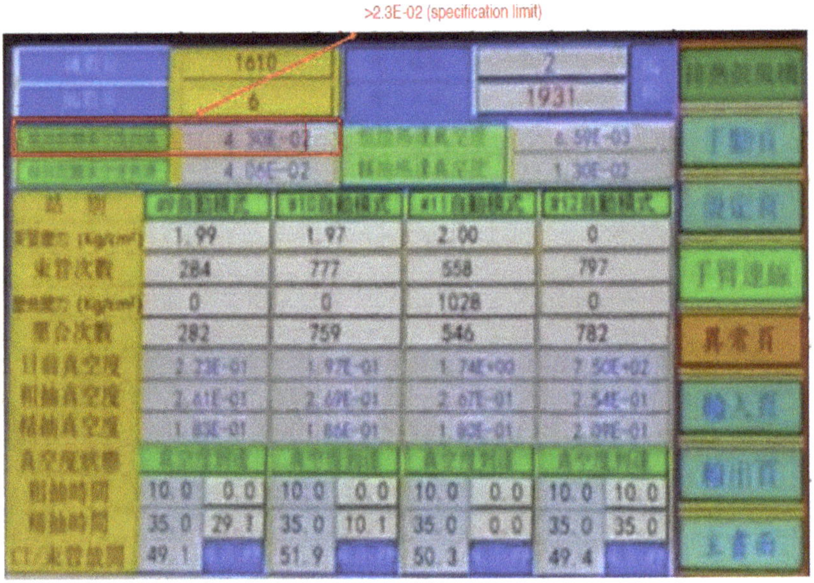

Figure 8. Sensor connected with internet of things technology on the machine interface.

An autonomous inspection has also been implemented in this company. This helps worker to do the monitoring, and in case there is a problem happened, the machine can be shut down automatically. Additive manufacturing process involves the creation of solid three-dimensional objects based on digital models. This company has developed various 3D printing technologies, which the common feature is to create layers by layers of a physical model. Making mistake-proof fixtures for the visually impaired is an innovative application of 3D printing technologies. A Digital Poka-Yoke and 3D printing system is used for blind-friendly fixtures that ensure the correct positioning of the assembly components. As a result, a person with visual impairments is now able to perform jobs that before were not available to them. Additionally, it improves the capability of autonomous inspection.

Figure 9 depicts the results of before and after the improvement of the water vapor residue in vacuum degassing equipment using Digital Poka-Yoke.

 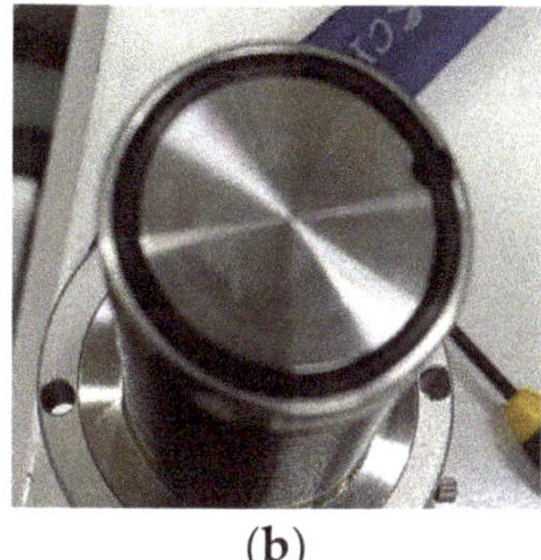

(a)   (b)

Figure 9. Comparison image (**a**) before and (**b**) after the improvement of the water vapor residue.

4.5. Control

During the control phase, the process capability index for new data is measured. After applying Digital Poka-Yoke, project team members collect up-to-date samples to identify new process capabilities and monitor improvements. A total of 1282 samples were collected, and the production yield was 100%. As shown in Figure 10, the process average improved significantly and moved closer to the specification target. The process proven to have a higher capability to meet customer specifications. The results of the analysis showed that the process average of the MA line decreased from 0.827 to 0.820. In addition, the process variation is also significantly reduced, especially for the MA-3 line, from 0.007 to 0.004. Consequently, PCR-Cpk was enhanced for all MA lines (whole and 1–4 lines) and was greater than the recommended minimum value of 1.33 (see Table 4). This means improved processes to meet customer requirements.

Table 4. Improved condition.

Storage Volume	MA Whole Line	MA-1	MA-2	MA-3	MA-4
Specification limits	0.82 ± 0.03	0.82 ± 0.03	0.82 ± 0.03	0.82 ± 0.03	0.82 ± 0.03
USL	0.85	0.85	0.85	0.85	0.85
LSL	0.79	0.79	0.79	0.79	0.79
Process mean	0.82	0.822	0.818	0.818	0.822
Standard deviation	0.005	0.006	0.004	0.005	0.004
Cpk	2	1.556	2.333	1.867	2.333

Furthermore, the adaptation of the Digital Poka-Yoke tool realizes this company achieve the SSMS that can be seen from three aspects. In relation to social impact, the working environment become safer with the presence of autonomous robots. The worker's load has decreased due to autonomous inspection. Additionally, the company has successfully created new job opportunity for visually impaired and involved them into work through the aid of additive manufacturing technology. From the economic perspective, the production yield rate has improved from 99.44% to 100% with the existence of IoT-enabled error-proofing devices. The direct savings of the project amounted to NT$68,000, mainly due to the high worker utilization rate and less rework process. Thus, it leads to high efficiency and increase profitability of this company. Lastly, the high level of operational efficiency has successfully reduced the excessive of material waste. This means that the Company-T contributes to the environmental sustainability.

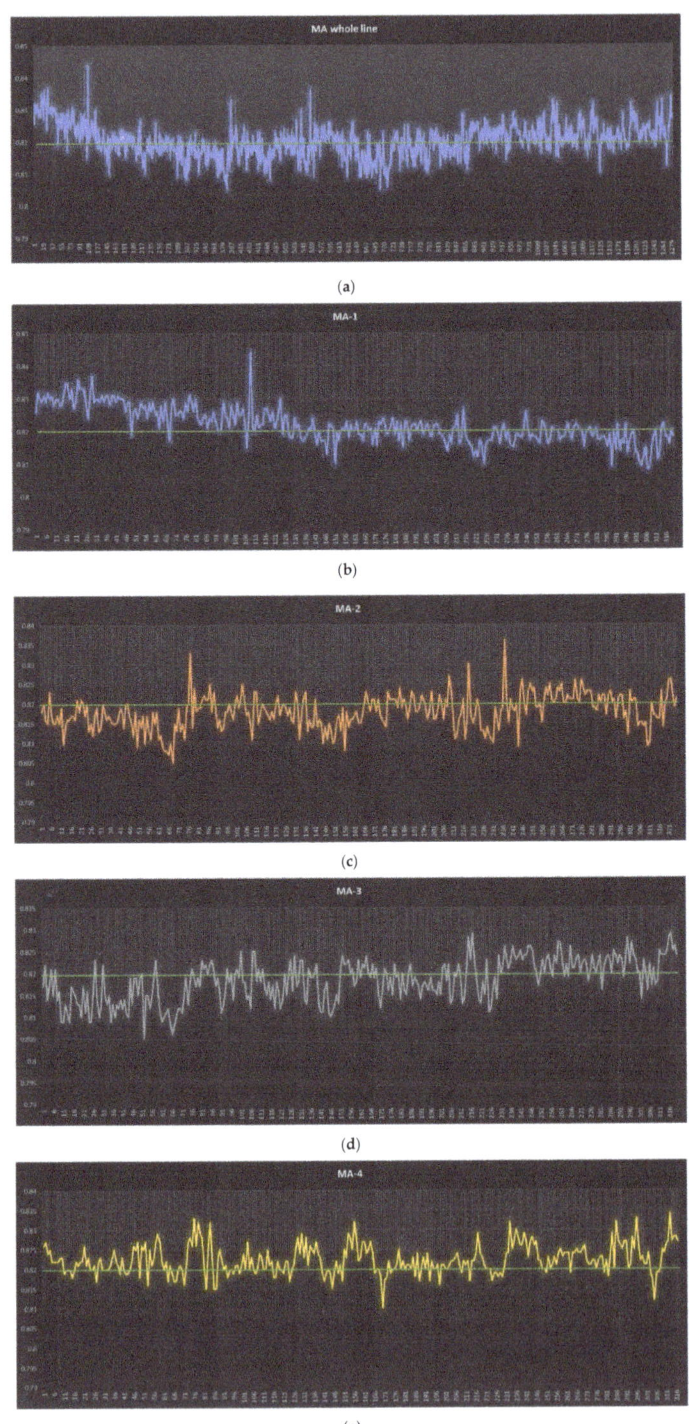

Figure 10. Improved state-run chart of storage volume on machines (**a**) MA whole line, (**b**) MA-1, (**c**) MA-2, (**d**) MA-3, and (**e**) MA-4.

5. Contribution and Practical Implications

This study offers valuable insights for scholars and practitioners. Similar to Antony et al.'s [16] studies on the integration of Lean Six Sigma and Industry 4.0, this paper discusses the development of lean principles in the digital environment of Industry 4.0 for the interest of researchers. More research is needed to understand the relationship of LM to Industry 4.0 technologies [17]. Therefore, this study aims to introduce a possible integration between several lean tools and emerging technologies, called Dynamic Lean 4.0 tools.

Although their relationship is still in its infancy, the literature review [18] found a synergistic need for each other. Cost competitiveness [19] and waste reduction [20] are expected to be the result of Industry 4.0 combined with lean. Furthermore, the impact of LM and Industry 4.0 on organizational performance is rarely investigated in real case studies [21]. Therefore, this paper fills the gap by introducing an SSMS framework that employs the concept of DMAIC methodology for continuous improvement of social impact, economic performance and environmental impact. It is worth noting that DMAIC provides an improvement roadmap. We modified the improvement phase by applying Dynamic Lean 4.0 tools and verify their effectiveness on organizational performance during the Control phase.

From a practical standpoint, this study provides valuable insights for all managers and leaders to better understand the application of lean tools in Industry 4.0. LM have been prominent manufacturing paradigms over the past decades, with the respective aim of reducing waste to achieve low cost in production processes. Industry 4.0 is transforming traditional manufacturing systems into smart ones. The current competitive market requires manufacturing companies to improve cost-efficiency with the application of Industry 4.0 technologies. LM implementation integrated with Industry 4.0 adoption leads to high operational performance improvement. Therefore, we introduce some novel tools, called Sustainable VSM, Extended SMED, and Digital Poka-Yoke, as extensions to popular lean tools combined with proper Industry 4.0 technologies. A case study is presented to verify the effectiveness of Digital Poka-Yoke tool. As a result, this novel tool successfully improves the process capability and production yield, resulting in a smart and sustainable manufacturing system.

6. Conclusions

This work aims to explore the possible relationship between Industry 4.0 technologies and useful LM tools on business performance. We propose a novel SSMS framework and some tools called Dynamic Lean 4.0 tools as an output of synergistic relationships to optimize production processes. In-depth exploration through project planning has been conducted to provide management insights into organizational performance and social, economic, and environmental impact.

In addition, this paper fills a gap in the literature on the combined impact of Industry 4.0 and LM principles on corporate performance, and the case study validating these impacts. A conceptual and theoretical discussion about LM and Industry 4.0 has been refined in this study to meet the research needs discussed in [11]. An application-oriented use case study is also presented to satisfy the shortfall of work described in [23]. Therefore, this study provides further advantages for researchers and practitioners.

We have demonstrated the improved effect of using the DMAIC method with Digital Poka-Yoke in the fabrication of vacuum degassing equipment. This project improvement program helped company-T successfully eliminate wasteful variations and improve business performance, increasing the process capability index, Cpk, from 1.278 Cpk to 2 Cpk and increasing the production yield from 99.44% to 100%. The direct savings of the project amounted to NTD 68,000, mainly due to the higher worker utilization rate. Additionally, safer working environments, new job opportunities, and excessive material waste are all improved to implement smart and sustainable manufacturing systems.

This paper is limited by the use of a single case study. More applications of Dynamic Lean 4.0 tools in SSMS should be explored. Future research may include implementing other Dynamic Lean 4.0 tools, such as Sustainable VSM and Extended SMED for SSMS framework. However, our approach provides a general learning perspective. A means of continuous improvement is demonstrated through the effective application of the DMAIC process. Therefore, this paper can serve as a unique roadmap for practitioners and academia to improve process capability to meet customer specifications, especially to improve social, economic, and environmental performance.

Author Contributions: Conceptualization, B.R. and F.-K.W.; methodology, B.R., F.-K.W. and Y.-P.C.; software, B.R.; validation, F.-K.W. and R.-H.Y.; resources, F.-K.W. and R.-H.Y.; data curation, Y.-P.C.; writing—original draft preparation, B.R. and F.-K.W.; writing—review and editing, F.-K.W. and R.-H.Y.; visualization, B.R. and Y.-P.C.; supervision, F.-K.W. and R.-H.Y.; project administration, F.-K.W. and R.-H.Y. All authors have read and agreed to the published version of the manuscript.

Funding: This research received no external funding.

Data Availability Statement: Not applicable.

Conflicts of Interest: The authors declare no conflict of interest.

References

1. Santos, A.C.O.; da Silva, C.E.S.; Braga, R.A.D.S.; Corrêa, J.É.; de Almeida, F.A. Customer value in lean product development: Conceptual model for incremental innovations. *Syst. Eng.* **2020**, *23*, 281–293. [CrossRef]
2. McDermott, O.; Antony, J.; Sony, M.; Healy, T. Critical Failure Factors for Continuous Improvement Methodologies in the Irish MedTech Industry. *TQM J.* **2022**, *34*, 18–38. [CrossRef]
3. Ratnayake, R.C.; Chaudry, O. Maintaining sustainable performance in operating petroleum assets via a lean-six-sigma approach: A case study from engineering support services. *Int. J. Lean Six Sigma* **2017**, *8*, 33–52. [CrossRef]
4. De Souza, J.P.E.; Alves, J.M. Lean-integrated management system: A model for sustainability improvement. *J. Clean. Prod.* **2018**, *172*, 2667–2682. [CrossRef]
5. Mathiyazhagan, K.; Gnanavelbabu, A.; Kumar, N.; Agarwal, V. A framework for implementing sustainable lean manufacturing in the electrical and electronics component manufacturing industry: An emerging economies country perspective. *J. Clean. Prod.* **2022**, *334*, 130169. [CrossRef]
6. Iranmanesh, M.; Zailani, S.; Hyun, S.S.; Ali, M.H.; Kim, K. Impact of lean manufacturing practices on firms' sustainable performance: Lean culture as a moderator. *Sustainability* **2019**, *11*, 1112. [CrossRef]
7. Maware, C.; Adetunji, O. Lean manufacturing implementation in Zimbabwean industries: Impact on operational performance. *Int. J. Eng. Bus. Manag.* **2019**, *11*, 1–12. [CrossRef]
8. Byrne, B.; McDermott, O.; Noonan, J. Applying lean six sigma methodology to a pharmaceutical manufacturing facility: A case study. *Processes* **2021**, *9*, 550. [CrossRef]
9. Trakulsunti, Y.; Antony, J.; Edgeman, R.; Cudney, B.; Dempsey, M.; Brennan, A. Reducing pharmacy medication errors using lean six sigma: A Thai hospital case study. *Total Qual. Manag. Bus. Excell.* **2022**, *33*, 664–682. [CrossRef]
10. Zhong, R.Y.; Xu, X.; Klotz, E.; Newman, S.T. Intelligent manufacturing in the context of industry 4.0: A review. *Engineering* **2017**, *3*, 616–630. [CrossRef]
11. Jamwal, A.; Agrawal, R.; Sharma, M.; Giallanza, A. Industry 4.0 technologies for manufacturing sustainability: A systematic review and future research directions. *Appl. Sci.* **2021**, *11*, 5725. [CrossRef]
12. Nara, E.O.B.; da Costa, M.B.; Baierle, I.C.; Schaefer, J.L.; Benitez, G.; do Santos, L.M.A.L.; Benitez, L.B. Expected impact of industry 4.0 technologies on sustainable development: A study in the context of Brazil's plastic industry. *Sustain. Prod. Consum.* **2021**, *25*, 102–122. [CrossRef]
13. Sony, M.; Antony, J.; Mc Dermott, O.; Garza-Reyes, J.A. An empirical examination of benefits, challenges, and critical success factors of industry 4.0 in manufacturing and service sector. *Technol. Soc.* **2021**, *67*, 101754. [CrossRef]
14. De Sousa Jabbour, A.B.L.; Jabbour, C.J.C.; Foropon, C.; Godinho Filho, M. When titans meet—Can industry 4.0 revolutionise the environmentally-sustainable manufacturing wave? the role of critical success factors. *Technol. Forecast. Soc. Chang.* **2018**, *132*, 18–25. [CrossRef]
15. Kamble, S.S.; Gunasekaran, A.; Sharma, R. Analysis of the driving and dependence power of barriers to adopt industry 4.0 in Indian manufacturing industry. *Comput. Ind.* **2018**, *101*, 107–119. [CrossRef]
16. Antony, J.; McDermott, O.; Powell, D.; Sony, M. The evolution and future of lean six sigma 4.0. *TQM J.* **2022**. *ahead-of-print*. [CrossRef]
17. Buer, S.-V.; Semini, M.; Strandhagen, J.O.; Sgarbossa, F. The complementary effect of lean manufacturing and digitalisation on operational performance. *Int. J. Prod. Res.* **2021**, *59*, 1976–1992. [CrossRef]

18. Felsberger, A.; Qaiser, F.H.; Choudhary, A.; Reiner, G. The impact of Industry 4.0 on the reconciliation of dynamic capabilities: Evidence from the European manufacturing industries. *Prod. Plan. Control* **2022**, *33*, 277–300. [CrossRef]
19. Ding, B.; Hernandez, X.; Jane, N. Combining lean and agile manufacturing competitive advantages through Industry 4.0 technologies: An integrative approach. *Prod. Plan. Control* **2023**. [CrossRef]
20. Tortorella, G.L.; Fettermann, D.C. Implementation of Industry 4.0 and lean production in Brazilian manufacturing companies. *Int. J. Prod. Res.* **2018**, *56*, 2975–2987. [CrossRef]
21. Vlachos, I.P.; Pascazzi, R.M.; Zobolas, G.; Repoussis, P.; Giannakis, M. Lean manufacturing systems in the area of industry 4.0: A lean automation plan of AGVs/IoT integration. *Prod. Plan. Control* **2023**. [CrossRef]
22. Calabrese, A.; Dora, M.; Ghiron, N.L.; Tiburzi, L. Industry's 4.0 transformation process: How to start, where to aim, what to be aware of. *Prod. Plan. Control* **2022**, *33*, 492–512. [CrossRef]
23. Ciano, M.P.; Dallasega, P.; Orzes, G.; Rossi, T. One-to-one relationships between Industry 4.0 technologies and lean production techniques: A multiple case study. *Int. J. Prod. Res.* **2021**, *59*, 1386–1410. [CrossRef]
24. Wang, F.-K.; Rahardjo, B.; Rovira, P.R. Lean six sigma with value stream mapping in industry 4.0 for human-centered workstation design. *Sustainability* **2022**, *14*, 11020. [CrossRef]
25. Ribeiro, M.A.S.; Santos, A.C.O.; de Amorim, G.D.F.; de Oliveira, C.H.; da Silva Braga, R.A.; Netto, R.S. Analysis of the implementation of the single minute exchange of die methodology in an agroindustry through action research. *Machines* **2022**, *10*, 287. [CrossRef]
26. Sundaramali, G.; Shankar, S.A.; Kummar, M.M. Non-conformity recovery and safe disposal by Poka Yoke and hallmarking in a piston unit. *Int. J. Product. Qual. Manag.* **2018**, *24*, 460–474. [CrossRef]
27. Haddud, A.; Khare, A. Digitalizing supply chains potential benefits and impact on lean operations. *Int. J. Lean Six Sigma* **2020**, *11*, 731–765. [CrossRef]
28. Borowski, P.F. Digitization, digital twins, blockchain, and industry 4.0 as elements of management process in enterprises in the energy sector. *Energies* **2021**, *14*, 1885. [CrossRef]
29. Vaidya, S.; Ambad, P.; Bhosle, S. Industry 4.0—A glimpse. *Procedia Manuf.* **2018**, *20*, 233–238. [CrossRef]
30. Guerra-Zubiaga, D.; Kuts, V.; Mahmood, K.; Bondar, A.; Nasajpour-Esfahani, N.; Otto, T. An approach to develop a digital twin for industry 4.0 systems: Manufacturing automation case studies. *Int. J. Comput. Integr. Manuf.* **2021**, *34*, 933–949. [CrossRef]
31. Ing, T.S.; Lee, T.C.; Chan, S.W.; Alipal, J.; Hamid, N.A. An overview of the rising challenges in implementing industry 4.0. *Int. J. Supply Chain. Manag.* **2019**, *8*, 1181–1188.
32. Horst, D.J.; Duvoisin, C.A.; de Almeida Vieira, R. Additive manufacturing at industry 4.0: A review. *Int. J. Eng. Tech. Res.* **2018**, *8*, 3–8.
33. Mourtzis, D.; Vlachou, E.; Zogopoulos, V.; Fotini, X. Integrated production and maintenance scheduling through machine monitoring and augmented reality: An industry 4.0 approach. In *IFIP International Conference on Advances in Production Management Systems*; Springer: Cham, Switzerland, 2017; pp. 354–362.
34. Park, J.; Bae, H. Big data and AI for process innovation in the industry 4.0 era. *Appl. Sci.* **2022**, *12*, 6346. [CrossRef]
35. Ejsmont, K.; Gladysz, B.; Corti, D.; Castaño, F.; Mohammed, W.M.; Lastra, J.L.M. Towards 'lean industry 4.0'—Current trends and future perspectives. *Cogent Bus. Manag.* **2020**, *7*, 1781995. [CrossRef]
36. Gallo, T.; Cagnetti, C.; Silvestri, C.; Ruggieri, A. Industry 4.0 tools in lean production: A systematic literature review. *Procedia Comput. Sci.* **2021**, *180*, 394–403. [CrossRef]
37. Langlotz, P.; Siedler, C.; Aurich, J.C. Unification of lean production and industry 4.0. *Procedia CIRP* **2021**, *99*, 15–20. [CrossRef]

Article

Data Science for Industry 4.0 and Sustainability: A Survey and Analysis Based on Open Data

Hélio Castro [1,2,*], Filipe Costa [1], Tânia Ferreira [3], Paulo Ávila [1,2], Manuela Cruz-Cunha [4], Luís Ferreira [4], Goran D. Putnik [5] and João Bastos [1,2]

1   ISEP—School of Engineering, Polytechnic of Porto, Rua Dr. António Bernardino de Almeida 431, 4249-015 Porto, Portugal
2   INESC TEC, Rua Dr. Roberto Frias, 4200-465 Porto, Portugal
3   Data CoLAB, Av. de Cabo Verde, 1, 4900-568 Viana do Castelo, Portugal
4   2Ai—Applied Artificial Intelligence Lab, School of Technology, IPCA, 4750-810 Barcelos, Portugal
5   School of Engineering, University of Minho, Campus de Azurém, 4800-058 Guimarães, Portugal
*   Correspondence: hcc@isep.ipp.pt

Abstract: In the last few years, the industrial, scientific, and technological fields have been subject to a revolutionary process of digitalization and automation called Industry 4.0. Its implementation has been successful mainly in the economic field of sustainability, while the environmental field has been gaining more attention from researchers recently. However, the social scope of Industry 4.0 is still somewhat neglected by researchers and organizations. This research aimed to study Industry 4.0 and sustainability themes using data science, by incorporating open data and open-source tools to achieve sustainable Industry 4.0. To that end, a quantitative analysis based on open data was developed using open-source software in order to study Industry 4.0 and sustainability trends. The main results show that manufacturing is a relevant value-added activity in the worldwide economy; that, foreseeing the importance of Industry 4.0, countries in America, Asia, Europe, and Oceania are incorporating technological principles of Industry 4.0 in their cities, creating so-called smart cities; and that the industries that invest most in technology are computers and electronics, pharmaceuticals, transport equipment, and IT (information technology) services. Furthermore, the G7 countries have a prevalent positive trend for the migration of technological and social skills toward sustainability, as it relates to the social pillar, and to Industry 4.0. Finally, on the global scale, a positive correlation between data openness and happiness was found.

Keywords: Industry 4.0; data science; sustainability; social sustainability open data; open-source tools

Citation: Castro, H.; Costa, F.; Ferreira, T.; Ávila, P.; Cruz-Cunha, M.; Ferreira, L.; Putnik, G.D.; Bastos, J. Data Science for Industry 4.0 and Sustainability: A Survey and Analysis Based on Open Data. *Machines* **2023**, *11*, 452. https://doi.org/10.3390/machines11040452

Academic Editor: Antonio J. Marques Cardoso

Received: 27 January 2023
Revised: 10 March 2023
Accepted: 30 March 2023
Published: 3 April 2023

1. Introduction

Sustainability and Industry 4.0 are probably the most important topics in the manufacturing field, both in academic and industrial environments. In the context of sustainability within Industry 4.0, as regards the triple bottom line, profit (economic) and planet (environmental) are being studied extensively by the research community, while people (social) are not receiving the same attention. The wider perspective of the social aspects of Industry 4.0 is neglected by the academic community [1]. Industry 5.0 emerged to overcome problems with the techno-centered approach of Industry 4.0 [2], namely the absence of the human being as the key player in an industrial digital environment and the influence on sustainability as it relates to the social pillar. Nowadays, the challenges are related to the way I4.0 is implemented and managed to achieve the desired outcomes—economic, environmental, and social [3].

Within the industrial digital environment, data act as an enabler or creator of value in Industry 4.0 [4], and most industrial organizations should have integrated low-cost analytical systems to collect and measure results [5] in order to stay competitive, especially small to medium-sized enterprises (SMEs). Industry 4.0 projects are perceived by SMEs as cost-driven initiatives [6], and lack of capacity to invest in advanced technologies and the need for specialist expertise [7] to implement them are constraints SMEs face in the adoption of Industry 4.0 and, probably as a consequence, sustainability practices.

Besides these limitations, those SMEs that are capable of incorporating technologies are unable to access sufficiently complex data due to limitations of scale. Leveraging data access, treatment, and analysis and creating intelligence are elements of a company's preparedness to improve its decision-making processes. For example, the relevance of big data and analytics to competitive intelligence has been demonstrated [8], and for this reason, data science is gaining relevance in contemporary society.

However, in small-scale organizations, such as SMEs, the cornerstone of data science is data availability, and a possible solution to overcome this problem is data access via an open approach, so-called open data. The open approach is a movement that tends to be focused on involving communities and leveraging social elements.

Due to this framework, and based on an inductive research method, the purpose of this work is to use the open approach to data science, by using open data, and the open approach in terms of analytical tools, by using open-source software, in the study of Industry 4.0 and sustainability concepts, filling a research gap in this domain. Therefore, a meta-analysis is presented in this study, based on quantitative data that is openly available and free of macro-indicators. The study will use induction analysis to generate conclusions and contributions that will lead to a better understanding of these topics relevant to in manufacturing, with a special emphasis on the technological elements of Industry 4.0 and the social aspects of sustainability.

This paper is organized as follows: Section 2 presents the literature review of the core topics in this work (data science, open data, Industry 4.0, and sustainability); Section 3 describes the research methodology; Section 4 demonstrates the analysis and the main results and discusses these; and, finally, the main conclusions are presented in Section 5.

2. Literature Review

This section provides an overview of the four main themes (data science, open data, Industry 4.0, and sustainability) that are explored across the research, and a brief bibliographic review of each of those themes.

2.1. Data Science and Open Data

In a world that constantly produces and consumes data, it is essential to understand the value that can be extracted from it. Mikalef et al. [9] consider data science and big-data domains as the next frontier for both practitioners and researchers as they embody significant potential in exploiting data to sustain competitive advantage. Data science is an interdisciplinary field that supports and guides the extraction of useful patterns from raw data by exploring advanced technologies, algorithms, and processes [10]. The actual extraction of knowledge from data is defined as data mining, and it can be applied to a broad set of business areas, such as marketing, customer relationship management, supply chain management, or product optimization [11]. Data science should be seen as the domain that originates from the merging of big-data technologies with data-management skills and behavioral disciplines [12]. Data science and big data can be combined with co-creation and data-sharing technologies to enable organizations to leverage creativity outside their organizational boundaries [13]. The development and operation of software have become increasingly dependent on data [14], and this data can be made more accessible to organizations and individuals through data-sharing and open-source technologies. Runeson [15] highlights the need for the adoption of co-creation and collaboration principles to harness innovation potential and manage costs in the age of data.

Today, data volumes are exploding, and not only is the rate of data generated per individual increasing but so also is the rate at which we share information. Lawmakers and organizations worldwide are trying to envision the future of data ownership. Information remains largely centralized, but the trend is shifting toward a distributed and open model of data sharing [16].

2.2. Open Data for Industry 4.0

As described by Tim Hall [17], one of the key drivers for the adoption of Industry 4.0 across the globe is the ability to use the power of data to revolutionize manufacturing. Open-data platforms provide innovative services with high impact on innovation [18], and data sources based on open data allow for the evaluation of Industry 4.0 readiness by regions [19].

However, the manufacturing sector has been slow to benefit from Industry 4.0 drivers evenly across different industries, enterprise sizes, and geographies. Since most Industry 4.0 technologies require substantial investment to be successfully implemented, the economic factor is undeniably crucial if they are to be adopted. Nevertheless, while differences in the economic situations of enterprises and countries have an obvious bearing on the speed and rate of success of Industry 4.0 adoption, they cannot be considered the only factor involved. Smart factories and smart cities are another relevant study theme, as technological advancements and digitalization are changing how companies operate their business and organizations reshape communities. All those changes and advancements require big R&D investments and qualified researchers and workers. Since there are many economic challenges as well as difficulties in recruiting the most qualified workers, the adoption of those technologies might be slow and unoptimized for SMEs, which need to adapt to technological changes in order to grow and compete.

Besides, the integration of open data is still oriented to applications, websites, and platforms [20], whereas it is necessary for it to be oriented toward product development. Only recently, a case-study project based on open data from academia and companies was applied in the development of Industry 4.0 technologies in additive manufacturing [21].

2.3. Sustainability in Industry 4.0

Wee et al. [22] reiterate that there is a need for deeper research on sustainability as it relates to Industry 4.0, since it has received little attention from academics and researchers. In Kamble et al.'s [23] framework for Industry 4.0 sustainability, the three sustainable outcomes that ideally should be accomplished from Industry 4.0 technologies and process integration are economic benefit, process automation, and safety and environmental protection. The sustainability pillars in manufacturing companies have a strong relationship with Industry 4.0 [24], and competitiveness and social and environmental advantages are potentialized in manufacturing companies that adopt Industry 4.0 [25].

Other models include open innovation and collaboration as guiding principles for sustainability in industry. In this research, analyzing the progress toward accomplishing sustainability goals using the open data available is considered an overall evaluation of sustainability across the three pillars. Since these are broad goals established not only in countries but also in organizations and companies, successful progress toward accomplishing them is also progress toward accomplishing sustainability in Industry 4.0. UN members need to collaborate across all established goals, even more so because Goal 17 itself—Partnerships for the Goals—focuses on evaluating members' progress toward economic, social, and environmental collaboration between them. For that reason, it is reasonable to assume that progressing in Goal 17 is essential to accomplish successful collaboration in the remaining goals. Social indicators are neglected within manufacturing companies [26], and Industry 4.0 still needs to improve intra-organization mechanisms for achieving social sustainability [1]. Within the social pillar of sustainability, one of the main social issues relating to the digitalization and automation of industry is how employment and skills requirements will be affected. The common understanding regarding this issue is that automation eliminates the need for human workers, which will bring unemployment and social dissatisfaction. However, researchers such as Shet and Pereira [27] believe that Industry 4.0 generates new job prospects in the emerging domains of science, technology, and engineering. Those domains usually require a higher level of skills and specialization than traditional jobs, leaving unskilled workers more vulnerable to the gradual increase in demand for qualified workers.

Recently, to evolve the concept of Industry 4.0 by stressing the social and human aspects, Industry 5.0 emerged. This approach combines the technologies of Industry 4.0, sustainability, and the human factor [28].

3. Research Methodology

The research methodology section aims to present the researchers' design, with all key design choices being detailed and justified logically, according to the research theme.

The appropriate research type for this study is the inductive and quantitative approach since the study aims to explore quantitative data relevant to the research themes and afterward form appropriate conclusions and contributions, instead of pre-establishing hypotheses or theories about those themes.

The research strategy was, from the starting point of the established research themes, to collect and aggregate data from different relevant organizations that had information relevant to the subject, namely the World Bank, Data World, Open Data Watch and Our World in Data. However, the datasets of these organizations could only be useful if their content was fully open for being downloaded, modified, and published, which is one of the main characteristics of open data. During the process of cleaning and organizing the data available within these institutions, some challenges were encountered regarding the datasets. For instance, specific information was quite hard to find, so it was necessary to select topics that by induction could lead us directly to the research themes and provide insights and conclusions. Another challenge was the different time horizon between datasets: data on the selected topics were provided by different institutions and had different objectives, and the time period for each topic was different. As a result, this study could not normalize a constant time period across all analyses, but instead used the longest period of time possible for each analysis made.

The datasets gathered contained data grouped according to regions, countries, industries, and enterprise size. For the data analysis and visualization, the most prevalent techniques across the study were frequency graphs and visualizations for inferential statistics that analyze correlations between selected variables, as correlation and regression analysis are techniques frequently used in inferential statistics [29]. In keeping with the open-approach emphasis of this study, the open-source software tools used to analyze the data were R and Python. R is a free open-source programming language that provides a statistical analytics computing environment. R provides a variety of statistical and graphical techniques that can be used by importing useful packages. These techniques can be used to handle raw data and to retrieve information in order to have a sense of how the data is distributed or whether there are patterns that are masked [30]. The R packages used were *arules* and *arulesViz* for rule association, and *RQDA* for quantitative analysis. Python is currently the fastest-growing programming language in the world, thanks to its open accessibility, ease of use, fast learning curve, and numerous high-quality packages for data science and machine learning. Together with R, Python is extremely useful for identifying correlations between variables and creating powerful visualizations such as graphs, matrixes, plots, or maps [31]. The main Python libraries used were *Matplotlib*, *Numpy*, and *Seaborn*. During this stage, the graphs produced were critically analyzed for the observations and conclusions that comprise the findings of this work.

Table 1 summarizes the research methods adopted during the present study, namely research type, research strategy, sampling strategy, data collection method, and data analysis tools.

Table 1. Research Methodology.

Research Design	Method
Research Type	Inductive and Quantitative
Research Strategy	1. Establishing the research themes 2. Collecting and Aggregating Open Data 3. Cleaning and Organizing Data 4. Data Analysis and Visualization 5. Results and Discussion
Sampling Strategy	Probability Sampling within groups such as regions, countries, industries, and enterprise size
Data Collection Method	Open Datasets
Data Analysis Tools	Open-Source software tools such as Python and R

4. Data Analysis, Results and Discussion

This section presents the data analysis (point 4 of the research strategy), followed by the results from the selected datasets and the discussion thereon (point 5 of the research strategy). Each research theme identified in the previous subsection is represented by several relevant visualizations and their respective critical analyses in the context of the research.

4.1. Open Data for Industry 4.0

The first part of the results and analysis obtained from the data treatment is the representation of the open data that was available for Industry 4.0 themes. This subsection deals with Industry 4.0 themes previously referred to, such as manufacturing value added to gross domestic product (GDP), smart cities, and R&D efforts for innovation.

4.1.1. Manufacturing Value Added to GDP

Manufacturing is one of the main sectors of industry around the world and one of the main adopters of Industry 4.0 [32]. By analyzing available open data and using it alongside other relevant variables that measure development, such as a country's GDP, this research intends to give a clearer perspective on the issues examined as part of the research model.

Since GDP characterizes the economic output of a country, according to the data on the global manufacturing value added to GDP in 2020 [33], Asia is the continent with the most countries in which manufacturing value added to GDP is greater than 30%, followed by Europe and America, with just one country each. In Oceania and Africa, no countries are in this situation. By country, China is one of the countries in the world with a big share of its GDP allocated to manufacturing—around 40%. Most countries appear to have manufacturing value added to GDP of between 10% and 20%. The continents with a larger share of countries that have less than 10% of their GDP value added from manufacturing are Africa and Oceania, while in Europe, America, and Asia there are few countries with less than 10% manufacturing value added to GDP.

With regards to the Industry 4.0 environment and the available open data, it can be observed that Asia is the leading continent in terms of countries with a high percentage of manufacturing value added to GDP, followed by Europe and America. China stands out as the country with the largest share of GDP derived from manufacturing.

4.1.2. Smart Cities

A smart city uses information and technology to improve operational efficiency, share information, and provide a better quality of life for its citizens and workers [34]. Implementing smart technologies and processes within factories and services through Industry 4.0 adoption also tends to promote economic growth, social integrity, and environmental sustainability in industrial sectors, creating new jobs in the high-tech and creative industries [17]. The dataset evaluates cities across six smart categories: mobility, environment, government, economy, people, and living. The aggregation of those scores translates to a city's ranking along IMD's Smart City Index [35]. Below, Figure 1a ranks countries by overall smart city scores and (b) presents a plot for correlations between all six variables and the overall score.

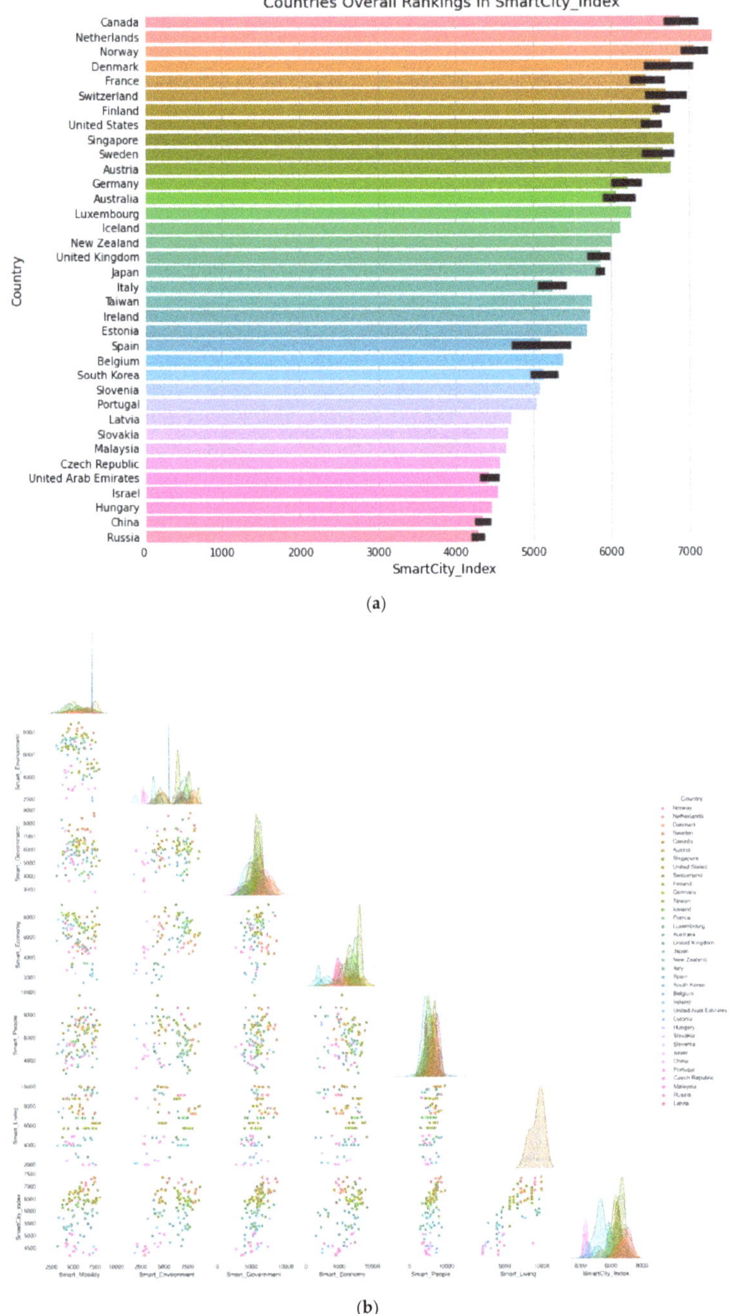

Figure 1. (**a**) Countries' Overall rankings in Smart City Index (**b**) Smart City Index categories and correlations with overall Smart City score.

Taking into account the first 20 countries of the overall ranking, the continent with most countries in this ranking is Europe, with 13 countries (the Netherlands, Norway, Denmark, France, Switzerland, Finland, Sweden, Austria, Germany, Luxembourg, Iceland, the United Kingdom, and Italy), followed by Asia, with three countries (Singapore, Japan, and Taiwan), and America and Oceania, with two each (Canada and the United States, and Australia and New Zealand, respectively).

The countries with the highest overall scores are Canada, the Netherlands, Norway, Denmark, and France. The lowest-scored countries are Russia, China, Hungary, Israel, and the United Arab Emirates. The fact that a country has a high number of smart cities does not necessarily mean that the country itself has a high smart score.

By analyzing each category within the overall Smart City Index, it looks as if the key factors that seem to correlate most strongly with the overall index are smart living and smart economy. Since Industry 4.0 is such a big driver for digitalization and automation in the global economy, it makes sense for developed and developing nations that seek to develop their cities technologically and sustainably to accelerate the transition to a smart economy and a smart way of living.

In terms of smart city rankings, therefore, the majority of the top 20 countries are located in Europe, followed by Asia, with Canada having the highest score.

4.1.3. R&D Efforts for Innovation

One of the main drivers of innovation, particularly in the technological and industrial fields, is the financing of research and development (R&D) by enterprises, academic researchers, and scientists [36]. However, because of the uncertainty around the level of return and the payback period, this kind of investment is not equally accessible to all countries, industries, and sizes of enterprises. Assessing which countries benefit the most from R&D investments by their enterprises and which industries allocate most expenditure to R&D from World Bank data [37] might provide a representation of the efforts to implement Industry 4.0. Figure 2 shows the expenditure in R&D and the patent share of various industries.

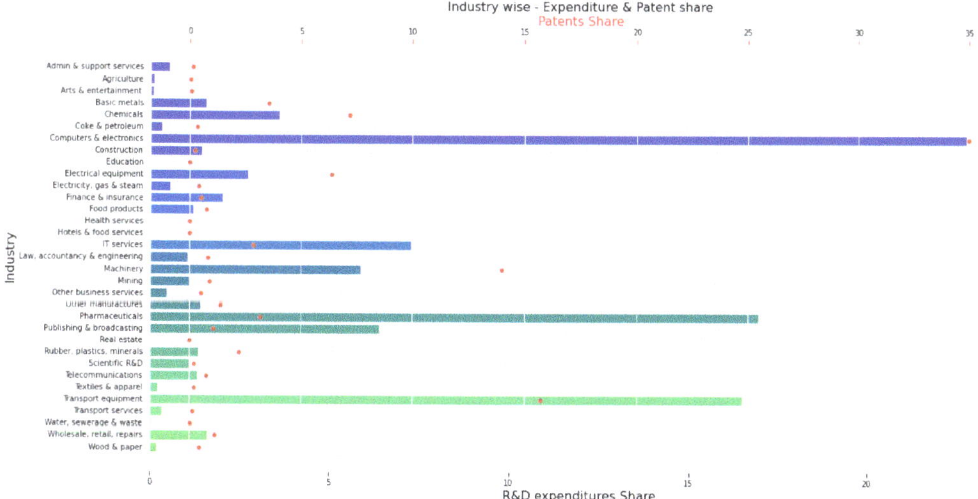

Figure 2. Industry expenditure in R&D and patent share.

It is possible to identify computers and electronics as the industry with the highest R&D expenditure share (close to 25%). As expected, this is also the industry with the highest patent share (35%). Pharmaceuticals appear as the second industry with the highest R&D expenditure, with around 17% of the share, followed by transport equipment, IT services, and publishing and broadcasting, with 16.5%, 7.5%, and 6% share, respectively. Surprisingly, the patent share does not follow the R&D expenditure distribution so closely in these industries. The pharmaceuticals sector ranks only seventh in the patent share even though it ranks second in the R&D expenditure share. This might be caused by other factors, such as regulation and difficulties in innovating existing solutions. IT services and transport equipment also reveal a much higher R&D expenditure share compared with patent share. On the other hand, machinery is the third-ranked sector for patent share, at almost 15%, even though it occupies sixth position in R&D expenditure. Electrical equipment, chemicals, and basic metals are other sectors with a much larger patent share compared with their R&D expenditure share.

4.2. Open Data for Sustainability

The second part of the results and analysis obtained from the data treatment is the representation of the open data that was available for sustainability. This subsection deals with sustainability themes such as skills migration and the relationship between openness and happiness.

Although sustainability is based on three known pillars, as was evidenced in the literature the social pillar is arguably one of the most pressing sustainability issues, and yet it is the one that receives the least attention from researchers. One of the main issues regarding the relationship between the adoption of Industry 4.0 and the future of work is job shortages. The increasing digitalization and automation of business and service tasks often leads to worries about the permanent replacement of the human labor force by machines. However, the literature shows that that can be a misconception concerning the future of work. Shet and Pereira [27] argue that Industry 4.0 can generate job prospects by creating new employment opportunities in emerging domains, such as science, technology, engineering, and mathematics. While technological advancements and automation tend to minimize employment prospects in some sectors, they also bring about the simultaneous emergence of new businesses and services linked with economic growth and new markets, which leads to a rise in new job opportunities. Shet and Pereira [27] also warn that those jobs created by digitalization and automation require high levels of skill, knowledge, competence, and specialization that are not required by traditional jobs, leaving unskilled workers more vulnerable to the gradual increase in the demand for a qualified workforce.

4.2.1. Skills Migration

To study sustainability from the social perspective, this work considered data on skill migration across different countries and industries, to compare supply and demand trends for skilled workers. Skills migration can be defined as the trends in both supply and demand for professional skills over a number of years. As economies and labor markets change, largely because of the evolution of consumer behavior and the adoption of new technologies, so do the skills that are demanded from enterprises and public services. Better education also means better qualified workers who migrate from traditional industries to more technological and digitalized ones [38]. This dynamic is also accelerated by Industry 4.0 adoption. However, since not all countries are equal in terms of economic growth, technology adoption, and industry digitalization, naturally skills migration varies not only between industries but also in terms of geography.

This study, based on data from the World Bank [39] analyzed three categories of skills migration, namely specialized industry skills, soft skills, and disruptive-tech skills. The specialized industry skills category could indicate the overall migration of workers. The soft skills category could provide information on perceptions of important social skills, such as problem-solving, leadership, teamwork, communication, time management, persuasion, and negotiation, which are essential skills for workers, independent of location or industry sector. The disruptive-tech skills category could demonstrate trends in high-tech jobs.

For this analysis, three groups were considered in both categories: Group a), comprising China and the United States (US), as these are two of the most influential economies in the world; Group b), the G7 (international Group of 7) countries, comprising seven advanced economies (Canada, France, Germany, Italy, Japan, the United Kingdom and the US); and Group c), BIC, comprising Brazil, China, and India due to their status as emerging economies in the BRIC bloc, with Russia's data not being available. The period of analysis was from 2015 to 2019. The purpose of the analysis of the graphs was to determine whether each skill saw an increase, a decrease, or a stabilization in demand, as plotted by their respective values.

For the specialized industry skills component, the analyses for the three groups (China and the US, the G7, and BIC) are represented in Figure 3a–c, respectively.

The plot in Figure 3a, which considers China and the US, presents China as having dynamic migration in relation to specialized industry skills, whereas the US presents a very stable graph, demonstrating that there was no migration of specialized industry skills. The overall dynamics of China showed that the dots are more located in negative that in positive values, suggesting an overall negative migration.

Excluding the US, which has been mentioned previously, for the G7 countries in Figure 3b, a stable-toward-positive migration trend for specialized industry skills was demonstrated, with very few exceptions. Among those exceptions, Germany and Japan are highlighted in the national security, army, and navy categories as having lost skilled workers in those categories over the years, revealing a possible antimilitaristic policy.

On the other hand, excluding China, which has been mentioned previously, the developing BIC economies of Brazil and India demonstrated an overall negative skills migration trend in specialized industries, which contrasts with the positive trend of the G7 countries.

For the soft skills component, the analysis for the three groups (China and the US, the G7, and BIC) are represented in Figure 4a–c, respectively.

Concerning the findings from the graph representing Group a), the first output is that the US was more stable as regards soft skills migration than China. For practically all skills during the studied period, values for the US are near zero and China showed different behavior. In China, some skills had become stable (active learning, communication, flexible approach, leadership, negotiation, problem solving, and writing), persuasion was floating from negative (2015, 2019) to positive (2016, 2018), social perception showed a negative trend throughout the period, and teamwork and time management displayed a positive trend.

Regarding the graph plotted for Group (b), besides the US, already analyzed in Group (a), there was a weak positive trend in the other G7 countries for active learning, communication, leadership, negotiation, social perception, and writing. In the flexible approach skill, most of the countries during the studied period showed fluctuation between a positive and a negative trend, but overall there was a positive trend (in 2019, most of the countries showed a positive value). The skill of persuasion also showed a positive trend in most of the countries, except Italy and the United Kingdom, where the trend was negative. Problem solving, teamwork, and time management skills showed a significant positive tendency, particularly in Germany and Japan.

For Group (c), besides China, already analyzed in Group (a), Brazil and India show a completely negative trend in all skills during the period 2015 to 2019. This tendency demonstrates a depreciation of the soft skills in Brazil and India, meaning a different approach compared with China and the G7 countries.

For the disruptive-tech skills component, the analyses for the three groups are characterized in Figure 5a–c.

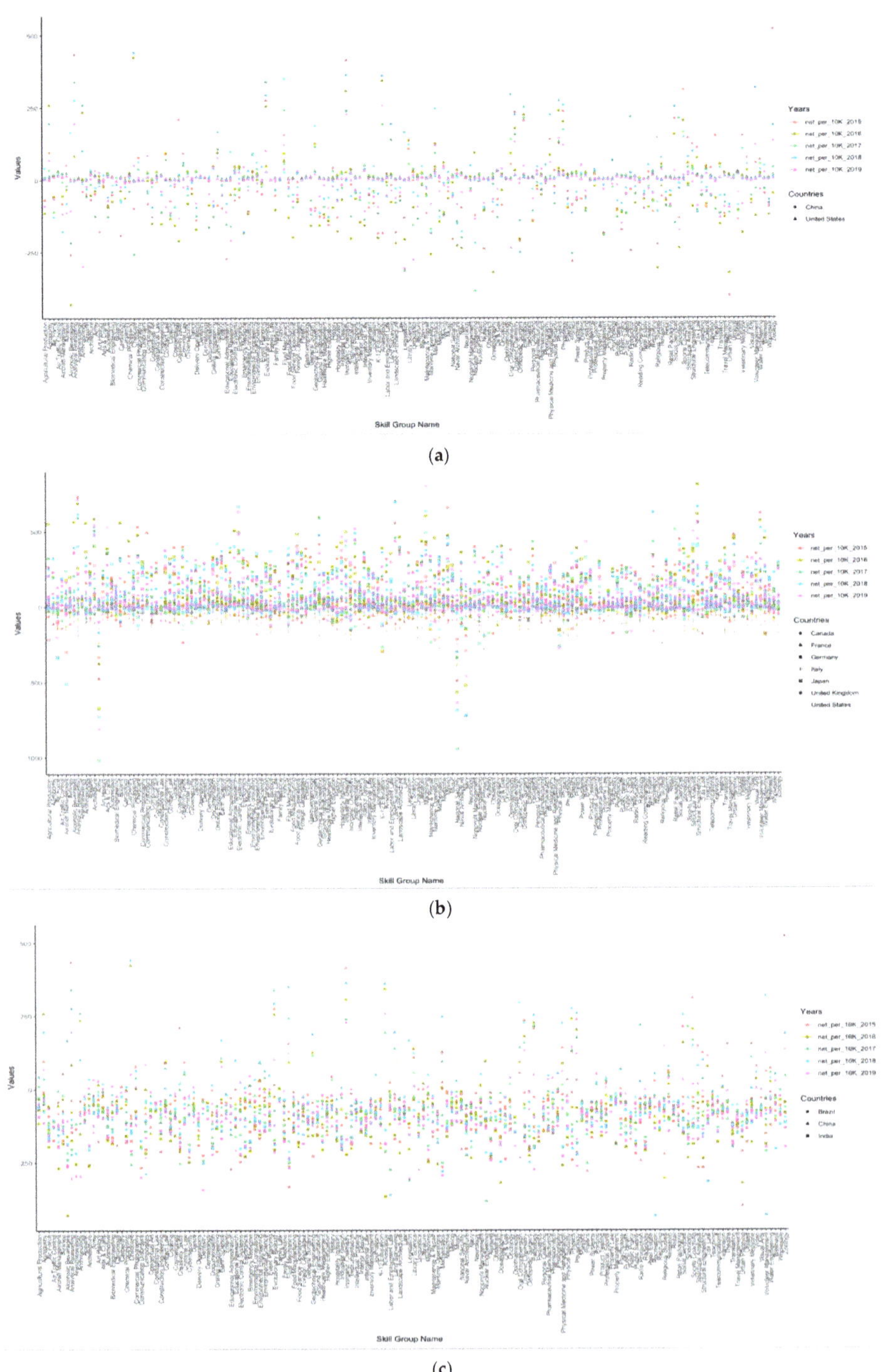

Figure 3. Specialized industry skills migration in (**a**) China and the US, (**b**) G7 countries, and (**c**) BIC countries.

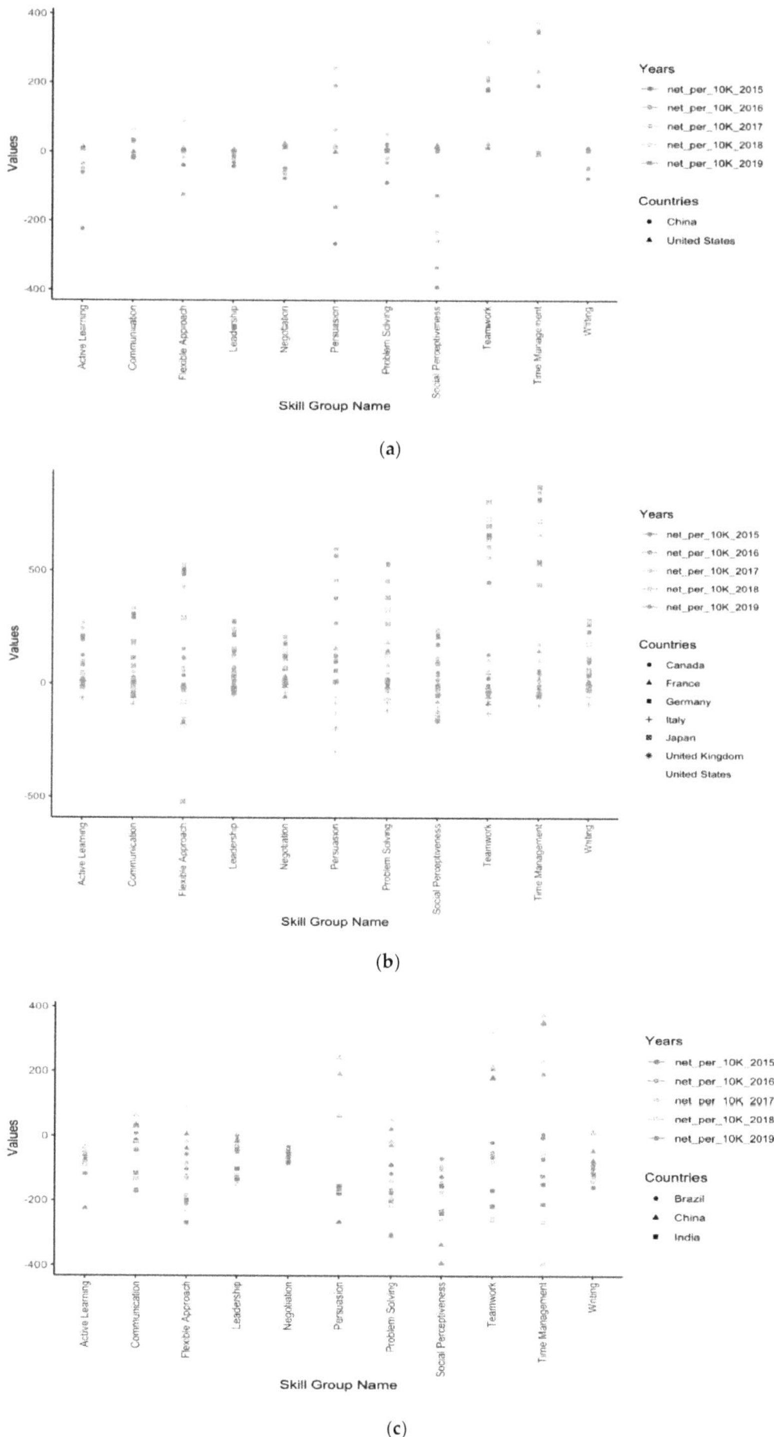

Figure 4. Soft skills migration in (**a**) China and the US, (**b**) G7 countries, and (**c**) BIC countries.

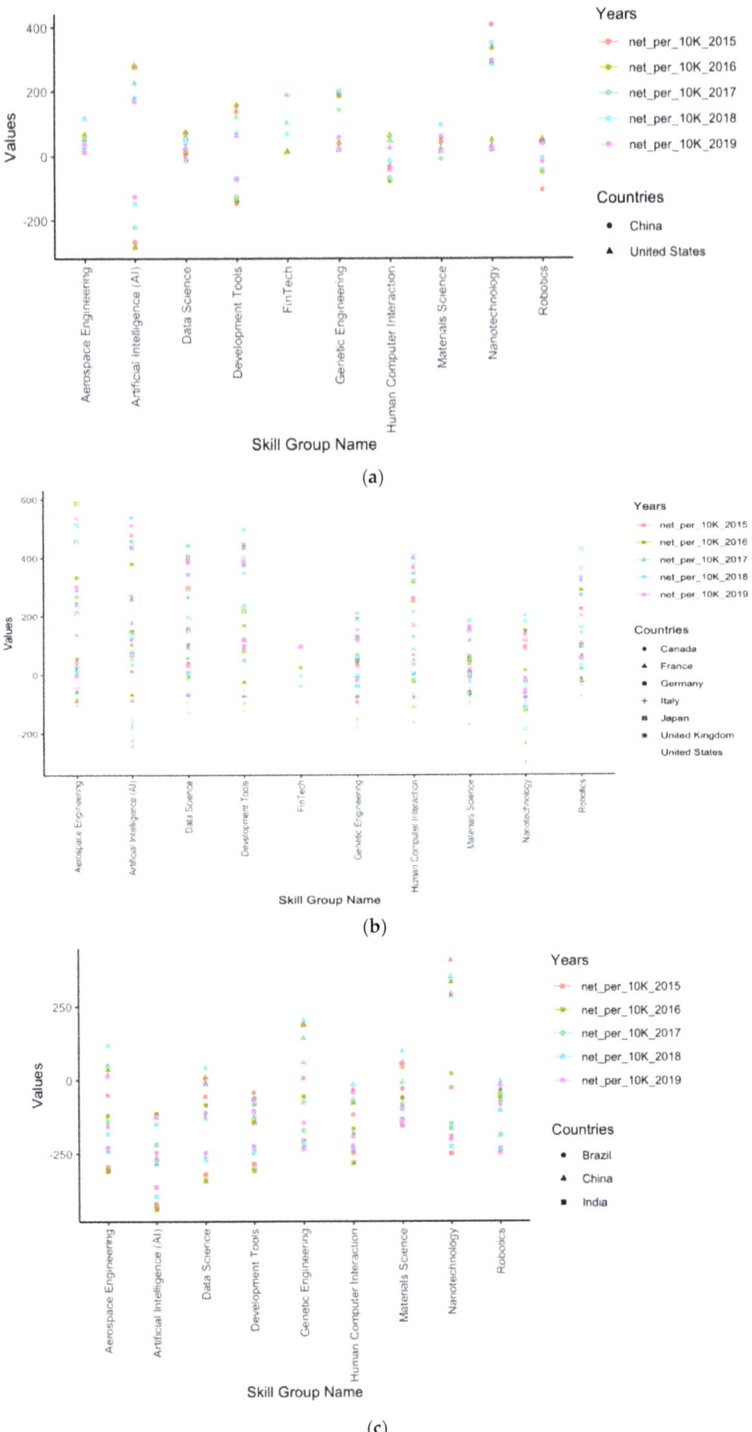

(**a**)

(**b**)

(**c**)

Figure 5. Disruptive-tech skills migration in (**a**) China and the US, (**b**) G7 countries, and (**c**) BIC countries.

The overall trend for disruptive-tech skills migration is positive for China and the US (Figure 5a). A weak positive tendency was found in aerospace engineering, data science, fintech (only in the US, Chinese data was not available), genetic engineering, and materials science. In the remaining skills (artificial intelligence (AI), development tools, human computer interaction, and robotics) a countercurrent was found between these countries, China having a negative migration trend and the US having a positive trend. For example, in China, AI showed a negative migration trend, while, in contrast, a positive migration trend was observed in the US.

Most of the G7 countries showed a very positive trend for migration of disruptive-tech skills, demonstrating the importance of this skill in high-tech jobs. The leading countries with higher values in most skills were Germany and Japan. Only Italy presented negative migration tendency in all skills. The fintech skill was presented only in the United Kingdom's data.

Similar to Italy, Brazil and India tended more to the negative position. When comparing the G7 countries with the BIC countries, the G7, except for Italy, had very high values for positive migration for these high-tech skills, while the BIC countries showed mixed results, tending more to the negative position, particularly Brazil and India, with China showing a mixed tendency, with both positive and negative skills migration.

Focusing on the disruptive-tech skills that are directly related to Industry 4.0, namely artificial intelligence, data science, development tools, human computer interaction, and robotics, the US and the G7 countries, except Italy, showed positive migration trends for these skills.

Table 2 presents a summary of the trends for the three categories of skills migration within the three defined groups.

Table 2. Summary of the skills migration trend analysis.

		Groups		
		China and the US	**G7**	**BIC**
Skills	Specialized Industry	China, US	G7, except the US	All BIC
	Soft	China and, US	G7, except the US	Brazil and India, China and
	Disruptive Tech	China and the US	G7, except Italy	Brazil and India, China
	Disruptive Tech in Industry 4.0	China, US	G7, except Italy	BIC

4.2.2. Openness and Happiness

As open data platforms are providers of innovative services, this work correlated the extent of available openness in data with people's happiness. Open-data-friendly countries can establish happier sustainable societies and serve as an example of social success for the rest of the globe. To study this topic, the Open Data Scoring from the ODIN [40] dataset was used to evaluate openness across different countries, with scores from 0 to 100 and considering the values for the year 2020. From the sustainability perspective of the social pillar, the World Happiness Report from 2020 [41] and its respective dataset were used to evaluate social happiness across different countries, with a score from 0 to 10. This report is a survey published by the UN that ranks 156 countries, reviewing their state of social happiness. To better understand how different regions of the globe experience this correlation, Figure 6 outlines the clusters of the openness and social happiness scores from different regions for 2020 and the respective trendlines.

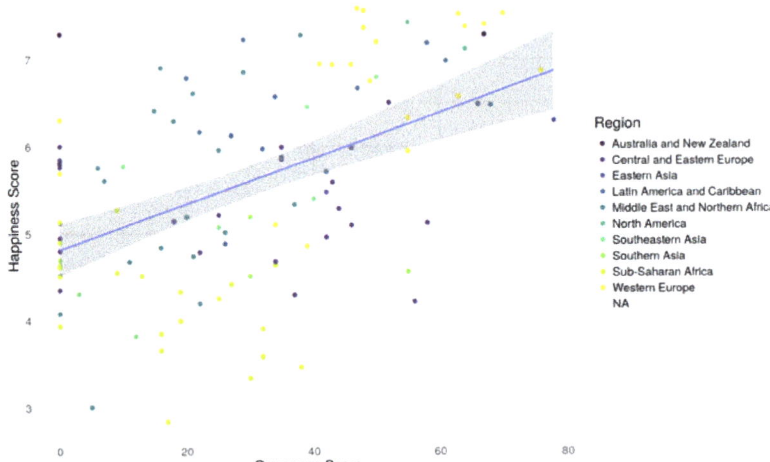

Figure 6. Openness and social happiness clustering and correlation in different regions.

Additionally, Figure 7 shows the correlation between openness and social happiness, for three established associations of countries: the G7 countries (Canada, France, Italy, Germany, Japan, the United Kingdom, and the US), the BRIC countries (Brazil, Russia, India, and China), and the Southeast Asian ASEAN countries (Brunei, Cambodia, Indonesia, Laos, Malaysia, Myanmar, the Philippines, Singapore, Thailand, and Vietnam).

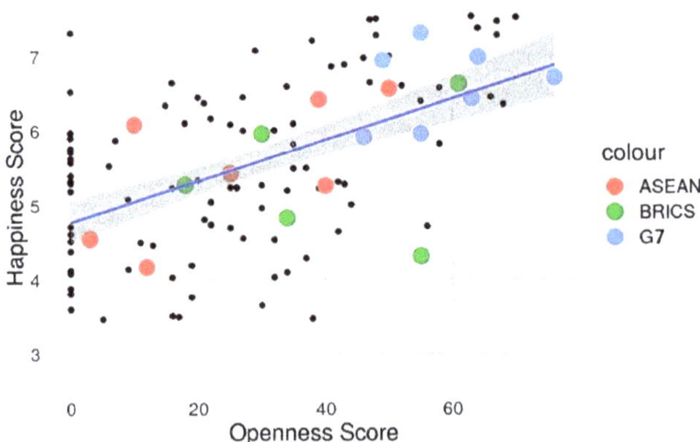

Figure 7. Clustering for the correlation between openness and happiness in the ASEAN, BRIC, and G7 countries.

The G7 countries have much higher openness in their data policies, as well as bigger indices of social happiness than the BRIC and ASEAN countries, which places the G7 clusters in the upper-right corner of the plot. The BRIC countries have a somewhat contradictory behavior, since the cluster with the second-highest openness score is also the one with the lowest happiness score, while the second lowest in openness is the second highest in happiness. Finally, ASEAN countries show clusters in the bottom-left corner, as well as clusters closer to the upper-right corner. Like the G7, this group closely matches the trend line, which means that countries in this group with high openness also tend to have high social happiness.

These visualizations allowed this study to conclude that a positive correlation exists between openness and happiness, both relevant aspect of Industry 4.0 and sustainability, especially the social pillar.

4.3. Relating the Open Data for Industry 4.0 and the Open Data for Sustainability

The third part of the analysis and results presentation involves establishing a connection between the findings from the results of the open data for Industry 4.0 and those from the open data for sustainability. To this end, Table 3 was developed to elucidate some links between the results obtained, namely the findings more relevant to the items considered in the scope of Industry 4.0 and those relevant to sustainability for the two economic blocs G7 and BIC. This table only considers the existence of simultaneous high significant values for both classes of parameters. This means that when G7 or BIC country(ies) appear in a specific square of the table, this indicates that for the country(ies) in these economic blocs significant values were verified for both entries of the table. When this apparent relationship was not verified, this is signified in the table by use of the term "No evidence".

Table 3. Matrix relating the Open Data for Industry 4.0 and Open Data for Sustainability.

		Industry 4.0		
		Manufacturing Value Added to GDP	**Smart City Index**	**R&D Efforts for Innovation**
Sustainability	Specialized Skills	Germany, Italy, Japan, and the US.	Canada, France, Germany, Italy, Japan, and the United Kingdom.	No evidence
	Soft Skills	Germany, Italy, Japan, and the US.	Canada, France, Germany, Italy, Japan, and the United Kingdom.	No evidence
	Disruptive Tech	Germany, Italy, Japan, and the US.	Canada, France, Germany, Japan, the United Kingdom, and the US.	No evidence
	Openness and Happiness	Germany, Italy, Japan, and the US.	Canada, France, Germany, Italy, Japan, the United Kingdom, and the US.	No evidence

According to Table 3, it can be seen that there are a considerable number of countries (belonging to the G7 and BIC) that demonstrated a positive tendency for both entries (Industry 4.0 and sustainability) of the table. However, this observation should be seen as a possible positive correlation between the entries of the table for these countries. At this moment, this work does not have the scope to evaluate the correlation, but the aim is to open up ideas for future research on the scope of the relationship between Industry 4.0 and sustainability. From the tables presented, data can be observed that seem to suggest the existence of some positive relationships between specialized skills, soft skills, and disruptive-tech skills, as well as openness and happiness, on the one hand and manufacturing value added to GDP, Smart City Index, and R&D efforts for innovation on the other hand.

5. Conclusions

In this study, the theme of Industry 4.0 and social sustainability was approached through the lens of openness. Data Science using open data, together with analytical open-source tools, provided the method for generating an inductive meta-analysis, with the resultant conclusions and contributions providing a better understanding of the technological and economical elements of Industry 4.0 and the social aspect of sustainability.

A broader conclusion from this work is that society, with its computational means, open-source tools, and data science know-how, when combined with an environment that offers open data, could facilitate the treatment and analysis of data that would offer new solutions for a happier knowledge society.

Manufacturing still carries significant weight in worldwide value added to GDP; countries in Asia, particularly China with a share around 40%, as well as Europe and the

United States, depend on this sector, and for this reason the concept of Industry 4.0 must be considered quite relevant. Due to the technological premise upon which Industry 4.0 is based, a relevant showcase of countries' willingness to embrace smart environments is the smart city. When considering the 20 leading countries in the overall rankings for 2020, Europe led, with a number of its countries in the leading group, followed by Asia, America, and Oceania, while Canada was the highest scoring smart city. Investment plays an important role in the successful implementation of Industry 4.0, and according to the data on industry expenditure on R&D and patent share, the industries investing most in Industry 4.0 are computers and electronics, pharmaceuticals, transport equipment, and IT (information technology) services.

In relation to the need for more attention to be paid to the social pillar within the concept of sustainability, the migration of technical and social skills was examined in this study. Most specialized industry skills revealed a positive migration tendency in most of the G7 countries, there was a particularly stable situation in the US, and a negative migration trend in the BIC countries. In soft skills migration, China and the G7 countries presented a positive situation, but Brazil and India maintained a negative position. The soft skills with the highest positive migration were problem solving, teamwork and time management. For all disruptive-tech skills, the BIC countries showed negative migration movement, China assumed a mix tendency depending on the skill, and almost all G7 countries revealed a moderate-to-high positive migration trend in all skills. Italy was the only exception to this trend. Germany and Japan showed a high migration rate in these high-technology jobs. Linking with the Industry 4.0 concept, five of these skills were identified: artificial intelligence, data science, development tools, human computer interaction, and robotics. A positive migration tendency was found in the US and the G7 countries, except Italy. The open approach is known as a movement that tends to become oriented toward involvement of communities and the leveraging of social elements as part of the concept of social sustainability, and in this study, based on two approaches in different worldwide regions, a positive correlation between openness and happiness is concluded to exist.

Some positive relationships might exist between specialized skills, soft skills and disruptive-tech skills, as well as openness and happiness, on the one hand, and manufacturing value added to GDP, Smart City Index, and R&D efforts for innovation on the other hand.

Regarding the Industry 4.0 and sustainability concepts, the above are the main theoretical contributions considering the limitations of an inductive study for this extremely complex context and topics. Concerning the practical and managerial implications of this study, it stresses the role of open data as used in data science for Industry 4.0, sustainability and their relationship, identifying which industries, countries, economic blocs, and continents provide an attractive environment for Industry 4.0, sustainability and their relationship. Researchers, policy makers, technological investors, and manufacturing industry owners and managers could make better research and business decisions based on the findings presented throughout this study.

In future work, the investigation of smart factories by industry, country, economic bloc, and world region might prove important, should this data become available.

Author Contributions: Conceptualization, H.C. and F.C.; methodology, H.C. and F.C.; software, F.C., T.F. and L.F.; validation, H.C., P.Á., M.C.-C., L.F., G.D.P. and J.B.; formal analysis, H.C.; investigation, H.C. and F.C.; resources, F.C.; data curation, F.C., T.F. and L.F.; writing—original draft preparation, H.C. and P.Á.; writing—review and editing, P.Á., M.C.-C., L.F., G.D.P. and J.B.; visualization, H.C.; supervision, H.C.; project administration, H.C.; funding acquisition, P.Á. All authors have read and agreed to the published version of the manuscript.

Funding: This work was financed by national funds through the Portuguese funding agency, FCT—Fundação para a Ciência e a Tecnologia, within project LA/P/0063/2020.

Data Availability Statement: The data presented in this study are openly available and presented along the paper.

Conflicts of Interest: The authors declare no conflict of interest.

References

1. Grybauskas, A.; Stefanini, A.; Ghobakhloo, M. Social Sustainability in The Age of Digitalization: A Systematic Literature Review on The Social Implications of Industry 4.0. *Technol. Soc.* **2022**, *70*, 101997. [CrossRef]
2. Sindhwani, R.; Afridi, S.; Kumar, A.; Banaitis, A.; Luthra, S.; Singh, P.L. Can Industry 5.0 Revolutionize the Wave of Resilience and Social Value Creation? A Multi-Criteria Framework to Analyze Enablers. *Technol. Soc.* **2022**, *68*, 101887. [CrossRef]
3. Putnik, G.D.; Ávila, P.S. Manufacturing System and Enterprise Management for Industry 4.0: Guest Editorial. *FME Trans.* **2021**, *49*, 769–772. [CrossRef]
4. Klingenberg, C.O.; Borges, M.A.V.; Antunes, J.A.V., Jr. Industry 4.0 as a Data-Driven Paradigm: A Systematic Literature Review on Technologies. *J. Manuf. Technol. Manag.* **2019**, *29*, 910–936. [CrossRef]
5. Boehmke, B.; Hazen, B.; Boone, C.A.; Robinson, J.L. A Data Science and Open Source Software Approach to Analytics for Strategic Sourcing. *Int. J. Inf. Manag.* **2020**, *54*, 102167. [CrossRef]
6. Moeuf, A.; Pellerin, R.; Lamouri, S.; Tamayo, S.; Barbaray, R. The Industrial Management of Smes in the Era of Industry 4.0. *Int. J. Prod. Res.* **2017**, *56*, 1118–1136. [CrossRef]
7. Ingaldi, M.; Ulewicz, R. Problems with the Implementation of Industry 4.0 in Enterprises from the SME Sector. *Sustainability* **2020**, *12*, 217. [CrossRef]
8. Ranjan, J.; Foropon, C. Big Data Analytics in Building the Competitive Intelligence of Organizations. *Int. J. Inf. Manag.* **2020**, *56*, 102231. [CrossRef]
9. Mikalef, P.; Boura, M.; Lekakos, G.; Krogstie, J. Big Data Analytics Capabilities and Innovation: The Mediating Role of Dynamic Capabilities and Moderating Effect of the Environment. *Br. J. Manag.* **2018**, *30*, 272–298. [CrossRef]
10. Provost, F.; Fawcett, T. Data Science and its Relationship to Big Data and Data-Driven Decision Making. *Big Data* **2013**, *1*, 51–59. [CrossRef] [PubMed]
11. Bilal, M.; Oyedele, L.O.; Qadir, J.; Munir, K.; Ajayi, S.O.; Akinade, O.O.; Owolabi, H.A.; Alaka, H.A.; Pasha, M. Big Data in the Construction Industry: A Review of Present Status, Opportunities, and Future Trends. *Adv. Eng. Inform.* **2016**, *30*, 500–521. [CrossRef]
12. Saritha, B.; Bonagiri, R.; Deepika, R. Open Source Technologies in Data Science and Big Data Analytics. *Mater. Today Proc.* **2021**, withdrawn. [CrossRef]
13. Runeson, P.; Olsson, T.; Linåker, J. Open Data Ecosystems—An Empirical Investigation into an Emerging Industry Collaboration Concept. *J. Syst. Softw.* **2021**, *182*, 111088. [CrossRef]
14. Gandomi, A.; Haider, M. Beyond the Hype: Big Data Concepts, Methods, and Analytics. *Int. J. Inf. Manag.* **2015**, *35*, 137–144. [CrossRef]
15. Runeson, P. Open Collaborative Data—Using OSS Principles to Share Data in SW Engineering. In Proceedings of the IEEE/ACM 41st International Conference on Software Engineering: New Ideas and Emerging Results (ICSE-NIER), Montreal, QC, Canada, 25–31 May 2019; pp. 25–28. [CrossRef]
16. Hickin, R.; Bechtel, M.; Golem, A.; Erb, L.; Buscalno, R. Technology Futures: Projecting the Possible, Navigating What's Next. April 2021. Available online: https://www3.weforum.org/docs/WEF_Technology_Futures_GTGS_2021.pdf (accessed on 26 December 2022).
17. Hall, T. The Role of Data in Industry 4.0. 20 May 2020. Available online: https://industrytoday.com/the-role-of-data-in-industry-4-0/ (accessed on 16 December 2022).
18. Cammarano, A.; Varriale, V.; Michelino, F.; Caputo, M. Open and Crowd-Based Platforms: Impact on Organizational and Market Performance. *Sustainability* **2022**, *14*, 2223. [CrossRef]
19. Czvetkó, T.; Honti, G.; Abonyi, J. Regional Development Potentials of Industry 4.0: Open Data Indicators of the Industry 4.0+ Model. *PLoS ONE* **2021**, *16*, e0250247. [CrossRef]
20. Sołtysik-Piorunkiewicz, A.; Zdonek, I. How Society 5.0 and Industry 4.0 Ideas Shape the Open Data Performance Expectancy. *Sustainability* **2021**, *13*, 917. [CrossRef]
21. Gronle, M.; Grasso, M.; Granito, E.; Schaal, F.; Colosimo, B.M. Open Data for Open Science in Industry 4.0: In-Situ Monitoring of Quality in Additive Manufacturing. *J. Qual. Technol.* **2022**, *55*, 1–13. [CrossRef]
22. Wee, D.; Kelly, R.; Cattel, J.; Breuning, M. *Industry 4.0-How to Navigate Digitization of the Manufacturing Sector*; Mckinsey & Company: New York, NY, USA, 2015; Volume 58.
23. Kamble, S.S.; Gunasekaran, A.; Gawankar, S.A. Sustainable Industry 4.0 framework: A Systematic Literature Review Identifying the Current Trends and Future Perspectives. *Process Saf. Environ. Prot.* **2018**, *117*, 408–425. [CrossRef]
24. Varela, L.; Araújo, A.; Ávila, P.; Castro, H.; Putnik, G. Evaluation of the Relation between Lean Manufacturing, Industry 4.0, and Sustainability. *Sustainability* **2019**, *11*, 1439. [CrossRef]
25. Kumar, V.; Vrat, P.; Shankar, R. Factors Influencing the Implementation of Industry 4.0 for Sustainability in Manufacturing. *Glob. J. Flex. Syst. Manag.* **2022**, *23*, 453–478. [CrossRef]

26. Contini, G.; Peruzzini, M. Sustainability and Industry 4.0: Definition of a Set of Key Performance Indicators for Manufacturing Companies. *Sustainability* **2022**, *14*, 11004. [CrossRef]
27. Shet, S.V.; Pereira, V. Proposed Managerial Competencies for Industry 4.0—Implications for Social Sustainability. *Technol. Forecast. Soc. Change* **2021**, *173*, 121080. [CrossRef]
28. Grabowska, S.; Saniuk, S.; Gajdzik, B. Industry 5.0: Improving Humanization and Sustainability of Industry 4.0. *Scientometrics* **2022**, *127*, 3117–3144. [CrossRef]
29. Hevner, A.R.; March, S.T.; Park, J.; Ram, S. Design Science in Information Systems Research. MIS Q. *Manag. Inf. Syst.* **2004**, *28*, 75–105. [CrossRef]
30. R Core Team. *R: A Language and Environment for Statistical Computing*; R Foundation for Statistical Computing: Vienna, Austria, 2019; Available online: https://cran.r-project.org/doc/manuals/r-devel/fullrefman.pdf (accessed on 30 May 2022).
31. Vallat, R. Pingouin: Statistics in Python. *J. Open Source Softw.* **2018**, *3*, 1026. [CrossRef]
32. Thames, L.; Schaefer, D. Industry 4.0: An Overview of Key Benefits, Technologies, and Challenges. In *Springer Series in Advanced Manufacturing*; Springer: Berlin/Heidelberg, Germany, 2017; pp. 1–33.
33. UN. *Goal 9—Industry, Innovation and Infrastructure, Sustainable Development Goals*; UN: New York, NY, USA, 2022.
34. Angelidou, M. Smart City Policies: A spatial approach. *Cities* **2014**, *41*, S3–S11. [CrossRef]
35. IMD. Smart City Observatory. Available online: https://www.imd.org/smart-city-observatory/home/ (accessed on 20 September 2022).
36. Mansfield, E.; Lee, J.-Y. The Modern University: Contributor to Industrial Innovation and Recipient of Industrial R&D Support. *Res. Policy* **1996**, *25*, 1047–1058. [CrossRef]
37. The World Bank. World Development Indicators. Available online: http://data.worldbank.org/data-catalog/world-development-indicators (accessed on 15 September 2022).
38. Kerr, S.P.; Kerr, W.; Özden, Ç.; Parsons, C. High-Skilled Migration and Agglomeration. *Annu. Rev. Econ.* **2016**, *9*, 201–234. [CrossRef]
39. The World Bank. Skills | LinkedIn Data. Available online: https://datacatalog.worldbank.org/search/dataset/0038027/Skills (accessed on 9 September 2022).
40. ODW. Open Data Inventory 2020/21 Annual Report. Available online: https://opendatawatch.com/publications/open-data-inventory (accessed on 11 May 2022).
41. Helliwell, J.F.; Layard, R.; Sachs, J.D.; Neve, J.-E.D. World happiness report 2020. Available online: https://worldhappiness.report/ed/2020/#appendices-and-data (accessed on 2 October 2022).

Article

Modelling of Determinants of Logistics 4.0 Adoption: Insights from Developing Countries

Shahbaz Khan [1], Rubee Singh [1], José Carlos Sá [2,3], Gilberto Santos [4,*] and Luís Pinto Ferreira [2,3]

[1] Institute of Business Management, GLA University, Mathura 281406, India
[2] School of Engineering, Polytechnic of Porto (ISEP), 4200-072 Porto, Portugal
[3] Institute of Science and Innovation in Mechanical and Industrial Engineering (INEGI), 4200-465 Porto, Portugal
[4] Design School, Polytechnic Institute Cavado and Ave, 4750-810 Barcelos, Portugal
* Correspondence: gsantos@ipca.pt

Abstract: With the emergence of industry 4.0, several elements of the supply chain are transforming through the adoption of smart technologies such as blockchain, the internet of things and cyber-physical systems. Logistics is considered one of the important elements of supply chain management and its digital transformation is crucial to the success of industry 4.0. In this circumstance, the existing logistics system needs to be upgraded with industry 4.0 technologies and emerge as logistics 4.0. However, the adoption/transformation of logistics 4.0 is dependent on several determinants that need to be explored. Therefore, this study has the prime objective of investigating the determinants of logistics 4.0 adoption in the context of a developing country, specifically, India. Initially, ten determinants of logistics 4.0 are established after a survey of the relevant literature and the input of industry experts. Further, a four-level structural model is developed among these determinants using the Interpretive Structural Modelling (ISM) approach. In addition, a fuzzy Matrix of Cross-Impact Multiplications Applied to Classification (MICMAC) analysis is also conducted for the categorization of these determinants as per their driving and dependence power. The findings show that top management supports, information technology infrastructure and financial investment are the most significant determinants towards logistics 4.0 adoption. This study facilitates the supply chain partners to focus on these high-level determinants for the effective adoption of logistics 4.0. Moreover, the findings lead to a more in-depth insight into the determinants that influence logistics 4.0 and their significance in logistics 4.0 adoption in emerging economies.

Keywords: determinants; industry 4.0; logistics 4.0; ISM; fuzzy MICMAC

Citation: Khan, S.; Singh, R.; Sá, J.C.; Santos, G.; Ferreira, L.P. Modelling of Determinants of Logistics 4.0 Adoption: Insights from Developing Countries. *Machines* **2022**, *10*, 1242. https://doi.org/10.3390/machines10121242

Academic Editor: Pingyu Jiang

Received: 28 October 2022
Accepted: 14 December 2022
Published: 19 December 2022

Publisher's Note: MDPI stays neutral with regard to jurisdictional claims in published maps and institutional affiliations.

1. Introduction

The pervasiveness of information and communication technologies provides the opportunity to incorporate "smartness" in factories and pushes the industry towards the next industrial revolution, i.e., the fourth industrial revolution. This industrial revolution focuses on the adoption of smart technologies to improve the system and process efficiency in order to achieve sustainability. In the current world, professionals, managers, and government representatives are becoming increasingly interested in Industry 4.0 as its implementation would boost national economies and corporate competitiveness [1–3]. To promote higher automation, industry 4.0 was first implemented in Germany [4,5]. This paradigm shift has been recognized by other countries, including the United States, China, United Kingdom, and India, through the adoption of several initiatives [5,6]. Although industry 4.0 originated in the manufacturing sector, it has significant implications for the supply chain [7]. The goal of industry 4.0 is to develop smart factories and supply chains by incorporating cutting-edge technologies such as analytics, big data, the Internet of Things (IoT) and Cyber-Physical Systems (CPS) [8,9]. These technologies are utilized to advance the many supply chain components, including planning, sourcing, manufacturing and logistics. It is widely recognized that logistics 4.0 is a substantial part of industry 4.0, and its adoption could benefit digitalization throughout the entire supply chain.

Logistics 4.0 is the logistical version of industry 4.0 and focuses on turning logistics activities into more effective activities with improved material and information flow, resulting in smart and intelligent logistics that are more precise, dependable, agile, and sustainable [10]. Some authors have recognized that logistics 4.0 is a new paradigm associated with industry 4.0. The logistics 4.0 concept entails integrating smart technologies with logistics subsystems, such as smart sensors, big data and IoT, and CPS. These technologies provide data-driven ecosystems that more successfully meet the customer's demand for customized products. Timm and Lorig [11] define logistics 4.0 as: " . . . a logistic system which consists of independent subsystems and behaviour of these subsystems depend on other surrounding subsystems". In addition to this, Wang [12] defined logistics 4.0 as " . . . a collective term for technologies and concepts of value chain organization. Within the logistics, CPS monitors physical processes creates a virtual copy of the physical world and makes decentralized decisions. Over the IoT, CPS communicate with machines and humans in real-time. Data mining discovers knowledge to support the decision-making process. Both internal and cross-organizational services are offered and utilized by participants of the value chain through the Internet of Services (IoS)".

In light of the above definitions, the combination of smart technology and logistics is referred to as "logistics 4.0" to address the demand for highly personalized goods and services. To implement logistics 4.0, more initiative and financial resources must be invested in the adoption of these cutting-edge technologies. This transformation is challenging because of the high cost of technology and infrastructure, complicated network of supply chains, global participation, and security and privacy issues.

The adoption of logistics 4.0 faces several challenges from the organizational, cultural and supply chain levels [13]. Despite these challenges, some factors are also present in the system that enable the adoption of logistics 4.0. In order to adopt logistics 4.0, these determinants need to be explored and analyzed in a comprehensive manner. The determinants of logistics 4.0 are the factors that facilitate its adoption. These determinants are the key factors that need to be addressed for the successful adoption of logistics 4.0. In addition, the shift from conventional logistics to logistics 4.0 is dependent on these determinates. To ensure a seamless and effective implementation of logistics 4.0, it is vital to explore these determinants. The adoption of logistics 4.0 is facilitated by several industry 4.0 technologies. The adoption of these technologies depends on external factors, including business environment, technological readiness, working culture and associated uncertainties. These factors differ between developed and developing countries. Therefore, the adoption of logistics 4.0 will not necessarily be identical because of their differing levels of infrastructure and economies of scale [14]. Due to this, the determinants of logistics 4.0 adoption differ for developing and developed countries. It is found that logistics 4.0 faces a wide range of business issues due to the lack of technological development, big data management, and market volatility [15]. In comparison with developed countries, such as the USA, United Kingdom and German, developing countries use relatively fewer sophisticated technological tools and techniques in the logistics sector [15]. Therefore, the adoption of logistics 4.0 needs to be addressed from the perspective of developed and developing countries [16]. The literature review reveals that the majority of current research focuses on logistics 4.0 adoption in developed countries. Only limited research has been conducted on logistics 4.0 adoption in emerging countries [10]. It is also important to note that the adoption of logistics 4.0 implementation has not been fully explored. Therefore, the disproportionate focus on developed countries in comparison to developing countries, the lack of guidance on the determinants of logistics 4.0 adoption, and the lack of sufficient evidence regarding logistics 4.0 adoption constitute compelling gaps in the existing literature. Hence, it is necessary to identify and analyse the determinants of logistics 4.0 in emerging economies in order to fill these research gaps. For efficient and effective adoption, managers also need to be aware of the structural relationships between the identified determinants in the context of developing countries. The structural relationship among the determinants of logistics 4.0 is rarely explored. Therefore, this study was conducted to identify and explore the structural relationship among determinants of logistics 4.0 adoption in the context of developing countries, and specifically, India. This study has the following research objectives to fill the knowledge gaps in the existing literature:

- To identify the key determinant of logistics 4.0 adoption in the context of developing countries
- To develop the structural relationship among the finalized determinant
- To categorize the determinant based on their driving and dependence power

We have identified the determinants to meet the aforementioned study objectives through a literature review and validated this with expert feedback. After the finalization of the logistics 4.0 adoption determinants, the structural relationship is developed between them using the ISM method. The results of this study will assist practitioners in implementing logistics 4.0 for incorporating sustainability into their supply chains. Logistics 4.0 and sustainability have a close relationship by reducing waste and carbon footprints, and creating new job prospects. The implementation of digital technologies in logistics improves the availability of items at the right place, time, and amount, hence reducing waste [17]. This study will also assist policymakers in formulating policies that support the systematic adoption of logistics 4.0. As the majority of the literature suggests, logistics 4.0 is not well adopted in developing countries and required further exploration. Therefore, the uniqueness of this paper lies in the identification of the determinants of logistics 4.0 in developing countries. With the findings of this study, supply chain partners that are located in developing countries will benefit in terms of logistics 4.0 adoption.

The remaining study is organized as follows: Section 2 gives a summary of studies relevant to logistic 4.0; the adopted solution methodology is provided in Section 3; Section 4 provides the data analysis; Section 5 discusses the findings; Section 6 provides the implications of the study and finally, Section 7 concludes the study and provides scope for future research.

2. Background of the Study

Several advancements and enhancements are introduced to the logistics industry under Industry 4.0. For instance, businesses are able to construct hyper-connected supply chains by implementing digital technologies at every stage of the planning process [18], to increase operational and financial limitations by capturing big data and operating at a different level of resilience and responsiveness [19]. In the framework of Industry 4.0, transportation and distribution networks have placed a significant focus on flexibility and resilience [20]. Moreover, as the supply chain partners become more integrated, logistics becomes an increasingly vital element for industry 4.0's success [21]. A digitalized supply chain and smart systems have enabled logistics to propose the name "logistics 4.0" as the equivalent of industry 4.0 [22].

Logistics 4.0's emergence from the industry 4.0 idea is crucial, not just for operational views such as sustainability, effectiveness, and customer responsiveness, but also for allowing advances in all basic business components [23]. Since logistics ensures the timely and appropriate availability of resources for manufacturing systems, logistics is ideally suited to exploring practical applications and reflecting industry 4.0 [24]. Therefore, logistics 4.0 adoption will significantly enhance the efficiency and effectiveness of the entire supply chain by optimizing the operations of logistics services. Tascón et al. [25] determine the logistics 4.0 service quality criteria and assess the impact of developing technologies on logistics sustainability. Their findings show that artificial intelligence, advanced robotics, blockchain and additive manufacturing are the most significant technologies that help in achieving logistics 4.0 sustainability. Similar to this, Parhi et al. [26] also develop a framework for the assessment of enabling factors that are responsible for the implementation of sustainable logistics 4.0 at various digitalization levels. They focused on the management perspective and found that technology infrastructure", "digital solutions", and "top management commitment", are the prime enablers for the adoption of logistics 4.0.

Sun et al. [27] focused on reverse logistics 4.0 and developed an integrated framework for smart reverse logistics by integrating Industry 4.0 enablers to achieve sustainable business objectives. Batz et al. [28] provide an overview of the maturity model developed for the assessment of Logistics 4.0.

The three primary characteristics of logistics 4.0 are vertical integration, horizontal integration and end-to-end engineering integration [23]. In vertical integration, different IT systems are integrated at different levels within a factory, whereas horizontal integration is about collaboration between firms, and end-to-end integration is about cross-linking stakeholders, products, and machines. In logistics 4.0, products are transported, and their information flow is controlled systemically from source to destination. With Logistics 4.0, customers can cost-effectively access logistics services by utilizing frontends and base technologies [29]. It primarily employs IoT, CPS, big data analytics, and cloud computing technologies [30], which help organizations to optimize their resources and provide enhanced value. The schematic diagram of logistics 4.0 is provided in Figure 1. Based on these technologies, advanced systems such as intelligent transportation systems, warehouse management systems, information security, and independent order processing through blockchain technologies and smart contracts are operated [21,24]. As a result of logistics 4.0 initiatives, organizations can reduce the costs of logistics (such as labour costs), boost productivity, and improve customer satisfaction [31–33].

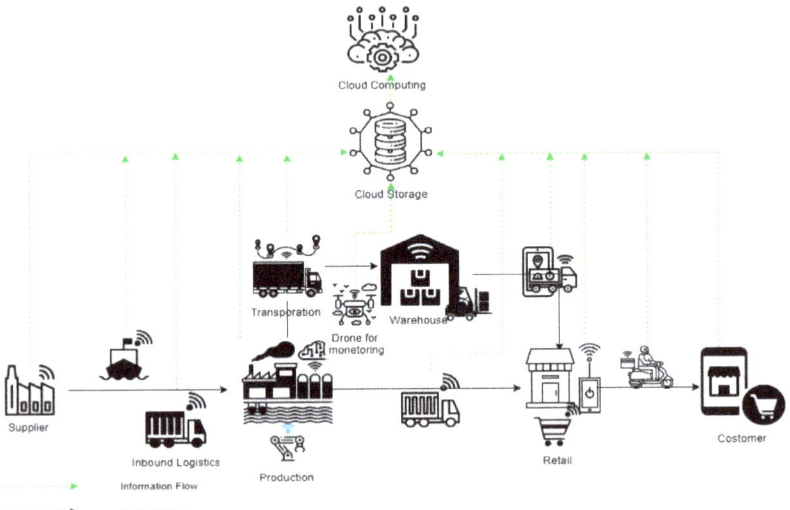

Figure 1. Conceptual structure of Logistics 4.0 (Source: Authors).

Despite these benefits, the adoption of logistics 4.0 is limited, particularly in developing countries. Studies pertaining to the adoption of logistics 4.0 are relatively insufficient. Typically, most of the studies focus on the technological side of logistics 4.0 and how these technologies facilitate adoption. For example, Atzeni et al. [34] claim that a robot is essential to logistics 4.0, assisting with picking, operations and placing a proposed idea of Cobots, or collaborating robots, can be used in logistics 4.0. Additionally, Markov and Vitliemov [35] investigated how blockchain technologies were applied to automobile supply chains and their logistics. They added that blockchain technologies have a significant positive impact on logistics 4.0. Further, they found potential supply chain and logistics 4.0 opportunities for creating sustainable supply chains and logistics.

From the operational perspective, some studies are available related to the adoption of logistics 4.0. A framework for logistics 4.0 was evaluated and purposed by Winkelhaus and Grosse [30]. They examine how this framework might be used to identify future logistics strategies and technological advancements that will enable sustainable logistics operations. They also proposed new technology solutions, such as IoT, CPS, and big data, to address current and future expectations. Kucukaltan et al. [36] observed from a multidimensional perspective to reflect the impact of Industry 4.0 on logistics. It may be impacted by the growth of industries when modifications are made to operational, financial, and human resource factors.

Several studies address the impact of logistics 4.0 on business performance [37,38]. To achieve sustainability, Torbacki and Kijewska [39] investigated the production methods that may be employed in logistics 4.0 in association with the performance characteristics of logistics. Additionally, it concentrated on manufacturing and logistics performance indicators, both of which can be used to gauge a business's performance. Additionally, Nantee and Sureeyatanapas [38] analyzed the impact of logistics 4.0 projects, such as automated warehouse systems, on sustainability performance in various industries. Kodym et al. [40] stated that throughout the logistics and manufacturing process, the supply chain may become more intelligent, efficient, and transparent with the help of digital transformation. They emphasized several cutting-edge technologies, including blockchain, IoT, big data, data mining, and machine learning, which professionals can utilize to identify the risks associated with logistics 4.0. Bag et al. [41] highlight three specific areas in which an organizations' performance is impacted by logistics 4.0 adoption: environmental, organizational, and technological. Planning and scheduling can help to save maintenance costs by utilizing some of these features. Additionally, the logistics 4.0 process can enhance businesses' manufacturing operations through sustained communication and visibility.

Some studies have evaluated the ways in which industry 4.0 technology may affect the operation of logistics [42,43]. Through integrating the lean 4.0 idea, these technologies are enhancing logistical processes by reducing waste. Industry 4.0 is not just a technological term, but rather an amalgamation of social and organizational circumstances [44]. However, Wagner et al. [43] suggested that a lean production system be utilized to estimate the first step in industry 4.0, particularly when a CPS-based Just-in-Time (JIT) method is being used. On the other hand, Rosin et al. [45] advise combining Jidoka and Industry4.0 for optimal results. They claim that IoT and simulation are the two technologies that are most frequently recommended for combining Lean 4.0 and Industry 4.0. For such integration, a cluster with the necessary knowledge base, IT solution expertise, robotics, and automation is required [46–49].

3. Materials and Methods

The three-stage framework is developed to fulfil the research objectives, as shown in Figure 2. An initial literature review, along with validation from multiple experts, has been used to identify the determinants of logistics 4.0. Further, these determinants of logistics 4.0 adoption are modelled using the ISM approach for the exploration of structural relationships. ISM has attracted researchers from different disciplines because it segregates complex relationships between elements of a system into hierarchical structures. It is widely used in the analysis of interrelationships among variables affecting a system, for example, barriers, critical success factors, determinants and drivers. The method has been successfully used in a variety of management studies [50,51]. For example, Khan et al. [52] applied the factors of remanufacturing adoption in the context of emerging economies to the model. Yadav, Luthra and Garg [53] applied the ISM to model the barriers of IoT integration in the agri-food supply chain. In the third stage, the fuzzy MICMAC analysis is used for the validation of the ISM model. The fuzzy MICMAC analysis also classifies the determinants into four groups based on driving and dependence power. The ISM–MICMAC methodology has been applied and recommended for use in management issues such as lean manufacturing [54,55], cold supply chain [56], digital supply chain [57], risk management [58], and industry 4.0 adoption. However, the ISM–MICMAC methodology has not been applied to logistics 4.0. Therefore, the ISM and fuzzy MICMAC is acknowledged by adopting it to solve the problems of logistics 4.0.

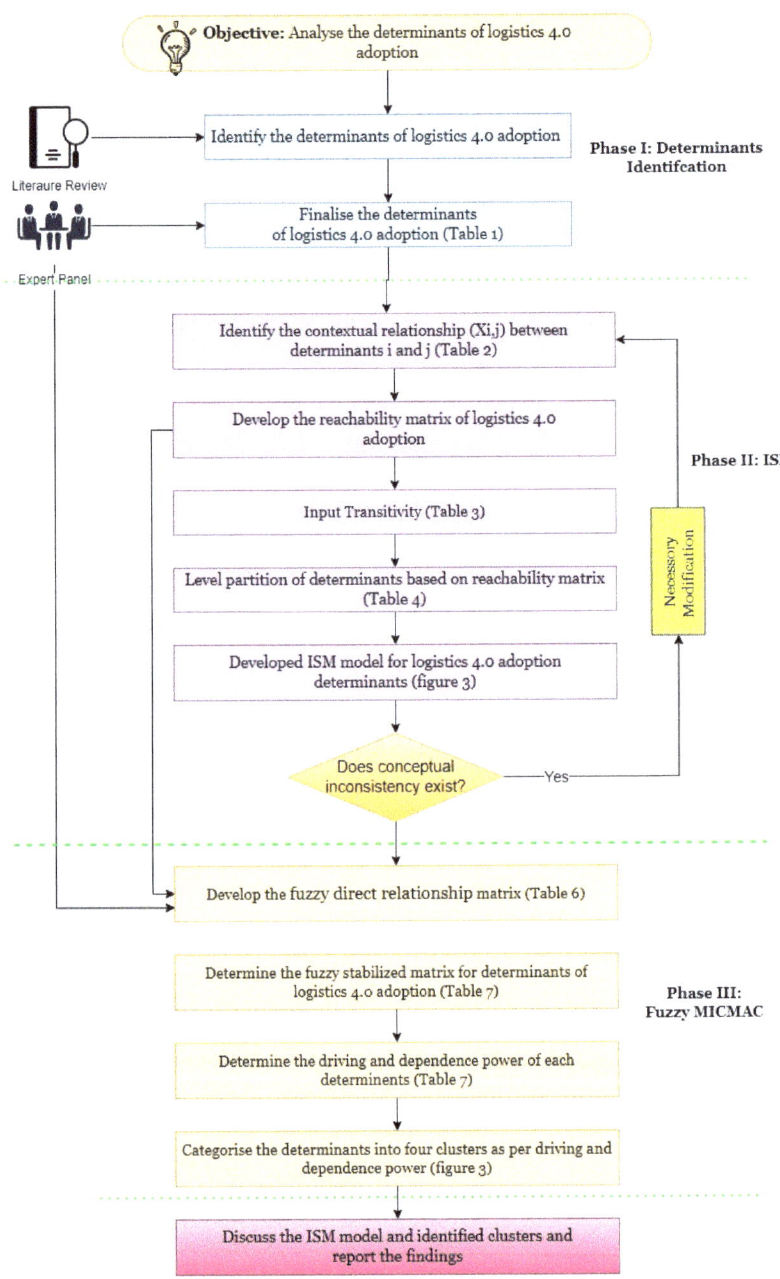

Figure 2. Proposed framework for modelling of determinants of logistics 4.0 adoption (Source: Authors).

3.1. ISM Method

The ISM approach was discovered by Warfield [59]. This approach has been frequently utilized in the past to build structural relationships between identified obstacles/factors/determinants. The steps to the ISM technique are as follows:

Step 1: The first step is to determine the factors that influence the system under study.

Step 2: Analyze the contextual interactions between the elements of the system to create a self-interaction matrix (SSIM).

Step 3: Converting the linguistic characters to binary values to establish an initial reachability matrix.

Step 4: By checking the transitive relationships, obtain the final reachability matrix.

Step 5: To determine the levels of variables, level partitioning is carried out.

Step 6: Create the ISM model by placing each variable at its appropriate level.

Step 7: The developed ISM model is reviewed to ensure that there are no theoretical irregularities, and modifications are made as necessary.

3.2. MICMAC Analysis

MICMAC was developed by Duperrin and Godet [60]. MICMAC enables the determination of the driving power and the dependence power of a variety of elements. The main purpose of MICMAC analysis is to examine how variables affect each other and how they are dependent on each other. Based on their driving and dependency power, these variables can be classified into four groups: autonomous variables, dependent variables, linkage variables, and driving variables. In order to evaluate the sensitivity of MICMAC, the fuzzy theory is combined with traditional MICMAC. Using fuzzy MICMAC, it is possible to measure the direct and indirect effects of relationships between variables. The driving power of a variable can be determined by summing the entries in its row in a fuzzy stabilized matrix. Similarly, the importance of a variable can be calculated by summing the entries in the columns of the fuzzy stabilized matrix.

4. Data Analysis

4.1. Stage 1: Determinants of Logistics 4.0

As a result of the literature review of relevant articles, the initial determinants of logistics 4.0 adoption have been identified. The articles for the literature survey were selected using the Scopus database, which is the largest collection of scientific journals with peer review. Next, the keywords for the literature search were finalized, which included "logistic 4.0", "smart logistics" "digital logistics", "supply chain 4.0", "drivers", and "determinates". The combinations (using a Boolean operator) of these keywords were used for the appropriate article identification. Further, the identified articles were reviewed in order to prepare the initial list of logistics 4.0 adoption determinants. An expert panel of eight members was formed, including five members from industry, two from academia and one policy planner. These industry professionals are well-versed in logistics 4.0 and supply chain digitalization. All the industrial participants in the study have management experience in the logistics industry of more than eight years. These professionals are working in a high-repute logistics company which have a minimum staff strength of 150 and a multinational market base, located in India. Additionally, two academic experts with specialized knowledge of logistics 4.0 operations and an understanding of industry 4.0 have been involved with this study. These two experts are also working in the Indian university at the professor and associate professor level. The details of the experts are provided in Table A1. As a result of the formation of the expert panel, the identified determinants list, which consists of twelve factors, is presented to the panel for its consideration. They were asked to check the relevance of the determinant in the context of the contemporary business environment of developing countries. After the discussion, the expert panel recommended eliminating two determinants due to their irrelevance in the contemporary business environment of developing countries. In this manner, ten determinants of logistics 4.0 adoption are finalized and depicted in Table 1.

Table 1. Determinants of logistics 4.0 adoption.

S. No	Determinants	Description	References
1.	IT Infrastructure	Logistic 4.0 requires IT infrastructure including IoT, big data, and CPS to meet the technological requirement of industry 4.0	[61–63]
2.	Mutual Trust	Logistics 4.0 is a sequence of interconnected activities with the integration of advanced technologies and the realization of logistics 4.0 depend on the mutual trust among the logistics partners	[41,64]
3.	Knowledge Management	Created a training program to foster continual learning which facilitates knowledge transmission, encouraging companies to adopt logistics. 4.0.	[65–67]
4.	Analytical competencies	Data is the main ingredient of Logistics 4.0 and their adoption relies on data analytics, which requires strong analytical skills.	[30,38]
5.	Digital work culture	Logistics 4.0 requires smart environments, new job descriptions, roles, and responsibilities that help to develop a smart work culture.	[63,68]
6.	Organizational strategies for logistics 4.0	Numerous logistics 4.0 initiatives must be integrated with organizational strategies for greater coordination to embrace logistics 4.0.	[69,70]
7.	Top management support	Logistics 4.0 demands several technology integrations, skill development and policy changes therefore, logistics 4.0 adoption requires top management commitment and support.	[6,71]
8.	Collaboration	For logistics 4.0 to succeed, it is essential to cultivate a collaborative relationship with logistics partners	[72]
9.	Financial investment	It is well-known that the adoption of innovative solutions, such as logistics 4.0, is strongly influenced by the amount of investment made into them.	[10,73]
10.	Skill development	As part of the logistics 4.0 initiative, seminars and workshops will be held to assist in the development of the essential analytical and technical capabilities needed	[74,75]

4.2. Stage 2: ISM Modelling

ISM methodology begins with the development of an initial structural self-interaction matrix (SSIM) showing relationships among variables. With the help of expert consultation, contextual relationships are developed between identified factors. To diagnose the interdependencies among the identified determinants of logistics 4.0 adoption, a contextual relationship of the type 'leads to' is considered. For instance, top management support leads to skill development. In the same manner, contextual relationships among determinants are developed, taking into account a determinant's contextual relationship with other determinants, the existence of any relation between two determinants (i and j), and their associated direction of relationship A between determinants (i and j) can be expressed using four symbols:

V: determinant i will lead to determinant j

A: determinant j will lead to determinant i

X: determinant i and j will lead to each other

O: determinant i and j are unrelated.

On the basis of contextual relationships, an SSIM was developed for the determinants of logistics 4.0 adoption and the same is shown in Table 2.

Table 2. SSIM for the determinants of logistic 4.0 adoption.

Determinants	10	9	8	7	6	5	4	3	2	1
1	V	A	V	X	V	O	V	O	O	
2	V	O	X	A	A	V	V	X		
3	V	A	X	A	A	V	V			
4	X	A	A	A	O	X				
5	X	A	A	A	A					
6	V	A	V	A						
7	V	X	V							
8	V	A								
9	V									
10										

By substituting V, A, X, and O with 1 and 0, the SSIM is transformed into a binary matrix, called the initial reachability matrix.

- if the (i, j) entry in the SSIM is V, then the (i, j) entry in the reachability matrix becomes 1 and the (j, i) entry becomes 0
- if the (i, j) entry in the SSIM is A, then the (i, j) entry in the reachability matrix becomes 0 and the (j, i) entry becomes 1
- if the (i, j) entry in the SSIM is X, then the (i, j) entry in the reachability matrix becomes 1 and the (j, i) entry becomes 1
- if the (i, j) entry in the SSIM is O, then the (i, j) entry in the reachability matrix becomes 0 and the (j, i) entry becomes 0.

As a result of applying the transitivity principle to the initial reachability matrix, the final reachability matrix can be constructed and is shown in Table 3. Furthermore, this table also depicts the driving and dependence power of each determinant. Driving power is defined as the number of determinants it influences, while dependence is the number of determinants that affect it.

Table 3. Final reachability matrix.

Determinants	1	2	3	4	5	6	7	8	9	10
1	1	1*	1*	1	1*	1	1	1	1*	1
2	0	1	1	1	1	0	0	1	0	1
3	0	1	1	1	1	0	0	1	0	1
4	0	0	0	1	1	0	0	0	0	1
5	0	0	0	1	1	0	0	0	0	1
6	0	1	1	1*	1	1	0	1	0	1
7	1	1	1	1	1	1	1	1	1	1
8	0	1	1	1	1	0	0	1	0	1
9	1	1*	1	1	1	1	1	1	1	1
10	0	0	0	1	1	0	0	0	0	1

1 shows the transitivity.*

The final reachability matrix could be partitioned into different levels with the help of the reachability and antecedent sets for each determinant. Each determinant has a reachability set consisting of the determinant itself and any other determinants that may contribute to achieving it, and an antecedent set consisting of the determinant itself and any other determinants that may support the achievement of it [60]. In other words, the reachability set consists of determinants that have one (row-wise) in the final reachability matrix.

For example, the reachability set of *determinant 2* is calculated as it has 0,1,1,1,1,0,0,1,0,1 (refer to the second row of Table 3) and their reachability set should be 2,3,4,5,8,10 (places where 1 exists). Similar to this, the antecedent set consists of the determinants that have one (column-wise) in the final reachability matrix. For instance, the antecedent set of *determinants 2* can be determined as it has 1,1,1,0,0,1,1,1,1,0 (refer to the second column of Table 3) and their antecedent set should be 1,2,3,6,7,8,9 (places where 1 exists). Further, the intersection (common elements in both sets) of reachability and the antecedent set are identified for all the determinants. Those determinants whose intersection set and reachability set are the same are labelled as Level I, and placed at the highest position in the ISM model. Following this, the top-level determinants are discarded, and subsequent iterations are repeated until all the determinants are placed. A digraph and ISM model were constructed based on the identified levels of variables.

Table 4 present the level partition that is used to construct the structural model of the determinants of logistics 4.0 adoption and based on the level of each determinant the digraph that is generated by removing transitivity in accordance with the ISM methodology. The digraph is transformed into the ISM model by labelling them appropriately, as shown in Figure 3.

Table 4. Shows the level of each determinant of logistics 4.0 adoption.

Factors	Reachability Set	Antecedent Set	Intersection	Level
1	1,2,3,4,5,6,7,8,9,10	1,7,9		
2	2,3,4,5,8,10	1,2,3,6,7,8,9		
3	2,3,4,5,8,10	1,2,3,6,7,8,9		
4	4,5,10	1,2,3,4,5,6,7,8,9,10	4,5,10	I
5	4,5,10	1,2,3,4,5,6,7,8,9,10	4,5,10	I
6	2,3,4,5,6,8,10	1,6,7,9		
7	1,2,3,4,5,6,7,8,9,10	1,7,9		
8	2,3,4,5,8,10	1,2,3,6,7,8,9		
9	1,2,3,4,5,6,7,8,9,10	1,7,9		
10	4,5,10	1,2,3,4,5,6,7,8,9,10	4,5,10	I
		Ittiration-2		
1	1,2,3,6,7,8,9,	1,7,9		
2	2,3,8,	1,2,3,6,7,8,9	2,3,8,	II
3	2,3,8,	1,2,3,6,7,8,9	2,3,8,	II
6	2,3,6,8,	1,6,7,9		
7	1,2,3,6,7,8,9,	1,7,9		
8	2,3,8,	1,2,3,6,7,8,9	2,3,8,	II
9	1,2,3,6,7,8,9,	1,7,9		
		Ittiration-3		
1	1,6,7,9,	1,7,9		
6	6,	1,6,7,9	6	III
7	1,6,7,9,	1,7,9		
9	1,6,7,9,	1,7,9		
		Ittiration-4		
1	1,7,9,	1,7,9	1,7,9,	IV
7	1,7,9,	1,7,9	1,7,9,	IV
9	1,7,9,	1,7,9	1,7,9,	IV

4.3. Stage 3: Fuzzy MICMAC Analysis

We used fuzzy MICMAC analysis to examine the driving and dependence power of the determinants of logistics 4.0. To apply fuzzy-MICMAC analysis, the final reachability matrix is used. In the final reachability matrix, binary digits are used, meaning zero or one, so that relationship strength is ignored. The fuzzy MICMAC analysis incorporates the relationship strength among determinants using triangular fuzzy numbers as provided in Table 5.

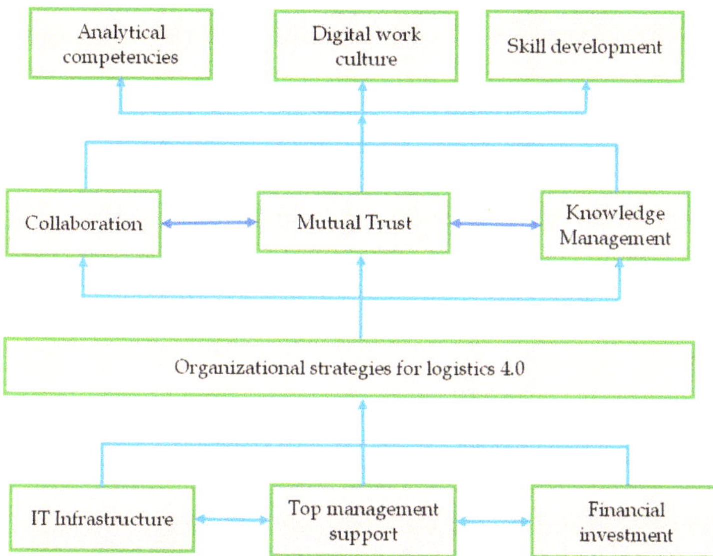

Figure 3. ISM model of determinants of logistics 4.0 adoption.

Table 5. Linguistic Scale and their associated TFNs.

Linguistic Variable	TFNs	Crips Value
No influence	(0, 0, 0)	0
Very low influence	(0, 0.1, 0.2)	0.1
Low influence	(0.2, 0.3, 0.4)	0.3
Medium influence	(0.4, 0.5, 0.6)	0.5
High influence	(0.6, 0.7, 0.8)	0.7
Very high influence	(0.8, 0.9, 1)	0.9
Complete influence	(1, 1, 1)	1

The experts assessed the relationship strength among the determinant using the linguistics scale. The influence of one determinant on the other is assessed by the experts and their linguistic response is converted into TFNs, and the resulting matrix is called the Fuzzy Direct Relationship Matrix (FDRM), as presented in Table 6.

Table 6. FDRM of determinants of logistics 4.0 adoption.

CSFs	1	2	3	4	5	6	7	8	9	10
1	0	0.3	0.3	0.7	0.3	0.7	0.5	0.9	0.3	0.7
2	0	0	0.9	0.7	0.9	0	0	0.5	0	0.9
3	0	0.7	0	0.5	0.7	0	0	0.5	0	0.3
4	0	0	0	0	0.5	0	0	0	0	0.5
5	0	0	0	0.5	0	0	0	0	0	0.3
6	0	0.9	0.7	0.9	0.7	0	0	0.9	0	0.7
7	0.7	0.7	0.3	0.7	0.7	0.7	0.7	0.5	0.3	0.5
8	0	0.7	0.6	0.7	0.3	0	0	0	0	0.3
9	0.7	0.1	0.5	0.1	0.7	0.9	0.7	0.5	0	0.5
10	0	0	0	0.9	0.7	0	0	0	0	0

The FDRM is multiplied repetitively until the driver and dependence power stabilize [51,76]. This stabilization infers that no significant changes in the value driving and dependence power occur if further multiplication is conducted. Through this process, the resultant stabilized matrix is obtained. The driving powers of each determinant are calculated by the row-wise summing of the data, whereas their dependence power is obtained by column-wise summation. Table 7 exhibits the resultant fuzzy stabilized matrix along with the driving and dependence power.

Table 7. Fuzzy stabilized matrix for determinants.

CSFs	1	2	3	4	5	6	7	8	9	10	Driving Power
1	0.5	0.7	0.7	0.7	0.7	0.5	0.5	0.5	0.3	0.7	5.8
2	0	0.7	0.5	0.7	0.7	0	0	0.5	0	0.5	3.6
3	0	0.5	0.7	0.7	0.7	0	0	0.5	0	0.7	3.8
4	0	0	0	0.5	0.5	0	0	0	0	0.5	1.5
5	0	0	0	0.5	0.5	0	0	0	0	0.5	1.5
6	0	0.7	0.7	0.7	0.7	0	0	0.5	0	0.7	4
7	0.7	0.7	0.7	0.7	0.7	0.7	0.7	0.7	0.3	0.7	6.6
8	0	0.6	0.7	0.7	0.7	0	0	0.5	0	0.7	3.9
9	0.7	0.7	0.7	0.7	0.7	0.7	0.7	0.7	0.3	0.7	6.6
10	0	0	0	0.5	0.5	0	0	0	0	0.5	1.5
Dependence Power	1.9	4.6	4.7	6.4	6.4	1.9	1.9	3.9	0.9	6.2	

The finalized determinants are classified into four clusters based on their driving and dependence power: autonomous, dependent, linkage, and driving. Figure 4 shows the four clusters of identified determinants of logistics 4.0 adoption.

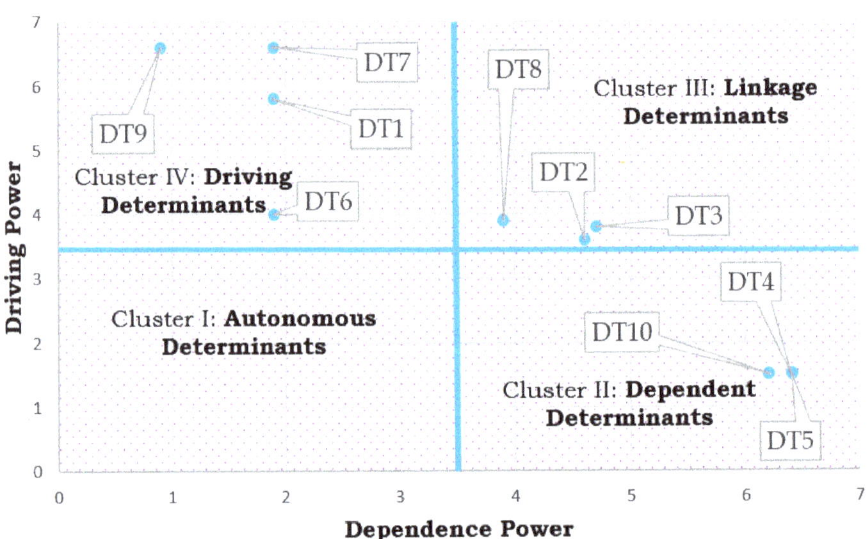

Figure 4. Fuzzy MICMAC analysis of determinants of Logistics 4.0.

5. Discussion

Logistics 4.0 is a relatively new and rapidly emerging topic for both academics and practitioners [10]. Following the literature, supply chain managers are concerned with factors/determinants that help the adoption of logistics 4.0. In this regard, an effort has been made to construct an integrated model for evaluating the linkages between determinants of logistics 4.0, which may be valuable for managers and practitioners. Through this study, ten significant determinants that are responsible for the adoption of logistics 4.0 have been identified. The structural relationship among these determinants is modelled through ISM.

The developed ISM model has four levels, as per their influence on each other. The ISM model reveals that top management support, IT infrastructure and financial investment are the most significant determinants for the adoption of logistics 4.0, asit is placed at the bottom of the model. This could be validated by Bag [41], who stated that dedication is required on a strategic level to implement any novel practices. These three determinants are highly effective for the adoption of logistics 4.0 among all of the identified determinants. As logistics 4.0 requires some strategic and operational transformation in conventional logistics activities, top management support is essential. Logistics 4.0 requires several advanced technological integrations such as big data analytics, IoT-enabled tracking and real-time data transfer. This integration is not possible without a robust IT infrastructure. With the support of top management and their financial investments, the IT infrastructure could be developed. The bottom-level determinants help to achieve the remaining determinants, which results in the effective adoption of logistics 4.0.

The third level of the ISM model contains one determinant: 'organizational strategies for logistics 4.0'. The top management's involvement and the presence of IT infrastructure help in the development of organizational strategies for the adoption of logistics 4.0. The organization could assess the impact of logistics 4.0 on its performance and formulate strategies to adopt logistics 4.0. Establishing a vision and mission to adopt logistics 4.0 help the working personnel to channelize their effort.

The second level of the ISM model contains three determinants, including collaboration, mutual trust, and knowledge management. These factors are influenced by 'organizational strategies for logistics 4.0'. The mutual trust among the logistics partners enables them to share the relevant data with concerned partners and resolve the initial conflict efficiently. This led to the collaboration among logistics/supply chain partners that help them in the long run. Knowledge management is also an essential component for the adoption of any novel practice. Efficient knowledge management helps the logistics partners to understand the logistics 4.0 practices and processes. This will help the organizations to effectively adopt logistics 4.0 in less time. Mutual trust, collaboration and knowledge management depend on the organizational strategies with regard to logistics 4.0 and influence the top-level determinants.

The top level of the ISM model contains three determinants: analytical competencies, digital work culture and skill development. These three determinants could be achieved through the successful adoption of the influential determinants that are placed on levels II, III and IV. The organization receives an outcome in terms of developing its analytical capabilities, digital work culture and skill development by effectively incorporating the above determinants. The analytical capabilities help the organizations to adopt logistics 4.0 and benefit in terms of competitive edge and sustainability. Further, skill development also supports the adoption of logistics 4.0 practices as well as enabling the digital work culture

Further, the fuzzy MICMAC analysis is conducted to categorize the determinants into four clusters based on their driving and dependence power. Based on this analysis, four clusters of determinates are identified; namely, autonomous, dependent, linkage and driving.

The first cluster of determinates, called autonomous determinants, have weak driving and dependence power. There is not much effect of the autonomous cluster factors on the system due to low driving and dependence power. As there are no determinants that fall into this cluster, it explains that all determinants are related, appropriate, and in control.

In the second cluster, there are dependent determinates that are highly dependent and weakly driven. Three determinants belong to this cluster- analytical competencies, digital work culture and skill development. These three determinants are important because their strong dependence points out that they need all the other determinants to adopt logistics 4.0. The organization's management need to focus on these determinants for the effective adoption of logistics 4.0.

The third cluster is made up of determinates that have both high driving power and high dependence power. These determinates are called linkage determinates and are inherently unstable. Changes in these linkage determinants have a positive or negative effect on the other determinates [60]. Three determinants fall into this cluster, including collaboration, mutual trust, and knowledge management. These determinants are highly unstable, so careful observation is required during the process of logistics 4.0 adoption. It should be noted that, at every stage of logistics 4.0 adoption, managers should continuously observe these determinants.

The fourth cluster is driving determinates that have strong driving power, and weak dependence on others. Four determinants associated with this category include top management support, IT infrastructure, financial investment and organizational strategies for logistics 4.0. Logistics 4.0 implementation relies heavily on these determinants, and logistics partners should give these determinants the highest priority. Any changes in these determinants may have an impact on the other determinants at all levels of the hierarchy. As a result, the highest priority must be given to these four determinants.

6. Implications of Research

This research provided significant implications and useful insights for decision-makers and practitioners to adopt logistics 4.0. As logistics 4.0 is an evolving practice and its adoption is a challenge for the logistics partners, in order to make decisions regarding the adoption of logistic 4.0, organizations are concerned about the factors that are responsible for the logistics 4.0 adoption. The findings of this study show that top management support, IT infrastructure and financial investment are the major component for the adoption of logistics 4.0. The structural changes are required for the adoption of logistics 4.0 are not possible without top management's involvement. Further, the findings claim that IT infrastructure is also an essential component to operationalizing logistics 4.0 practices. The findings of this study corroborate previous research by Khan et al. [10], which indicated that top management involvement has a perceiving influence on the adoption of logistics 4.0. In addition, the organizational strategies need to align with logistics 4.0 to make the adoption process more effective. The current organizational strategies need to be revised and aligned with logistics 4.0 through the incorporation of industry 4.0 technology adoption. Through adopting the determinants of logistics 4.0, organizations can develop its analytical capabilities, digital work culture and skill development. With these characteristics, the organization can smoothly adopt logistics 4.0 and develop a competitive edge. Further, the fuzzy MICMAC analysis also supports the contextual relationship that is established through the ISM model. The fuzzy MICMAC analysis shows that organizations need to primarily focus on the driving determinants that include top management support, IT infrastructure, financial investment and organizational strategies for logistics 4.0. Moreover, it also suggests that the management needs to keep a close look at the linkage factors, collaboration, mutual trust, and knowledge management, at every stage of logistics 4.0 adoption. Based on the interplay among the determinants of logistics 4.0, policy planners could formulate their plans and strategies for the adoption of logistics 4.0.

7. Conclusions, Limitations and Future Scope

As the consequences of the fourth industrial revolution begin to emerge, the necessity for many industries to adopt new technology becomes increasingly evident. Several supply chain solutions are proposed to meet the consumer preferences in the present business environment. Logistics are a crucial component of the supply chain, and supply chain managers gives a lot of attention to them. The customer gains more from the implementation of logistics 4.0 due to enhanced transparency, reduced lead times, traceability, condition monitoring, etc. Therefore, the adoption of logistic 4.0 is the principal focus of this study through the adoption of its determinants. An integrated approach of a literature review and input from experts in the field of logistics is used to first identify the determinants of the adoption of logistics 4.0. These determinants are modelled through ISM for exploring the structural relationships among them. The developed structural model has four levels that show the relationship among these determinants. The structural relationship is helpful for management to align their efforts. Further, fuzzy MICMAC analysis is also used to categorize the determinants into four clusters based on their driving power. The finding shows that top management support, IT infrastructure and financial investment are the most significant determinants towards the adoption of logistics 4.0. It is important to consider that changes to the driving factors leading to improvements in the other determinants. Managers, decision-makers, and experts should instantly concentrate on these aspects to adapt logistics 4.0. Furthermore, based on the driving and dependence power, the MICMAC analysis shows that these three determinants, along with organizational strategies for logistics 4.0 factors, have high driving power. Therefore, the logistics partners should make the strategies for logistics 4.0 at the organizational and supply chain levels.

Similar to other studies, this work has some limitations. The first limitation is that, because there has been so little research conducted on logistics 4.0, it may be possible to overlook some determinants. Second, the expert's input, which is based on the finalization of the uncovered determinants, may be biased in favor of their managerial position, location, and organization. Thirdly, the ISM method is not capable of quantifying the strength of the relationship among the determinants of logistics 4.0. Future research could address these limitations. For identifying the determinants in the subsequent studies, a systematic literature review involving a larger number of documents, along with grey literature, might be conducted. With the help of multiple case studies, the findings of this study will be generalized. Other modelling approaches, such as structural equation modelling, system dynamics, and modified TISM, could be employed to establish the causal link.

Author Contributions: S.K. and R.S. conceived the idea. R.S. surveyed the literature and found determinants. S.K. collected the responses and contributed to the analysis, and J.C.S. and L.P.F. analyzed the results. S.K. and R.S. wrote the paper, and L.P.F. edited the original draft and administration the project. G.S. validated the results. All authors have read and agreed to the published version of the manuscript.

Funding: This research received no external funding.

Data Availability Statement: Not applicable.

Conflicts of Interest: The authors declare no conflict of interest.

Appendix A

Table A1. Details of the participated experts.

S. No	Position	Experience	Qualification	Gender	Specialisation(s)	Country
1.	Professor	28 Years	Doctorate	Male	Industry 4.0, Logistics 4.0, Technology transfer	India
2.	Logistics Manager	16 Years	Postgraduate	Female	Logistics Solution Provider, Logistics 4.0	India
3.	Warehouse managers	12 Years	Graduate	Male	Warehousing	India
4.	Technology Transfer Head	14 Years	Postgraduate	Male	Automation and Industry 4.0 adoption	India
5.	Associate Professor	15 Years	Doctorate	Male	Supply chain management 4.0	India
6.	Research and Development Head	18 Years	Doctorate	Female	Innovation and Technology transfer,	India
7.	Transport Manager	12 Years	Postgraduate	Male	Logistics Management	India
8.	Deputy Director	10 Years	Doctorate	Female	Industrial infrastructure Development	India

References

1. Sony, M.; Antony, J.; Mc Dermott, O.; Garza-Reyes, J. An empirical examination of benefits, challenges, and critical success factors of industry 4.0 in manufacturing and service sector. *Technol. Soc.* **2021**, *67*, 101754. [CrossRef]
2. Masood, T.; Sonntag, P. Industry 4.0: Adoption challenges and benefits for SMEs. *Comput. Ind.* **2020**, *121*, 103261. [CrossRef]
3. Culot, G.; Nassimbeni, G.; Orzes, G.; Sartor, M. Behind the definition of Industry 4.0: Analysis and open questions. *Int. J. Prod. Econ.* **2020**, *226*, 107617. [CrossRef]
4. Yin, Y.; Stecke, K.; Li, D. The evolution of production systems from Industry 2.0 through Industry 4.0. *Int. J. Prod. Res.* **2017**, *56*, 848–861. [CrossRef]
5. Liao, Y.; Deschamps, F.; Loures, E.; Ramos, L.F.P. Past, present and future of Industry 4.0—A systematic literature review and research agenda proposal. *Int. J. Prod. Res.* **2017**, *55*, 3609–3629. [CrossRef]
6. Khan, S.; Singh, R.; Kirti. Critical Factors for Blockchain Technology Implementation: A Supply Chain Perspective. *J. Ind. Integr. Manag.* **2021**, *7*, 479–492. [CrossRef]
7. Ding, B. Pharma Industry 4.0: Literature review and research opportunities in sustainable pharmaceutical supply chains. *Process. Saf. Environ. Prot.* **2018**, *119*, 115–130. [CrossRef]
8. Khan, M.; Khan, S.; Khan, U.; Haleem, A. Modeling the Big Data challenges in context of smart cities—an integrated fuzzy ISM-DEMATEL approach. *Int. J. Build. Pathol. Adapt.* 2021, *ahead-of-print*. [CrossRef]
9. Javaid, M.; Haleem, A.; Singh, R.P.; Rab, S.; Suman, R.; Khan, S. Exploring relationships between Lean 4.0 and manufacturing industry. *Ind. Robot. Int. J. Robot. Res. Appl.* **2021**, *49*, 402–414. [CrossRef]
10. Khan, S.; Singh, R.; Haleem, A.; Dsilva, J.; Ali, S.S. Exploration of critical success factors of logistics 4.0: A DEMATEL approach. *Logistics* **2022**, *6*, 13. [CrossRef]
11. Timm, J.; Lorig, F. Logistics 4.0—A Challenge for Simulation. In Proceedings of the 2015 Winter Simulation Conference, Huntington Beach, CA, USA, 6–9 December 2015; Yilmaz, L., Chan, W.K.V., Moon, I., Roeder, T.M.K., Macal, C., Rossetti, D., Eds.; IEEE Press: Piscataway, NJ, USA, 2015; pp. 3118–3119.
12. Wang, K. Logistics 4.0 Solution-New Challenges and Opportunities. In Proceedings of the 6th International Workshop of Advanced Manufacturing and Automation, London, UK, 27–28 October 2016.
13. Queiroz, M.M.; Fosso Wamba, S.; Chiappetta Jabbour, C.J.; Lopes de Sousa Jabbour, A.B.; Machado, M.C. Adoption of industry 4.0 technologies by organizations: A Maturity Levels Perspective. *Ann. Oper. Res.* **2022**. [CrossRef] [PubMed]
14. Imran, M.; Hamid, S.N.B.A.; Aziz, A.B.; Hameed, W.-U. The contributing factors towards e-logistic customer satisfaction: A mediating role of Information Technology. *Uncertain Supply Chain Manag.* **2019**, 63–72. [CrossRef]
15. Tamvada, J.P.; Narula, S.; Audretsch, D.; Puppala, H.; Kumar, A. Adopting new technology is a distant dream? The risks of implementing Industry 4.0 in emerging economy SMEs. *Technol. Forecast. Soc. Chang.* **2022**, *185*, 122088. [CrossRef]
16. Horváth, D.; Szabó, R.Z. Driving forces and barriers of industry 4.0: Do multinational and small and medium-sized companies have equal opportunities? *Technol. Forecast. Soc. Chang.* **2019**, *146*, 119–132. [CrossRef]
17. Sharma, H.P.; Kumar, K. Developing and implementing environment management practices in small and medium size manufacturing companies in India. *IOP Conf. Ser. Earth Environ. Sci.* **2021**, *795*, 012022. [CrossRef]
18. Bányai, T. Real-Time Decision Making in First Mile and Last Mile Logistics: How Smart Scheduling Affects Energy Efficiency of Hyperconnected Supply Chain Solutions. *Energies* **2018**, *11*, 1833. [CrossRef]
19. Kayikci, Y. Sustainability impact of digitization in logistics. *Procedia Manuf.* **2018**, *21*, 782–789. [CrossRef]
20. Teschemacher, U.; Reinhart, G. Ant Colony Optimization Algorithms to Enable Dynamic Milkrun Logistics. *Procedia CIRP* **2017**, *63*, 762–767. [CrossRef]
21. Herter, J.; Ovtcharova, J. A Model based Visualization Framework for Cross Discipline Collaboration in Industry 4.0 Scenarios. *Procedia CIRP* **2016**, *57*, 398–403. [CrossRef]
22. Barreto, L.; Amaral, A.; Pereira, T. Industry 4.0 implications in logistics: An overview. *Procedia Manuf.* **2017**, *13*, 1245–1252. [CrossRef]
23. Strandhagen, J.; Vallandingham, L.R.; Fragapane, G.; Stangeland, A.; Sharma, N. Logistics 4.0 and emerging sustainable business models. *Adv. Manuf.* **2017**, *5*, 359–369. [CrossRef]
24. Hofmann, E.; Rüsch, M. Industry 4.0 and the current status as well as future prospects on logistics. *Comput. Ind.* **2017**, *89*, 23–34. [CrossRef]
25. Tascón, D.C.; Mejía, G.; Rojas-Sánchez, D. Flexibility of operations in developing countries with industry 4.0. A systematic review of literature. *Production* **2022**, *32*, e20210055. [CrossRef]
26. Parhi, S.; Joshi, K.; Gunasekaran, A.; Sethuraman, K. Reflecting on an empirical study of the digitalization initiatives for sustainability on logistics: The Concept of Sustainable Logistics 4.0. *Clean. Logist. Supply Chain* **2022**, *4*, 100058. [CrossRef]
27. Sun, X.; Yu, H.; Solvang, W.D. Towards the smart and sustainable transformation of Reverse Logistics 4.0: A conceptualization and research agenda. *Environ. Sci. Pollut. Res.* **2022**, *29*, 69275–69293. [CrossRef] [PubMed]
28. Batz, A.; Oleśków-Szłapka, J.; Stachowiak, A.; Pawłowski, G.; Maruszewska, K. Identification of Logistics 4.0 Maturity Levels in Polish Companies—Framework of the Model and Preliminary Research. In *Sustainable Logistics and Production in Industry 4.0*; Springer: Cham, Switzerland, 2019; pp. 161–175. [CrossRef]
29. Frank, A.; Dalenogare, L.; Ayala, N. Industry 4.0 technologies: Implementation patterns in manufacturing companies. *Int. J. Prod. Econ.* **2019**, *210*, 15–26. [CrossRef]

30. Winkelhaus, S.; Grosse, E. Logistics 4.0: A systematic review towards a new logistics system. *Int. J. Prod. Res.* **2019**, *58*, 18–43. [CrossRef]
31. Ghadge, A.; Kara, M.E.; Moradlou, H.; Goswami, M. The impact of Industry 4.0 implementation on supply chains. *J. Manuf. Technol. Manag.* **2020**, *31*, 669–686. [CrossRef]
32. Castro, C.; Pereira, T.; Sá, J.C.; Santos, G. Logistics reorganization and management of the ambulatory pharmacy of a local health unit in Portugal. *Evaluation Program Plan.* **2020**, *80*, 101801. [CrossRef]
33. Choudhury, A.; Behl, A.; Sheorey, P.; Pal, A. Digital supply chain to unlock new agility: A TISM approach. *Benchmarking Int. J.* **2021**, *28*, 2075–2109. [CrossRef]
34. Atzeni, G.; Vignali, G.; Tebaldi, L.; Bottani, E. A bibliometric analysis on collaborative robots in Logistics 4.0 environments. *Procedia Comput. Sci.* **2021**, *180*, 686–695. [CrossRef]
35. Markov, K.; Vitliemov, P. Logistics 4.0 and supply chain 4.0 in the automotive industry. *IOP Conf. Ser. Mater. Sci. Eng.* **2020**, *878*, 012047. [CrossRef]
36. Kucukaltan, B.; Saatcioglu, O.Y.; Irani, Z.; Tuna, O. Gaining strategic insights into Logistics 4.0: Expectations and impacts. *Prod. Plan. Control* **2020**, *33*, 211–227. [CrossRef]
37. Di Nardo, M.; Clericuzio, M.; Murino, T.; Sepe, C. An Economic Order Quantity Stochastic Dynamic Optimization Model in a Logistic 4.0 Environment. *Sustainability* **2020**, *12*, 4075. [CrossRef]
38. Nantee, N.; Sureeyatanapas, P. The impact of Logistics 4.0 on corporate sustainability: A performance assessment of automated warehouse operations. *Benchmarking Int. J.* **2021**, *28*, 2865–2895. [CrossRef]
39. Torbacki, W.; Kijewska, K. Identifying Key Performance Indicators to be used in Logistics 4.0 and Industry 4.0 for the needs of sustainable municipal logistics by means of the DEMATEL method. *Transp. Res. Procedia* **2019**, *39*, 534–543. [CrossRef]
40. Kodym, O.; Kubáč, L.; Kavka, L. Risks associated with Logistics 4.0 and their minimization using Blockchain. *Open Eng.* **2020**, *10*, 74–85. [CrossRef]
41. Bag, S.; Gupta, S.; Luo, Z. Examining the role of logistics 4.0 enabled dynamic capabilities on firm performance. *Int. J. Logist. Manag.* **2020**, *31*, 607–628. [CrossRef]
42. Javaid, M.; Haleem, A.; Singh, R.P.; Khan, S.; Suman, R. Blockchain technology applications for Industry 4.0: A literature-based review. *Blockchain Res. Appl.* **2021**, *2*, 100027. [CrossRef]
43. Wagner, T.; Herrmann, C.; Thiede, S. Industry 4.0 Impacts on Lean Production Systems. *Procedia CIRP* **2017**, *63*, 125–131. [CrossRef]
44. Beier, G.; Ullrich, A.; Niehoff, S.; Reißig, M.; Habich, M. Industry 4.0: How it is defined from a sociotechnical perspective and how much sustainability it includes—A literature review. *J. Clean. Prod.* **2020**, *259*, 120856. [CrossRef]
45. Rosin, F.; Forget, P.; Lamouri, S.; Pellerin, R. Impacts of Industry 4.0 technologies on Lean principles. *Int. J. Prod. Res.* **2019**, *58*, 1644–1661. [CrossRef]
46. Silva, F.J.G.; Kirytopoulos, K.; Ferreira, L.P.; Sá, J.C.; Santos, G.; Nogueira, M.C.C. The three pillars of sustainability and agile project management: How do they influence each other. *Corp. Soc. Responsib. Environ. Manag.* **2022**, *29*, 1495–1512. [CrossRef]
47. Götz, M.; Jankowska, B. Clusters and Industry 4.0—Do they fit together? *Eur. Plan. Stud.* **2017**, *25*, 1633–1653. [CrossRef]
48. Sousa, G.; Sá, J.C.; Santos, G.; Silva, F.J.G.; Ferreira, L.P. The Contribution of Obeya for Business Intelligence. *Des. Appl. Maint. Cyber-Phys.* **2021**, 244–269. [CrossRef]
49. Santos, G.; Sá, J.; Félix, M.; Barreto, L.; Carvalho, F.; Doiro, M.; Zgodavová, K.; Stefanović, M. New Needed Quality Management Skills for Quality Managers 4.0. *Sustainability* **2021**, *13*, 6149. [CrossRef]
50. Sharma, H.P.; Chaturvedi, A. Adoption of smart technologies: An Indian perspective. In Proceedings of the 2021 5th International Conference on Information Systems and Computer Networks (ISCON) [Preprint], Mathura, India, 22–23 October 2021. [CrossRef]
51. Khan, S.; Haleem, A.; Khan, M.; Abidi, M.; Al-Ahmari, A. Implementing Traceability Systems in Specific Supply Chain Management (SCM) through Critical Success Factors (CSFs). *Sustainability* **2018**, *10*, 204. [CrossRef]
52. Khan, S.; Haleem, A.; Fatma, N. Effective adoption of remanufacturing practices: A step towards circular economy. *J. Remanufacturing* **2022**, *12*, 167–185. [CrossRef]
53. Yadav, S.; Luthra, S.; Garg, D. Internet of things (IoT) based coordination system in Agri-food supply chain: Development of an efficient framework using DEMATEL-ISM. *Oper. Manag. Res.* **2020**, *15*, 1–27. [CrossRef]
54. Lakra, A.; Gupta, S.; Ranjan, R.; Tripathy, S.; Singhal, D. The Significance of Machine Learning in the Manufacturing Sector: An ISM Approach. *Logistics* **2022**, *6*, 76. [CrossRef]
55. Sharma, A.; Abbas, H.; Siddiqui, M.Q. Modelling the inhibitors of cold supply chain using fuzzy interpretive structural modeling and fuzzy MICMAC analysis. *PLoS ONE* **2021**, *16*, e0249046. [CrossRef] [PubMed]
56. Deepu, T.S.; Ravi, V. An ISM-MICMAC approach for analyzing dependencies among barriers of supply chain digitalization. *J. Model. Manag.* **2022**. preprint. [CrossRef]
57. Zekhnini, K.; Cherrafi, A.; Bouhaddou, I.; Benghabrit, Y.; Belhadi, A. Supply Chain 4.0 risk management: An interpretive structural modelling approach. *Int. J. Logist. Syst. Manag.* **2020**, *41*, 171–204. [CrossRef]
58. Ferreira, N.; Santos, G.; Rui Silva, R. Risk level reduction in construction sites: Towards a computer aided methodology—A case study. *Appl. Comput. Inform.* **2019**, *15*, 136–143. [CrossRef]
59. Warfield, J.N. *Structuring Complex Systems*; Battelle Memorial Institute: Columbus, OH, USA, 1974; p. 4.

60. Duperrin, J.C.; Godet, M. Méthode de hiérarchisation des éléments d'un système: Essai de prospective du système de l'énergie nucléaire dans son contexte sociétal. Centre national de l'entrepreneuriat(CNE), CEA: Paris, France, 1973.

61. Silva, N.; Barros, J.; Santos, M.Y.; Costa, C.; Cortez, P.; Carvalho, M.; Gonçalves, J. Advancing Logistics 4.0 with the Implementation of a Big Data Warehouse: A Demonstration Case for the Automotive Industry. *Electronics* **2021**, *10*, 2221. [CrossRef]

62. Cimini, C.; Lagorio, A.; Pirola, F.; Pinto, R. Exploring human factors in Logistics 4.0: Empirical evidence from a case study. *Ifac-Papersonline* **2019**, *52*, 2183–2188. [CrossRef]

63. Sá, J.C.; Vaz, S.; Carvalho, O.; Lima, V.; Morgado, L.; Fonseca, L.; Doiro, M.; Santos, G. A model of integration ISO 9001 with Lean six sigma and main benefits achieved. *Total Qual. Manag.* **2022**, *33*, 218–242. [CrossRef]

64. Joshi, S.; Sharma, M. Digital technologies (DT) adoption in agri-food supply chains amidst COVID-19: An approach towards food security concerns in developing countries. *J. Glob. Oper. Strat. Sourc.* **2021**, *15*, 262–282. [CrossRef]

65. Jagtap, S.; Bader, F.; Garcia-Garcia, G.; Trollman, H.; Fadiji, T.; Salonitis, K. Food Logistics 4.0: Opportunities and Challenges. *Logistics* **2020**, *5*, 2. [CrossRef]

66. Santos, G.; Mandado, E.; Silva, R.; Doiro, M. Engineering learning objectives and computer assisted tools. *Eur. J. Eng. Educ.* **2019**, *44*, 616–628. [CrossRef]

67. Mathivathanan, D.; Mathiyazhagan, K.; Rana, N.P.; Khorana, S.; Dwivedi, Y.K. Barriers to the adoption of blockchain technology in business supply chains: A Total Interpretive Structural Modelling (TISM) approach. *Int. J. Prod. Res.* **2021**, *59*, 3338–3359. [CrossRef]

68. Rejeb, A.; Keogh, J.; Zailani, S.; Treiblmaier, H.; Rejeb, K. Blockchain Technology in the Food Industry: A Review of Potentials, Challenges and Future Research Directions. *Logistics* **2020**, *4*, 27. [CrossRef]

69. Facchini, F.; Oleśków-Szłapka, J.; Ranieri, L.; Urbinati, A. A Maturity Model for Logistics 4.0: An Empirical Analysis and a Roadmap for Future Research. *Sustainability* **2019**, *12*, 86. [CrossRef]

70. Rahman, M.; Kamal, M.M.; Aydin, E.; Haque, A.U. Impact of Industry 4.0 drivers on the performance of the service sector: Comparative study of cargo logistic firms in developed and developing regions. *Prod. Plan. Control* **2020**, *33*, 228–243. [CrossRef]

71. Kumar, A.; Choudhary, S.; Garza-Reyes, J.A.; Kumar, V.; Khan, S.A.R.; Mishra, N. Analysis of critical success factors for implementing Industry 4.0 integrated circular supply chain—moving towards sustainable operations. *Prod. Plan. Control* **2021**, 1–15. [CrossRef]

72. Ozkan-Ozen, Y.D.; Kazancoglu, Y.; Kumar Mangla, S. Synchronized Barriers for Circular Supply Chains in Industry 3.5/Industry 4.0 Transition for Sustainable Resource Management. *Resour. Conserv. Recycl.* **2020**, *161*, 104986. [CrossRef]

73. Wang, B.; Ha-Brookshire, J.E. Exploration of Digital Competency Requirements within the Fashion Supply Chain with an Anticipation of Industry 4.0. *Int. J. Fash. Des. Technol. Educ.* **2018**, *11*, 333–342. [CrossRef]

74. Hou, T.; Cheng, B.; Wang, R.; Xue, W.; Chaudhry, P.E. Developing Industry 4.0 with systems perspectives. *Syst. Res. Behav. Sci.* **2020**, *37*, 741–748. [CrossRef]

75. Diabat, A.; Govindan, K. An analysis of the drivers affecting the implementation of Green Supply Chain Management. *Resour. Conserv. Recycl.* **2011**, *55*, 659–667. [CrossRef]

76. Raji, I.O.; Shevtshenko, E.; Rossi, T.; Strozzi, F. Modelling the relationship of digital technologies with lean and agile strategies. *Supply Chain. Forum Int. J.* **2021**, *22*, 323–346. [CrossRef]

Article

Implementation of a Lean 4.0 Project to Reduce Non-Value Add Waste in a Medical Device Company

Ida Foley [1], Olivia McDermott [1,*], Angelo Rosa [2] and Manjeet Kharub [3]

1 College of Science and Engineering, University of Galway, H91 TK33 Galway, Ireland
2 Department of Management Studies, Università LUM Jean Monnet, 70010 Casamassima, Italy
3 Department of Mechanical Engineering, CVR College of Engineering, Hyderabad 500039, India
* Correspondence: olivia.mcdermott@nuigalway.ie or olivia.mcdermott@universityofgalway.ie

Abstract: The fourth industrial revolution, also referred to as Industry 4.0, has resulted in many changes within the manufacturing industry. The purpose of the study is to demonstrate how an Industry 4.0 project was scoped and deployed utilising Lean tools to reduce non-value add wastes and aid regulatory compliance. A case study research approach was utilised to demonstrate how the Lean Industry 4.0 project was implemented in a Medtech company to enhance Lean processes while increasing digitalisation. This research demonstrates that Industry 4.0 can enhance Lean, improve flow, reduce nonvalue add waste, and facilitate product lifecycle regulatory compliance to reduce defects, enhance quality, improve cycle time, and minimise reworks and over-processing. Lean and Industry 4.0 combined offer many benefits to the MedTech Industry. This research will support organisations in demonstrating how digital technologies can synergistically affect Lean processes, positively impact product lifecycle regulatory compliance, and support the industry in building a business case for future implementation of Industry 4.0 technologies.

Keywords: Industry 4.0; medical device; Medtech; regulatory compliance; engineering change management; product lifecycle management; Regulatory 4.0; Lean 4.0

Citation: Foley, I.; McDermott, O.; Rosa, A.; Kharub, M. Implementation of a Lean 4.0 Project to Reduce Non-Value Add Waste in a Medical Device Company. *Machines* **2022**, *10*, 1119. https://doi.org/10.3390/machines10121119

Academic Editor: Mosè Gallo

Received: 17 October 2022
Accepted: 17 November 2022
Published: 26 November 2022

Publisher's Note: MDPI stays neutral with regard to jurisdictional claims in published maps and institutional affiliations.

1. Introduction

The Medical Device Industry is one of the largest growing industries in the world. This growth is driven by ageing populations, advancing technologies and new innovations to meet clinical needs [1]. In order to reduce costs, improve manufacturing productivity, and reduce cycle times, the Medtech industry, along with other industries, has embraced Lean [2]. However, with the advent of Industry 4.0 and increased digitalisation, the MedTech industry can improve efficiencies, reduce operational costs, and support organisational decisions through big data analytics [3,4]. Some studies have investigated Lean 4.0, the combination of Lean and Industry 4.0, and concluded that there is a synergistic effect between Lean and Industry 4.0 [2,5]. A recent Boston Consulting Group (BCG) study showed that states have a multiplier effect when lean and Industry 4.0 are combined. The study found that Lean can reduce operational costs by 15–20%, and digitalisation can reduce costs by 10–15% but combine both, and you get up to 40% cost reduction [6].

The impact of digital technologies on product lifecycle regulatory compliance has also not been widely researched. Product lifecycle regulatory compliance or regulatory compliance is how medical device manufacturers comply with the different statutory, mandatory, and voluntary regulatory requirements to ensure their organisations deliver a safe and effective product and meet various regulatory jurisdictions' specific regulatory requirements [7,8]. Increasingly changing regulations in Europe related to medical devices and other jurisdictions and staying compliant with technological advances have increased regulatory compliance complexity [9].

There have been few practical case studies of a Lean Industry 4.0 application [10], and neither one specifically focused on the Medtech organisation nor Lean Industry 4.0's impact on regulatory compliance [11,12]. This study will utilise a case study of a multinational medical device manufacturer with several global sites. This research aims to investigate the impact of Lean practices combined with Industry 4.0 on regulatory compliance and enhancing Lean processes within the MedTech Industry using the case study organisation as a reference. This research will address the following research questions:

RQ1: What impact can Lean 4.0 have on a Medical Device manufacturer's Total Product Lifecycle and Regulatory Compliance?

RQ2: How can Industry 4.0 enhance and enable Lean in a Medical Device manufacturer?

Section 2 reviews the published literature that is currently available on Lean Industry 4.0 in Medtech and how Lean and digitalisation can support regulatory compliance. Section 3 discusses the research methodology, while Section 4 documents the findings and results. Finally, the discussion and conclusion are outlined in Sections 5 and 6.

2. Literature Review

2.1. Lean 4.0

According to Antony et al. [11,12], in studies on Lean and Six Sigma combined with Industry 4.0, Lean 4.0 has emerged as a topic of researcher interest only from 2017 onwards. A systematic literature review found that Lean and Industry 4.0 (or Lean 4.0) combined, while still a nascent area, have many symbiotic and synergistic needs for each other [13,14]. Lean has become digitally enabled [15]. Integrating Industry 4.0 with Lean can enhance cost-competitiveness [16] and can generate reduced waste [17]. Several Lean concepts can be improved by integrating I4.0 technologies [13]. I4.0 can increase data availability which will enable Lean and aid in measuring, monitoring, and improving key performance indicators (KPI's) in organisations [18]. Thus, the synergistic effect between Lean and Industry 4.0 in Lean processes by improving flows and reducing bottlenecks [19].

Within a Lean value stream, integration of Industry 4.0 technologies benefits the Lean approach by combining the simplicity and efficiency of Lean with the agility of the I4.0 technologies [17]. Antony et al. [11] argued that there is a bidirectional relationship between Lean and Industry 4.0. Some studies have argued that, while Lean is an enabler for I4.0 or a pre-requisite for its introduction, there still needs to be more studies and guidance on its integration [13,20].

2.2. How Is Lean and Industry 4.0 Impacting Medtech Regulatory Compliance?

The Medtech sector is by its very nature, highly regulated with many different regulatory requirements globally, from the European Medical Device Regulation (MDR) and in vitro diagnostic medical device Regulation (IVDR) to the American Food & Drug Administration (FDA)'s Code of Federal Regulations (CFR) in the United States of America (USA), the Pharmaceutical and Medical Device Act (PMD Act) in Japan, the Regulation on the Supervision and Administration of Medical Devices, Order 739 in China, and the Therapeutic Goods (Medical Devices) Regulations 2002 in Australia, to name just a few. Global Regulations set out the regulatory requirements, including pre and post-market requirements, to ensure that medical devices are produced which are safe and effective [21].

Many Medical device companies have deployed Lean, with one recent study by McDermott et al. [2] on the Irish Medtech sector highlighting that over 95% of Irish Medtech companies had a Lean program. Lean in the medical device industry, as in other industries, has enabled waste reduction, particularly non-value add activity and improved process flow [22]. However, medical device regulatory compliance involves manual tracking and surveillance of multiple databases, leading to over-processing.

While Industry 4.0, Quality 4.0, Supply Chain 4.0, and even Healthcare 4.0 are studied in academic literature, Regulatory 4.0 or Industry 4.0's impact on regulatory affairs digitalisation is not a term that has been widely used [23–25]. In particular, the Quality (QA) function is more advanced on its digital transformation path than the Quality Assurance and Regulatory Affairs (QARA) partner function Regulatory Affairs (RA) [26]. Industry 4.0, in particular, can support Regulatory compliance using tools such as Regulatory information management systems (RIMs) [27]. RIMs provide secure access to real-time regulatory data and visibility across regions. A challenge for device manufacturers is to remain current with global regulations and changes in achieving regulatory compliance throughout a product's life cycle [28]. IMs aid the RA function in quicker regulatory submission times and product registrations, resulting in faster access to markets in organisations across global sites.

RA functions must access several regulatory databases; for example, the European database on medical devices (EUDAMED) is used to access medical device-related data to understand how a device is performing in the market, its risks and benefits, and if post-market surveillance corrective actions are required based information that has been inputted into the system on individual devices as required by the European Union Medical Device Regulations (EU-MDR). To adhere to the MDR, manufacturers must register devices, sites, unique device identification (UDI), notified bodies' information and certificates, clinical investigation results data, device performance studies, vigilance, and post-market surveillance (PMS) information [29].

Many regulatory functions utilise Excel for tracking and trending, which is not Lean. Regulatory intelligence can be obtained and managed using digital technology, removing data inventory, defects or errors, waiting, delays, and over-processing [30]. Several types of information must be tracked, including Regulatory Impact Assessments (RIAs), change notifications (CN), licenses, submissions, and device registrations [31]. Industry 4.0 technologies can help aid RIMs to be more efficient and Leaner. The digitalisation of an organisation's regulatory data is key in supporting RA moving forward on its Lean journey. The digitalisation of RIMs will drive flow, a reduction in non-value add activities, and ensure standardised, efficient systems. Industry 4.0 digitisation ensures RA functions know when regulators have made changes to guidance documents, standards, and regulations, reducing the non-value add waste of checking global regulatory websites to keep abreast of the latest changes and other systems that can manage regulatory information [22]. It is key that manufacturers are aware of changes to standards or regulations as they occur, as they need to demonstrate regulatory compliance and have access to the latest revisions in a more automated manner [32]. Much of an RA professional's time is spent waiting and searching for regulatory information in a non-value add manner.

Within the medical device regulatory world, several global jurisdictions have put in place legislation to protect patient data, enhance cybersecurity in relation to smart devices, and implement standards and guidance documents that can support their implementation [33]. Industry 4.0 can aid device manufacturers' data security, implementation of digital signatures, transaction time stamping and data encryption, which enhance traceability and increase cybersecurity [33]. However, there are many regulators and standards organisations, such as the International Organization for Standardization (ISO), American National Standards Institute (ANSI), European Committee for Standardization (CEN), the American Society for Testing and Materials (ASTM), and European Telecommunications Standards Institute (ETSI); there must be a more effective technological method of keeping abreast of all relevant regulatory requirements [34,35].

2.3. Challenges to Lean Industry 4.0 Deployment

Implementing Lean 4.0 is impacted by many factors, including management support, organisational vision, and investment [36]. The difficulty in implementing Industry 4.0 systems can be off-putting due to the technical complexity and resources involved, as well as the time required [37]. In particular, for the medical device industry, new European medical device regulations have provided severe resource challenges in preparing for more stringent regulatory requirements [38]. While this EU-MDR is not precluding Industry 4.0 deployment, it has stifled the MedTech Industry from implementing it [39]. System changes in device manufacturers need regulatory authority approvals [40]. These regulatory approvals consume time and resources [2]. Many studies have highlighted the importance of management support and leadership commitment in both Lean and

Industry 4.0 deployment [41,42]. However, given the costly nature of digitalisation, it is very important that the right technology is chosen and understood and the cost benefits analysed [43,44]. In addition, the technology chosen needs to be aligned with the organisation's strategic vision so that the technology can be integrated across the organisation and multisite functions [45]. The timing of when Industry 4.0 is adopted can also affect organisations. According to Antony, Sony, and McDermott [43], late adopters benefit from cost reductions in technology and can benchmark tried and tested solutions, while early adopters pay more but can achieve market share through increased competitive advantages. Table 1 summarises the Lean 4.0 opportunities from the literature.

Table 1. Lean Industry 4.0 Opportunities and Challenges.

Technology	Opportunities	Challenges
Cybersecurity	Reduced waste	Creating automated
Cloud Computing	Improved flow	waste
Mobile Technologies	Available data	Resources
Machine to Machine	Accurate data	Resistance to change
3D Printing	Data Analytics	Timing of adoption
Advanced Robotics	Quicker decisions	Data security
Big Data/Analytics	Flexibility	Data protection
Internet of Things	Connectivity	Change Management
RFID Technologies	Reduced errors	Lack of digital data
	New markets	Costly
	New products	Time-consuming
	New customers	Location
	New regulations	Management support
	New standards	Alignment with strategy
	Flexible working	Choosing the right technology
Cognitive Computing	Faster	Cost–benefit analysis
	Cheaper	Lack of communication of
	Innovation	strategy
	Increased productivity	
	Revenue growth	
	Predictive maintenance	
	Regulatory Compliance	

3. Methodology

This research aims to demonstrate, through a case study on a Lean Industry 4.0 project, that digitisation positively affects both Lean processes and regulatory compliance. The case study approach allows the researcher to focus on just one instance rather than multiple instances, supporting an in-depth review that can provide insight that may not be visible using multiple cases. A case study can also help the reader better understand the researched topic [46]. This study uses a single case to support the research. The case study will concentrate on one of the organisation's Industry 4.0 projects currently in implementation. Using a case study is a means by which the researcher can explore the subject in-depth, understanding how and why the subject is being implemented and how it is received by the organisation [47]. Data for the case study was gathered using local documents and having Microsoft Teams meetings with the case study organisations project lead to understand how the project progressed through to implementation, what challenges there were and why this particular Industry 4.0 project was chosen.

This research focuses solely on Company X, a medium-sized MedTech company in the early stages of its digital transformation. The case study will review and demonstrate how regulatory compliance has been impacted through detailed planning and execution of one of the organisation's Industry 4.0-type projects. Data was collected throughout the project with data collected beforehand to justify and quantify the need for the project.

Company X has over 23,000 products in its portfolio, employs over 14,000 people globally, generates just under $3 billion dollars in revenue, and has over 120,000 customers.

Company X products are used in over 24,000 surgical procedures in the United States, and its products are used in Intensive Care Units (ICU), Cardiology, Radiologists, Vascular Surgeons, and Emergency Responders. Therefore, Company X must continue to deliver products that meet customer and regulatory requirements. With the organisation's growth, its use of digital technology has also expanded. Due to how Company X has grown, through acquisition, multiple management systems manage its data, including product data, complaints, records, documentation, and the supply chain. Multiple systems have led to complex, difficult-to-manage processes, inefficiencies, a lack of global processes, and interconnectivity between IT systems, non-conformances, and recalls. Because of these issues, Company X is currently working on having one platform, system, and data source across all sites to enhance its production, reporting capabilities and compliance. While company X has had a mature Lean program for many years, it is considered a late adopter in terms of its Industry 4.0 deployment. Antony et al. [43] defined late adopters of Industry 4.0 as those organisations which delay the implementation of enhanced technology and adopt a more cautious approach to investing in such technologies.

The project this case study will focus on is internally referred to as "Project Impact". Project Impact is the organisation's Enterprise Change Management (ECM) program. ECM is the cornerstone program that will support the organisation's roadmap for the rollout of future Industry 4.0 initiatives. The project is a strategic initiative that aims to deliver a best-in-class ECM process for Company X's product data. Managing change in organisations is a laborious task that consumes value-added time in various segments of the product lifecycle, including design and development, production, delivery, and product disposition [48]. ECM and Product lifecycle management play an important role in minimising the time required for managing engineering changes [49].

4. Results

4.1. What Were The Industry 4.0 Tools Implemented?

The ECM program focuses on two key elements, Product Lifecycle Management (PLM) and Master Data Management (MDM). PLM is the process of managing the entire lifecycle of a product from inception, through engineering design and manufacture, to service and disposal of manufactured products and product end of life. ("Product Life Cycle Management System for the PLM Process") PLM is the business activity that effectively manages and supports Company X products throughout their lifecycle; refer to Figure 1 for an overview of PLM. The new PLM will use Oracle Agile, cloud-based software that will manage the following electronically: the Design History File, Registrations, Device Master Record, Change Process and Sustaining.

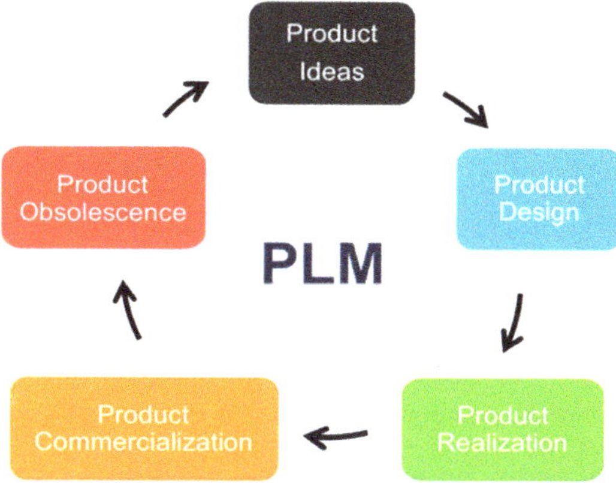

Figure 1. PLM Structure.

PLM impacts all aspects of Company X's business, including people, culture, technology, and processes, as seen in Figure 1 above.

MDM is a combination of systems and processes that link, manage and process key product data. MDM is a comprehensive method enabling an enterprise to link all its critical data to one file, called a master file or golden record. Refer to below Figure 2 for an overview of MDM. MDM uses Systems Applications and Products (SAP) Master Data Governance (MDG) software that will be the following for Company X's master data: a data hub, a golden Record, a Gatekeeper, and a workflow, which automates and defines data ownership.

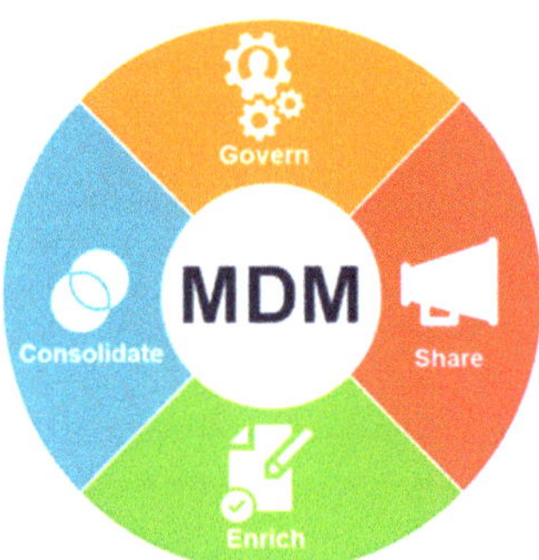

Figure 2. MDM Structure.

MDM will provide Company X employees with clear roles and responsibilities; it will deliver end-to-end metrics so that decisions can be made based on accurate data; it will simplify the current complex processes using technology and implementing one global system. The interface between PLM and MDM is a Business-to-Business (B2B) interface; MDG, in turn, consolidates and shares data with SAP. The two systems were chosen as both provide different functionalities. PLM will be used for managing product design and engineering specifications, change control, product lifecycle, workflow and task management, registrations, training, and document management. MDM will be the central repository for consolidated and clean data containing rules for integration and synchronisation of data that will be shared across both systems through workflow and task management using an interface.

As well as the two systems, another important element of Project Impact is Organisational Change Management (OCM) which is key in any project but even more so when implementing such a transformational change across the organisation. Anticipating and managing changes to people and process is critical to mitigating risk and enabling success [45]. Effective change management is more than training and communications; it also includes having and sharing the organisation's vision, having leadership support, bringing people on the journey as it happens, encouraging and enabling behavioural change, managing stakeholders, and continuously analysing and assessing on a daily, weekly, monthly, annual basis how the goals and objectives of the project and the team are progressing. Having a governance model in place to help and support the team in their decision-making gives the team the autonomy it needs to be successful and deliver per the agreed-upon timelines. Having the support of the Steering Committee, Project Leaders, and Project Team helps to ensure that decisions are made in a timely manner so that timelines are not impacted. It is about managing the change so that the people, processes, and technology are aligned, which ensures a successful outcome, benefiting all involved.

To deliver such a project, the team worked on obtaining buy-in from the Senior Leadership Team, which enabled them to build the team required to plan and execute the deliverables. The team includes a Steering Committee, Program Leadership/Advisors, and Project Management Office (PMO) Leadership who offer knowledge and support to each workstream, including PLM, MDM, Transformational Change, and IT. In addition, each workstream is supported by a core team and extended teams across the organisation.

4.2. Why Implement Industry 4.0 Tools?

The reasons for embarking on this transformational journey include product quality and compliance, recall reduction, revenue growth, improved time to market, operational efficiency, re-registration cost savings, effort during quality and regulatory audits, cycle time reduction for product management, and cost of goods sold (COGS) reduction, including scrap reduction and acceleration of cost improvement projects (CIPS). The team first built a strong business case to obtain Senior Leadership buy-in and support to support this project. The business case included reasons and examples of why and what could be achieved through implementing the PLM and MDM technologies and what the benefits are including customer, internal, and financial benefits. Table 2 gives an example of the importance of these technologies from a customer and internal point of view, including the issue, risk, and impact. Having an effective PLM/MDM prevents the type of error and consequences.

Table 2. Example of Internal and Customer Impact scenario that PLM/MDM can prevent.

Customer Story
Product: A medical device kit designed to support the most urgent clinical needs of the critical care patient.
Issue: Incorrect component listed on Bill of Materials (BOM)
Risk: The issue could have resulted in patient irritation during kit use, thereby complicating an already compromised patient during use.
Impact: Recall, Regulatory non-compliance, Business Impact
Internal Impact Story
Site to Site (Transfer issue)
Issue: Component manufactured in a new facility not cleared for release in EMEA by a regulatory agency. Insufficient controls for containment between regulatory plans, change control process, and product release at finished goods, semi-finished, or component level.
Risk: Finished goods/components were distributed without required regulatory clearance
Impact: Possible Recall, Regulatory non-compliance, Business Impact

The Teams vision enables Company X's accelerated growth by driving excellence in managing product creation and change through a unified global process. Thus the ECM will support the organisation's vision by standardising and deploying a global PLM process to reduce the risk of quality-related issues and introduce an MDM system for the management of product-related data for consistency and accuracy.

Other non-value add wastes specifically related to compliance identified by Company X as part of project brainstorming sessions and Value Stream Mapping (VSM) demonstrated the need to implement such technologies. The challenges included safety risks, quality risks, compliance risks, recall risks, impact on brand equity, highly manual work, and increased costs:

- Product changes implemented without adequate review/approval;
- Lack of verification of requirements to ensure the design meets the intended functionality;
- Discrepancies between product specifications, BOM, and commercial labels;
- Manufacturing processes not updated in coordination with product design updates;
- Inaccurate and non-compliant label information being released;
- Poor management of global label variations (language, metadata).

In addition to the above challenges, Company X has many disparate processes to manage product master data and document changes across the organisation. These result in very complicated workflows that can be challenging to manage and control, resulting in manual processes with many resources required to maintain them. Value Steam Mapping was utilised to map the current process and identify all non-value add (NVA) wastes [50]. Figure 3 demonstrate schematics based on the high-level VSMs before and after implementation of the Industry 4.0 project (Note: a schematic has been included rather than the original VSMs for legibility purposes). The new system creates pull and flow, adds value and can aid continuous improvement. The new system is more "Lean" and has a less complicated process resulting in reduced overprocessing and more streamlined processes for change management and master data maintenance.

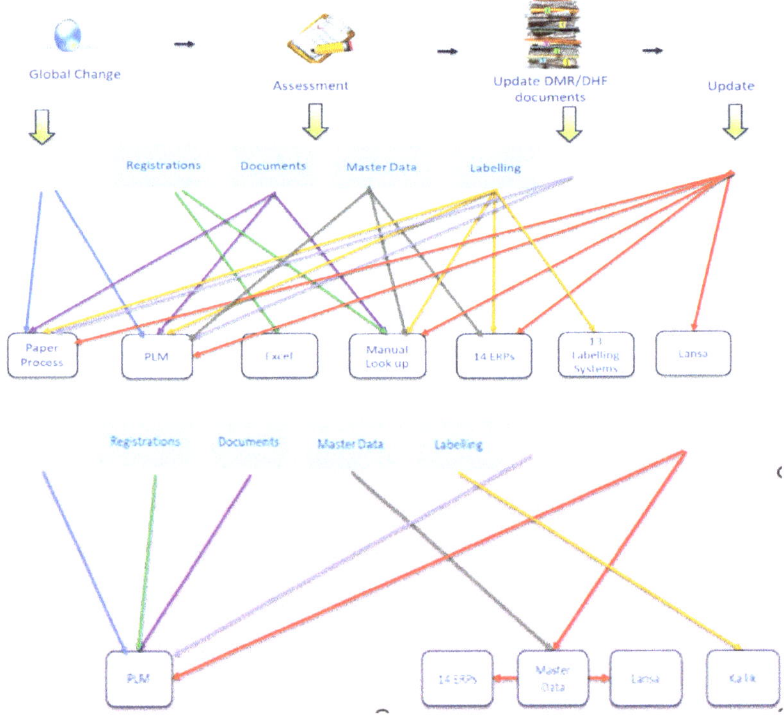

Figure 3. High-Level VSM before and after implementation.

Note: The diagram is intended to show visually the removal of non-value add systems and steps (represented as differently coloured lines in the diagram) via the research project implementation. Also the 22 types of system interactions before project versus the 8 systems after (PLM, ERP, Lansa and Kata) are represented.

4.3. Benefits

In addition to the above, Table 3 below lists the additional benefits that were gained post-full implementation of PLM and MDM.

The benefits directly impact the people, processes, and systems at Company X. Many of the benefits have a positive impact on regulatory compliance, ensuring that Company X products, processes, and services deliver a product that is safe and effective, meets customer requirements and expectations and meets regulatory requirements by delivering a harmonised global change management process with access to accurate and reliable master data.

Table 3. Benefits of the Lean 4.0 project.

Benefits	Lean Non-Value Add Waste Reduction
Reduction in Recalls	Defects/Transport/Over-processing
Reduced effort for compliance audits (internal and external)	Over-processing/Waiting/Over-production
Cycle Time Reduction	Waiting
Reduction in re-registration efforts	Over-processing
Reduction in Scrap	Defects/Over-processing
New Product Introductions (NPI) are delivered to the customer faster	Waiting/Over-processing
Better access and visibility to manage change internally leads to a streamlined, efficient process.	Over-processing/Waiting/Defects
Scalable process to assist how Company X can grow in the integration of future Mergers and Acquisitions (M&A)	Waiting/Transport/Inventory
Harmonised processes, data standards, and direct access to a source of true information	Defects/Over-processing/Inventory
Centralised, digital design documentation	Over-processing/Inventory/Defects/Waiting
Globally consistent Change Management process	Waiting/Inventory/Defects/Over-processing/ Over-production
Product management from conception to termination	All 7 wastes
Correct decision ownership	Under-utilisation of employee skillset
Process and data ownership defined with end-to-end metrics	Under-utilisation of employee skillsets
Aligning documentation	Inventory/Defects/Over-processing
70% Reduction in user interfaces	Over-processing/Inventory/Defects/ Over-production/ Waiting

4.4. Detailed Examples of ECM Impact on Regulatory Compliance

The following section takes a more in-depth look at some of the benefits associated with implementing ECM and how they will positively impact regulatory compliance. One common element across all areas of the ECM is the reduction in non-value add wastes in terms of man-hours and human interaction across each process. Reducing the number of people involved in any process reduces the number of opportunities for human error. Human error is one of the main sources of non-conformances across Company X; therefore, reducing human interaction directly impacts regulatory compliance by reducing non-conformances and defects.

4.5. Reduction in Non-Conformance Investigations and Recalls Related to ECM

Based on initial figures, 15% of Non-Conformance Reports (NCRs) were due to ECM activities (17 out of 110). Implementing an ECM program will reduce the number of ECM-related NCRs, positively impacting regulatory compliance and patient safety. Less NCRs result in fewer recalls and reduced effort in processing both NCRs and recalls freeing the ECM team up to work on other tasks, such as continuous improvement projects. ECM will deliver a 50% improvement in the number of ECM-related NCRs and recalls. Refer to Table 4 below for improvements relating to NCRs / recalls.

Table 4. Reduction in NCRs and Recalls related to ECM.

Reduction in Recalls	
% Related to ECM	55%
% Improvement with ECM	50%
Reduction in NCR	
% Improvement with ECM	50%

4.6. Reduced Effort for Quality Audits

The introduction of an ECM program means having all the product and master data available electronically. Having data that is readily available and easily accessed during audits/inspections reduces the number of people involved in pulling data manually and having to copy or scan documents to provide to an auditor/inspector. In addition, it ensures that documents, when requested, are available to the auditor/inspector promptly and without undue delay. This is particularly useful where there are many actors within an organisation working across many different time zones who are required to support audits and inspections across multiple sites depending on their actor statuses such as manufacturer, sub-contract manufacturer, component supplier, importer, distributor, authorised representative, or specification developer. As a result, ECM will deliver a 20% improvement in the effort it takes to manage a quality audit/inspection. Refer to Table 5 below for improvements relating to quality audits/inspections.

Table 5. Reduced effort for Quality Audits.

Reduced Efforts for Quality Audits	
# of audits	42 External
	300 Internal
Resources	15 people over 3 days
Total Effort	123,120 h
% Improvement with ECM	20%
Total Current Effort	123,120 man-hours
20% Inefficiency due to lack of ECM	24,624 man-hours
Post-ECM deployment Effort (hours)	11,650
% man-hour reduction	60%

4.7. Reduction in Re-Registration Efforts

As stated above, introducing an ECM program means having all product and master data available electronically. Having data that is readily available and easily accessed supports the registration process. When it is time to re-register products, rather than reaching out to different business units that must pull documents manually, scan them and arrange them for submission, ECM will support this process and make it easier and less time-consuming. As a result, ECM will deliver a 20% improvement in the effort it takes re-register the product. Refer to Table 6 below for improvements relating to re-registration.

Table 6. Reduction in re-registration efforts (Source: Project Impact Lead).

Reduction in Re-Registration Efforts	
Annual # of registrations	616
Re-Registration Effort	10.5 days/registration
Average Time to Support Each registration	51,744 h
% Improvement with ECM	20%
Risk Factor of 40% factored in	
Total Current effort for Re-Registrations	51,7444 man-hours
% Improvement due to ECM (20%)	10,349 man-hours
Post ECM Deployment Effort	4140 man-hours
Man-hour reduction	6209 man-hours
% man-hour reduction	60%

4.8. Cycle Time Reduction

Implementing an ECM will deliver a 48% reduction in the cycle time. Reducing cycle time means faster time to market, so customers and patients will have access to devices. Refer to Table 7 below for improvements relating to cycle time.

Table 7. Cycle Time Reduction (Source: Project Impact Lead).

Cycle Time Reduction	
Total # of changes	11,344
# of changes requiring rework	10% or 1134
Average Approval and Creation Time	
Approval	2 h
Creation	3.04 h
Total	5.04 h
Enterprise # of changes	7631
Average hours spent per change	16.03 h
% improvement by PLM	48%
Total No of hours	58,715 h
Total current effort	64,432 man-hours
% man-hour reduction	38,659 man-hours (60%)

4.9. Faster Time to Market

Based on the project's complexity, the time to market based on the implementation of ECM differs from between 5% and 15% improvement in the time it takes to get a device to market post-implementation. Having products on the market faster means customers can access life-changing and life-saving devices quicker, as seen below in Table 8.

Table 8. Faster Time to Market Benefits.

Device Product Complexity	Average Time to Market (Months)	% Improvement by ECM	Adjusted Time to Market (Months)	Saving in Months
High	27	15%	22.95	4.05
Medium	16	10%	14.40	1.60
Low	9	5%	8.55	0.45

The case study was performed on Company X, a medium-sized medical technologies (MedTech) manufacturer that provides medical devices and technologies globally. Company X has grown through acquisition resulting in its many management systems. The case study provides an overview of Company X's history, which includes why the organisation has started implementing some Industry 4.0 tools to aid its Lean processes. These include simplifying processes, improving Lean flow, realising efficiencies, and reducing the number of errors and recalls across the organisation through implementing a global system for managing changes and product data. In addition, as a case study, company X provides examples of how Lean Industry 4.0 tools have a more positive than negative impact on regulatory compliance.

5. Discussion

This research met its objectives to define the impact Lean 4.0 can have on the Total Product Lifecycle and Regulatory Compliance in a Medical Device manufacturer (RQ1) and to demonstrate how Industry 4.0 enhance and enable Lean (RQ2).

Lean processes, combined with Industry 4.0 technology as an enabler, can aid in regulatory compliance by optimising processes, reducing non-value add work and over-processing, and enabling ease of vigilance and aces to regulatory information. Improved Industry 4.0 technology can aid process flow and prevent errors that can result in missed compliance deadlines and errors that could result in recalls. Lean 4.0 is an enabler for enhanced Lean processes and reduces manual tasks [11,17]. The synergistic effects between the two concepts ensure a more successful and symbiotic relationship and implementation of Lean 4.0 [44].

Many studies on Lean, Industry 4.0, and Lean 4.0 combined discussed the benefits of an enhanced product and process quality, improved compliance, faster time to market and product cycle times, improved profits and revenue, and increased market share [12,23,41]. In addition, there have been improvements in the case study organisation in the following areas.

- Product Quality and Compliance Recall reduction: Patient safety and shrinkage in costs related to recalls caused by product management issues. Effort for Quality Audits: Reduction in man-hours related to searching and finding necessary data from across Company X sites and providing documents more efficiently and timely during audits.
- Time to Market Acceleration of initial launch: Products made available to the end user quicker and increased revenue achieved based on an acceleration of average time to market.
- Cost of goods sold reduction Scrap reduction: Minimising scrap cost related to preventable issues based on accurate product definition. Acceleration of Cost Improvement programs (CIPs): Accelerated time to adoption of cost improvement projects leading to increased cost-saving duration.
- Operational Efficiencies Cycle Time reduction: Streamline the change approval process to eliminate non-value add activities. Effort for re-registrations: Cutback on required man-hours per registration based on ease of visibility to data.

Other benefits included brand equity, faster integration of Mergers and acquisitions (M&A), procurement efficiencies and inventory optimisation. To achieve these benefits, Company X chose two well-established technologies, Agile for PLM and SAP for MDM. Choosing the right technology and understanding how it can be integrated into an organisation is a critical success factor for Industry 4.0 [51]. These technologies are the foundation for the organisation's digital transformation journey. These technologies will provide the organisation with the infrastructure needed to execute the organisation's Strategic Vision and what is also considered the organisation's Industry 4.0 roadmap. A strategic plan and map for Industry 4.0 implementation is key to the success of the initiative [52].

From the project's initiation, Company X's leadership team were fully invested, involved and supportive of the strategic plan for Industry 4.0 deployment. While it took some time to gain approval from the Senior Leadership Team (just under 2 years), significant investment was approved by the organisation in terms of resources, both people and finances. Many Industry 4.0 projects can fail without this level of management support and involvement to understand the alignment of digitalisation with strategy [53]. The project team had to provide the Senior Leadership Team with the evidence they needed in terms of benefits and return on investment before committing to the project. A detailed cost–benefit analysis and understanding of the need for such technology are key to the success of such deployments [24].

Another key aspect of the project was driving change within the organisation and the requirement for effective communication and training. People need to understand what is being changed and why it is being changed so they can buy into and support the project [4].

Digital transformation involves significant effort, time, and money [54]. However, the benefits the project could bring to make the organisation Leaner and enhance its regulatory compliance, the project needed Senior Leadership to buy in given its significance and for it to be successful and deliver the benefits to the organisation. The data presented in this case study demonstrates how Industry 4.0 and Lean combined can have a synergistic effect.

6. Conclusions

The research met its research objectives to demonstrate that Lean and Industry 4.0 can improve and enhance Lean processes, reduce waste and improve productivity and quality while enhancing digitalisation. Integrating Lean and Industry 4.0 can enhance regulatory compliance to ensure that organisations adhering to global regulations and legislation can deliver safe and effective products. A limitation of this research was that it was a single case study. Using similar or different-sized companies (small or large) would have provided another perspective on how and why other companies are implementing Lean 4.0 and at what stage they are in their journey so that a comparison could be made. The case study organisation used was only in the early implementation of its strategic plan for digitalisation to enhance Lean. Therefore, while it is possible to review the first stages of the project's success, further research could focus on the ongoing deployment across company X. Further research should be taken post-implementation to gain more long-term data on the digital technologies implemented, their effects on Lean, and their impact on regulatory compliance. In addition, future studies should consider including other MedTech companies to make a comparison.

Author Contributions: Conceptualisation, I.F. and O.M.; methodology, I.F. and O.M.; validation, I.F., O.M., A.R. and M.K.; formal analysis, O.M., I.F. and A.R.; investigation, M.K. and I.F.; resources, I.F. and A.R.; writing—original draft preparation, I.F., O.M., A.R. and M.K.; writing—review and editing, O.M., A.R. and M.K.; supervision, O.M. and M.K.; project administration, O.M. and I.F. All authors have read and agreed to the published version of the manuscript.

Funding: This research received no external funding.

Data Availability Statement: Not applicable.

Conflicts of Interest: The authors declare no conflict of interest.

References

1. Bhamra, R.; Hicks, C.; Small, A.; García-Villarreal, E. Value, product delivery strategies and operational performance in the medical technology industry. *Int. J. Prod. Econ.* **2022**, *245*, 108399. [CrossRef]
2. McDermott, O.; Antony, J.; Sony, M.; Healy, T. Critical Failure Factors for Continuous Improvement Methodologies in the Irish MedTech Industry. *TQM J.* **2022**, *34*, 18–38. [CrossRef]
3. Sony, M.; Antony, J.; Mc Dermott, O.; Garza-Reyes, J.A. An empirical examination of benefits, challenges, and critical success factors of industry 4.0 in manufacturing and service sector. *Technol. Soc.* **2021**, *67*, 101754. [CrossRef]
4. Antony, J.; McDermott, O.; Sony, M. Quality 4.0 conceptualisation and theoretical understanding: A global exploratory qualitative study. *TQM J.* **2021**. *ahead of print.* [CrossRef]
5. McDermott, O.; Antony, J.; Sony, M.; Swarnakar, V. Mapping the Terrain for the Lean Supply Chain 4.0. In *Journey to Sustainable Supply Chains*; Centre for Concurrent Enterprise, Cork University Business School: Cork, Ireland, 2022; pp. 167–179.
6. Kupper, D.; Heidemann, A.; Strohle, J.; Knizek, C. When Lean Meets Industry 4.0. Available online: https://www.bcg.com/publications/2017/lean-meets-industry-4.0 (accessed on 16 October 2022).
7. Maci, J.; Marešová, P. Critical Factors and Economic Methods for Regulatory Impact Assessment in the Medical Device Industry. *Risk Manag. Healthc Policy* **2022**, *15*, 71–91. [CrossRef]
8. Maganga, D.P.; Taifa, I.W.R. Quality 4.0 transition framework for Tanzanian manufacturing industries. *TQM J.* **2022**. *ahead-of-print.* [CrossRef]
9. Melvin, T.; Torre, M. New Medical Device Regulations: The Regulator's View. *EFORT Open Rev.* **2019**, *4*, 351–356. [CrossRef]
10. Vlachos, I.P.; Pascazzi, R.M.; Zobolas, G.; Repoussis, P.; Giannakis, M. Lean manufacturing systems in the area of Industry 4.0: A lean automation plan of AGVs/IoT integration. *Prod. Plan. Control* **2021**, 1–14. [CrossRef]
11. Antony, J.; McDermott, O.; Powell, D.; Sony, M. The Evolution and Future of Lean Six Sigma 4.0. *TQM J.* **2022**. *ahead-of-print.* [CrossRef]
12. Antony, J.O.; McDermott, O.; Powell, D.; Sony, M. Mapping the Terrain for Lean Six Sigma 4.0. Learning in the Digital Era. In *7th European Lean Educators Conference (ELEC) 2021 Trondheim*; Powel, D.J., Alfnes, E., Holmemo, M.D.Q., Reke, E., Eds.; Springer: Cham, Switzerland, 2021; pp. 193–204.
13. Buer, S.-V.; Semini, M.; Strandhagen, J.O.; Sgarbossa, F. The complementary effect of lean manufacturing and digitalisation on operational performance. *Int. J. Prod. Res.* **2021**, *59*, 1976–1992. [CrossRef]
14. Felsberger, A.; Qaiser, F.H.; Choudhary, A.; Reiner, G. The impact of Industry 4.0 on the reconciliation of dynamic capabilities: Evidence from the European manufacturing industries. *Prod. Plan. Control* **2022**, *33*, 277–300. [CrossRef]
15. Calabrese, A.; Dora, M.; Ghiron, N.L.; Tiburzi, L. Industry's 4.0 Transformation Process: How to Start, Where to Aim, What to Be Aware Of. *Prod. Plan. Control* **2022**, *33*, 492–512. [CrossRef]

16. Ding, B.; Hernandez, X.; Jane, N. Combining lean and agile manufacturing competitive advantages through Industry 4.0 technologies: An integrative approach. *Prod. Plan. Control* **2021**, 1–17. [CrossRef]

17. Tortorella, G.L.; Fettermann, D. Implementation of Industry 4.0 and Lean Production in Brazilian Manufacturing Companies. *Int. J. Prod. Res.* **2018**, *56*, 2975–2987. [CrossRef]

18. Hughes, L.; Dwivedi, Y.K.; Rana, N.P.; Williams, M.D.; Raghavan, V. Perspectives on the future of manufacturing within the Industry 4.0 era. *Prod. Plan. Control* **2022**, *17*, 138–158. [CrossRef]

19. Moeuf, A.; Pellerin, R.; Lamouri, S.; Tamayo-Giraldo, S.; Barbaray, R. The industrial management of SMEs in the era of Industry 4.0. *Int. J. Prod. Res.* **2018**, *56*, 1118–1136. [CrossRef]

20. Chiarini, A.; Kumar, M. Lean Six Sigma and Industry 4.0 integration for Operational Excellence: Evidence from Italian manufacturing companies. *Prod. Plan. Control* **2021**, *32*, 1084–1101. [CrossRef]

21. World Health Organization. *Medical Device Regulations: Global Overview and Guiding Principles*; World Health Organization: Geneva, Switzerland, 2003.

22. Anna, T.; Manto, D.; McDermott, O. A Review of Lean Adoption in the Irish MedTech Industry. *Processes* **2022**, *10*, 391. [CrossRef]

23. Antony, J.; Sony, M.; Furterer, S.; McDermott, O.; Pepper, M. Quality 4.0 and Its Impact on Organizational Performance: An Integrative Viewpoint. *TQM J.* **2021**. *ahead-of-print*. [CrossRef]

24. Antony, J.; Douglas, J.A.; McDermott, O.; Sony, M. Motivations, barriers and readiness factors for Quality 4.0 implementation: An exploratory study. *TQM J.* **2021**. *ahead-of-print*. [CrossRef]

25. Antony, J.; Sony, M.; McDermott, O.; Jayaraman, R.; Flynn, D. An exploration of organisational readiness factors for Quality 4.0: An intercontinental study and future research directions. *Int. J. Qual. Reliab. Manag.* **2021**. *ahead-of-print*. [CrossRef]

26. Veeva Medtech Modernizing Regulatory Affairs: Veeva MedTech Regulatory Benchmark Study. 2021. Available online: Veeva.com (accessed on 12 October 2022).

27. Arden, N.S.; Fisher, A.C.; Tyner, K.; Yu, L.X.; Lee, S.L.; Kopcha, M. Industry 4.0 for pharmaceutical manufacturing: Preparing for the smart factories of the future. *Int. J. Pharm* **2021**, *602*, 120554. [CrossRef]

28. Nick, G.; Kovács, T.; Kő, A.; Kádár, B. Industry 4.0 readiness in manufacturing: Company Compass 2.0, a renewed framework and solution for Industry 4.0 maturity assessment. *Procedia Manuf.* **2021**, *54*, 39–44. [CrossRef]

29. Bianchini, E.; Francesconi, M.; Testa, M.; Tanase, M.; Gemignani, V. Unique device identification and traceability for medical software: A major challenge for manufacturers in an ever-evolving marketplace. *J. Biomed. Inform.* **2019**, *93*, 103150. [CrossRef]

30. Badreddin, G.O.; Mussbacher, D.; Amyot, S.A.; Behnam, R.; Rashidi-Tabrizi, E.; Braun, M.; Alhaj, G. Richards Regulation-Based Dimensional Modeling for Regulatory Intelligence. In Proceedings of the 2013 6th International Workshop on Requirements Engineering and Law (RELAW), Rio de Janeiro, Brazil, 16 July 2013; pp. 1–10.

31. Koshechkin, K.; Lebedev, G.; Tikhonova, J. Regulatory Information Management Systems, as a Means for Ensuring the Pharmaceutical Data Continuity and Risk Management. In *Intelligent Decision Technologies 2019*; Czarnowski, I., Howlett, R.J., Jain, L.C., Eds.; Springer: Singapore, 2020; pp. 265–274.

32. Kazlovich, K.; Mishra, S.R.; Behdinan, K.; Gladman, A.; May, J.; Mashari, A. Open ventilator evaluation framework: A synthesised database of regulatory requirements and technical standards for emergency use ventilators from Australia, Canada, UK, and US. *HardwareX* **2022**, *11*, e00260. [CrossRef]

33. Kwon, J.; Johnson, M.E. Security practices and regulatory compliance in the healthcare industry. *J. Am. Med. Inform Assoc* **2013**, *20*, 44–51. [CrossRef]

34. Kumar, R.; Sharma, R. Leveraging blockchain for ensuring trust in IoT: A survey. *J. King Saud Univ. Comput. Inf. Sci.* **2021**, *34*, 8599–8622. [CrossRef]

35. Morrison, R.J.; Kashlan, K.N.; Flanagan, C.L.; Wright, J.K.; Green, G.E.; Hollister, S.J.; Weatherwax, K.J. Regulatory Considerations in the Design and Manufacturing of Implantable 3D-Printed Medical Devices. *Clin. Transl. Sci.* **2015**, *8*, 594–600. [CrossRef] [PubMed]

36. Czvetkó, T.; Honti, G.; Abonyi, J. Regional Development Potentials of Industry 4.0: Open Data Indicators of the Industry 4.0+ Model. *PLoS ONE* **2021**, *16*, e0250247. [CrossRef] [PubMed]

37. Ramírez-Durán, V.J.; Berges, I.; Illarramendi, A. Towards the implementation of Industry 4.0: A methodology-based approach oriented to the customer life cycle. *Comput. Ind.* **2021**, *126*, 103403. [CrossRef]

38. Sorenson, C.; Drummond, M. Improving medical device regulation: The United States and Europe in perspective. *Milbank Q* **2014**, *92*, 114–150. [CrossRef]

39. Malvehy, J.; Ginsberg, R.; Sampietro-Colom, L.; Ficapal, J.; Combalia, M.; Svedenhag, P. New regulation of medical devices in the EU: Impact in dermatology. *J. Eur. Acad. Derm. Venereol* **2022**, *36*, 360–364. [CrossRef] [PubMed]

40. Niemiec, E. Will the EU Medical Device Regulation help to improve the safety and performance of medical AI devices? *Digit. Health* **2022**, *8*, 20552076221089079. [CrossRef] [PubMed]

41. Antony, J.; Sony, M.; McDermott, O. Conceptualizing Industry 4.0 Readiness Model Dimensions: An Exploratory Sequential Mixed-Method Study. *TQM J.* **2021**. *ahead-of-print*. [CrossRef]

42. Duggan, J.; Cormican, K.; McDermott, O. Lean Implementation: Analysis of Individual-Level Factors in a Biopharmaceutical Organisation. *Int. J. Lean Six Sigma* **2022**. *ahead-of-print*. [CrossRef]

43. Antony, J.; Sony, M.; McDermott, O.; Furterer, S.; Pepper, M. How Does Performance Vary between Early and Late Adopters of Industry 4.0? A Qualitative Viewpoint. *Int. J. Qual. Reliab. Manag.* 2021. *ahead-of-print*. [CrossRef]

44. McDermott, O.; Nelson, S. *Readiness for Industry 4.0 in West of Ireland Small and Medium and Micro Enterprises*; College of Science and Engineering, University of Galway: Galway, Ireland, 2022.
45. Eltaief, A.; Ben Makhlouf, A.; Ben Amor, S.; Remy, S.; Louhichi, B.; Eynard, B. Engineering Change Risk Assessment: Quantitative and Qualitative Change Characterisation. *Comput. Ind.* **2022**, *140*, 103656. [CrossRef]
46. Denscombe, M. *The Good Research Guide: For Small-Scale Social Research Projects*; McGraw-Hill Education: London, UK, 2014.
47. Crowe, S.; Cresswell, K.; Robertson, A.; Huby, G.; Avery, A.; Sheikh, A. The case study approach. *BMC Med. Res. Methodol.* **2011**, *11*, 100. [CrossRef]
48. Brian, B.; McDermott, O.; Kinahan, N.T. Manufacturing control system development for an in vitro diagnostic product platform. *Processes* **2021**, *9*, 975.
49. Habib, H.; Menhas, R.; McDermott, O. Managing Engineering Change within the Paradigm of Product Lifecycle Management. *Processes* **2022**, *10*, 1770. [CrossRef]
50. Brian, B.; McDermott, O.; Noonan, J. Applying Lean Six Sigma Methodology to a Pharmaceutical Manufacturing Facility: A Case Study. *Processes* **2021**, *9*, 550. [CrossRef]
51. Yadav, N.; Shankar, R.; Singh, S.P. Hierarchy of Critical Success Factors (CSF) for Lean Six Sigma (LSS) in Quality 4.0. *JGBC* **2021**, *16*, 1–14. [CrossRef]
52. Sony, M.; Antony, J.; Mc Dermott, O. How Do the Technological Capability and Strategic Flexibility of an Organization Impact Its Successful Implementation of Industry 4.0? A Qualitative Viewpoint. *Benchmarking Int. J.* **2022**. *ahead-of-print*. [CrossRef]
53. Sony, M. Pros and Cons of Implementing Industry 4.0 for the Organisations: A Review and Synthesis of Evidence. *Null* **2020**, *8*, 244–272. [CrossRef]
54. Frère, E.; Zureck, A.; Röhrig, K. Industry 4.0 in Germany—The Obstacles Regarding Smart Production in the Manufacturing Industry. *SSRN Electron. J.* **2018**. Available online: https://papers.ssrn.com/sol3/papers.cfm?abstract_id=3223765 (accessed on 16 October 2022).

Article

Competitive Priorities and Lean–Green Practices—A Comparative Study in the Automotive Chain' Suppliers

Geandra Alves Queiroz [1,*], Alceu Gomes Alves Filho [2] and Isotilia Costa Melo [3]

[1] Engineering Department, Production Engineering, Universidade do Estado de Minas Gerais (UEMG), Passos 37900-106, Minas-Gerais, Brazil
[2] Industrial Engineering Department, Federal University of São Carlos, São Carlos 13565-905, São Paulo, Brazil
[3] Escuela de Ingeniería de Coquimbo (EIC), Universidad Católica del Norte (UCN), Coquimbo 25803-35, Chile
* Correspondence: geandraquelroz@gmail.com

Abstract: For organizations to remain competitive, they must now adapt to sustainability requirements, which have become performance criteria for supplier selection for most original Equipment manufacturers (OEMs). In this sense, environmental performance is now included as a competitive priority throughout the supply chain. Therefore, this study aims to verify, through two case studies, the competitive priorities of two first-tier suppliers from the automotive chain that have adopted lean and green practices. The findings show that the quality priority is the main source of competitive advantage and the focus of the operations that are analyzed here, while the environmental priority is not considered the most important by the companies. However, it is still included as a priority. Furthermore, it is demonstrated that lean practices could generate compatibility for the environmental priority, even indirectly, while trade-offs can arise between priorities. Therefore, the integration between lean and green practices can facilitate the inclusion of the environmental priority into the operations strategy and management systems.

Keywords: lean–green; sustainability; competitive priority; operations strategy; supplier

Citation: Queiroz, G.A.; Filho, A.G.A.; Costa Melo, I. Competitive Priorities and Lean–Green Practices—A Comparative Study in the Automotive Chain' Suppliers. *Machines* **2023**, *11*, 50. https://doi.org/10.3390/machines11010050

Academic Editor: Dan Zhang

Received: 24 October 2022
Revised: 28 November 2022
Accepted: 7 December 2022
Published: 1 January 2023

1. Introduction

1.1. Contextualization and Research Objective

As shown by Skinner (1969) [1] in his seminal article, manufacturing balances consumer demands and production function resources. Thus, industrial operations play a crucial role in achieving these goals. Remarkably, operations strategies that address environmental issues have been adopted by organizations through the inclusion of environmental performance as a competitive priority, referred to here as the "environmental priority" [2–4].

In this context, to be competitive, organizations have to establish long-term strategies to achieve environmental sustainability, and all supply chain members have an essential role in supporting this [5]. Besides managers, researchers in operations management also face a significant challenge in including an environmental priority [6]. In this context, equalizing the cost, quality, delivery, flexibility, and service priorities in a stable trade-off become an urgent necessity [7]. Along these lines, organizations and researchers have sought solutions that promote the integration and alignment of practices, enabling operational (lean) and environmental (green) gains [8,9].

The literature discusses the integration of lean manufacturing, known as the production management philosophy, and green manufacturing, an approach to reduce environmental impacts in manufacturing. This integration is named lean–green manufacturing and has been understood as a key to improving the competitiveness of organizations as a way to balance the environmental priority with the other competing priorities [3,7,10,11].

The literature also points out several compatible aspects between these approaches. Especially regarding the reductions in waste generated by lean processes that lead to the efficient use of resources, this approach can indirectly lead to the removal of negative environmental impacts caused by production flows [12]. Additionally, there are studies such as the paper by Jamali et al. (2017) [13] that argue that competitive strategies can implement lean and green practices. However, there are studies such as the one by Suifan (2019) [7] that point out that the competitive priorities can differ in each approach. However, these studies still do not provide a wide understanding of the relationships between lean–green and operations strategies.

Furthermore, a literature review [14] showing the current state of the trade-offs between the competitive priorities of lean and green processes found a few studies that have developed and validated a conceptual framework that seeks to put into context these practices. Additionally, there are discussions about digital innovations and their impacts on improving environmental performance, as presented by Yin et al. (2022) and Queiroz et al. (2022) [15,16].

Therefore, although synergies between lean and green processes and issues related to new technologies have been presented, these studies have not provided a discussion from a strategic perspective while presenting and understating the competitive priorities of such integration [4]. Considering the relevance of the operations for an organization's global strategy and the lack of studies about the lean–green approach from the operations strategy perspective, an analysis of the competitive priorities can provide an update about how these practices relate to the competitive strategy.

In this fashion, the following question arises, "What are the competitive priorities in companies that adopt lean and green practices?".

Given this background, this paper aims to identify the competitive priorities of two first-tier supplier companies of the automotive chain that have adopted lean and green practices, specifically to understand how these priorities are ordered and how the environmental priority is considered. This industry is considered a reference case for lean manufacturing. Consequently, most of the lean–green models are developed in organizations in the automotive industry [17].

The remainder of this paper is organized as follows. Section 2 presents the fundamental concepts about the competitive priorities of operations strategy and the lean and green approach. Next, Section 3 describes the research method. Section 4 presents the results and a discussion. Finally, the final considerations of this study are also given.

1.2. Theoretical Background

1.2.1. Competitive Priorities

The corporate strategy gives rise to the functional designs of a company, among them the operations strategy, which will be the focus of this study and is considered the primary source of competitive advantage over the last 40 years and until today, meaning it deserves attention [18,19]. Additionally, the operations strategy seeks to define an organization's business operations and decisions regarding the acquisition and allocation of resources. It is aimed at the entire organization [20].

Skinner's (1969) [1] study was the pioneer in highlighting and conceptualizing operations strategy, showing the importance of incorporating and aligning the operational elements of the production function into the corporate strategy. According to this author, production should be considered strategic and a source of competitive advantage. In this way, companies must recognize and establish a relationship between corporate and operations strategies so that the production systems are competitive and collaborate to achieve the organization's goals [21]. Hence, the operations play a decisive role in achieving a favorable competitive position [22].

In this sense, the operations strategy is defined as a sequence of decisions that over time allow a business unit to achieve a structure, the desired production infrastructure, and a set of specific resources, i.e., a consistent pattern of decision-making in the production function, aligned to the business strategy [20].

The content of the operations strategy is related to the company's decisions around the corporate system's effectiveness [23]. This is a set of competitive priorities related to the operations and decisions in the structural and infrastructural areas of production [24]. The competitive priorities are related to the performance objectives that the production function adopts to align with the company's competitive strategy [1]. In other words, as the organization outlines a strategy to meet the market requirements, it is determined how the operations need to be performed [25].

The competitive priorities of production, also called performance objectives, competitive dimensions, and production missions, should be part of the priorities that will guide the programs to be implemented by the production function of a company. This means competitive priorities represent how the company will meet customer needs concerning the production function's performance targets. In other words, they define how the company intends to compete in the market to meet the needs of its customers [26].

In this article, the competitive priorities proposed by Garvin (1993) [27] and Slack and Lewis (2011) [25] are adopted, which represent the most recurrent studies on operations strategy. Hence, the adopted priorities for this research are the cost, quality, delivery, service, flexibility, and environment. The cost priority may be related to the objectives of reducing the costs of acquisition, production, and distribution and the price to customers. The quality priority involves aiming to produce goods according to specifications, aesthetics, perceived quality, and performance. Flexibility is the ability to react to changes in the volume and mix of products and in the production schedule. The delivery priority is related to reductions in lead time between the beginning and the end of the operations. Additionally, it is related to the availability and quicker delivery of the product and meeting the agreed delivery deadline. Finally, the environmental priority is the search to reduce the environmental impacts from energy consumption, the use of materials, gas emissions, and waste generation from certain processes, as presented in the lean and green literature [3,28–31].

The inclusion of the environmental priority might make operations management even more complex, given that this impacts the company's performance in a multidimensional manner [32]. Thus, when the environment is considered a competitive priority, it is essential to consider the environmental issues in the operations strategy. Consequently, modifications or redesigns of the operations strategy are required [33].

Furthermore, it is crucial to consider that there are trade-offs between the competitive priorities, as argued by Skinner (1969) [1]. For this author, the organization must prioritize only one or another competitive priority, seeking to be better than its competitors. The operations must be focused once it is not possible to obtain low cost and quality at the same time. To this effect, Skinner (1969) [1] explains that organizations must make certain decisions regarding the size of the manufacturing unit, whether to have high stocks or low stocks, the types of equipment to be used, and the level of standardization. In summary, Skinner (1974) [34] introduced the concept of a focused factory, which concerns the impossibility of a factory working well for all competing priorities.

However, Sarmiento, Thurer, and Whelan (2016) [35] consider it is possible to focus on more than one priority. However, choices need to be made and trade-offs are inevitable, since a production system must be excellent to meet all criteria to create a competitive advantage. In a context where competitiveness increases and it becomes necessary to meet more than one customer need, the trade-off model proposed by Skinner (1969) [1] is questioned. Such questioning culminated in the proposal of a cumulative capabilities model that simultaneously implies high performance in more than one competitive priority [36].

In one of the first studies on cumulative production capabilities, as pointed out by Boyer and Lewis (2002) [36], Japanese organizations developed productive capabilities based on a previously established order, and the practices adopted allowed cost reductions and the production of quality products simultaneously. Ferdows and De Meyer (1990) [37] propose the "sand cone model", which establishes that the organization can achieve all competitive priorities over time and that there is an adequate sequence for their construction, with quality being considered the basis for the implementation of other improvements. In this way, Ferdows and De Meyer (1990) [37] also argue that it is important to focus on avoiding failures in the system, and that in this way the costs could be reduced by means of other capabilities, such as via better quality in the processes. The authors also point out that improvements obtained through good production practices are more lasting and stable.

However, Flynn and Flynn (2004) [38] noted no evidence for the sequence of priorities presented in the sand cone model. The authors argue that the development of cumulative capabilities is complex and not limited to a specific sequence, as several factors influence it.

Regarding the inclusion of the "environment" as a new competitive priority, the literature emphasizes possible trade-offs that may arise between the environment and the other priorities [4,39]. Vargas-Berrones, Sarmiento, and Whelan (2020) [39] show for small- and medium-sized enterprises (SMEs) that the implementation of some green initiatives may be extraordinarily costly, and this scenario could discourage businesses from pursuing them. However, according to Porter and Linde (1995) [40], it is possible to meet the economic and environmental objectives of products and processes, since the preservation of resources generates greater process efficiency.

1.2.2. Lean–Green

Lean manufacturing emerged in the 1950s. It is considered a management philosophy and has been one of the most widely used approaches to managing operations [41]. Lean manufacturing has been considered a great solution to improve all kinds of processes, in the production of both goods and services [42]. Lean manufacturing is a set of principles and practices that seek to eliminate all forms of waste from processes. The main lean practices are 5S, Kaizen, value stream mapping, just-in-time (JIT), rapid tool change, total productive maintenance, standardization work, visual management, 5 whys or Ishikawa diagram (fishbone), and Kanban (pull production) [43]. In addition, the focus on the implementation of lean systems is via delivery (lead time reductions), quality, and cost reduction [7,42,44]. Organizations from various industries around the world are adopting lean practices to become more competitive [45].

On the other hand, green manufacturing emerged in the 1990s as an operational and philosophical approach to reducing the adverse environmental effects of products and processes [3]. In short, this approach aims to reduce the impacts generated by operations and deals with the search for reduced pollution, energy consumption, and emissions of toxic substances through the development of new processes in manufacturing [46].

Green manufacturing is composed of different practices. These practices seek to reduce the environmental impacts generated by production processes, such as via the environmental accreditation of suppliers and the use of product life cycle analyses, reverse logistics, environmental management systems, waste management policies, and effluent treatments, as well as via programs for water conservation, energy, recycling, materials consumption, and environmental education [47].

The pioneering study on lean and green manufacturing was presented in 1996 by Florida (1996) [11] and discussed the integration of these approaches, exploring how organizations could include the environmental issue in manufacturing through the "lean–green" approach, arguing that the waste reduction generated by lean manufacturing contributes to environmental performance. Based on the previous literature, the lean–green concept is understood as an approach that supports the search for sustainable development in the economic, environmental, and social pillars of a production system [48] and focuses on waste reductions and the efficient use of resources [3,12,49].

Some studies [50–52] have shown that lean manufacturing can bring environmental benefits and that this can be attributed to the more efficient use of resources (such as water and other inputs). In lean implementation cases, it is possible to note many improvement efforts to reduce variation or waste from operations [53,54]. Thus, the congruent aspect of lean and green manufacturing is waste reduction [49].

The lean–green literature [55,56] suggests that implementing lean practices can offer significant advantages and synergy with a firm's environmental performance, without compromising other competing priorities. Lean and green processes are considered complementary [29,57,58]. Moreover, according to these studies, the organizational structure and lean culture facilitate the development of environmental management and the formation of a "green" company. Additionally, it was demonstrated that the vital link between operational excellence and the lean–green approach enables the achievement of competitive advantage in the sustainability era [59].

However, generally lean manufacturing does not directly cover environmental impacts. In the literature, surveys about the lean implementation model [60] did not find a framework for lean implementation that considers environmental impacts. Therefore, there are blind spots in lean manufacturing concerning the environment, such as the environmental risks of the improvements and practices [61]. Along these lines, organizations need to use green tools to fill this gap [29] and make the "environment" a competitive priority. Consequently, since lean and green processes have different objectives, trade-offs may arise between competitive priorities [4,7]; it is an urgent necessity to integrate green processes into lean manufacturing explicitly by considering the environmental aspects of lean performance indicators and practices [55,62,63]. Some studies point out that digitalization can be an enabler in supporting this integration and solving these trade-offs [8,16].

The lean–green literature emphasizes that the main trade-offs are between delivery and the environment, because JIT can increase emissions [4,64,65]. Additionally, the trade-off between the environment and quality is relevant because of the utilization of raw materials to achieve the best product quality [66]. Other trade-offs pointed out in the literature are related to flexibility and the environment, since small batches allow more product variety, but they may increase the number of setups [12,67]. Additionally, the cost can present a trade-off to green implementation [39]. Some studies point out that cost can be a motivation for lean organizations to reduce their environmental impacts [55,68]. Additionally, environmental performance has been considered an important criterion for supplier selection [69].

Therefore, based on the literature, the use of lean–green practices can be understood as an approach that supports the pursuit of sustainable development in the economic, environmental, and social pillars of a production system [48] and focuses on reducing waste and focusing on the efficient use of resources [70,71].

The lean paradigm was created in the automotive industry. Additionally, considering the concept of lean–green manufacturing, it is possible to find studies that sought to understand these practices in the automotive industry. In Iran, it was identified that the lean–green efforts are focused on packaging materials and concentrated on increasing the useful life of recyclable materials. At the operational level, the focus is on reducing pollution and waste. Finally, the strategy is seen as the basis for enhancing operational and environmental efficiencies [72]. Another example of lean–green manufacturing in the automotive chain [73], taking the concept of waste "Muda" and based on lean tools, is another case study in Iran that concluded that the assembly body and paint rooms are the areas, in this order, where the lean practices impact the green performance more.

The study also investigated how the integration between agility and lean manufacturing led to enhanced sustainability in the Indian automotive industry. The main conclusion of this investigation was that the legislation represents a driver for automotive companies to improve the ecological aspects of their business operations. Ecological aspects are seen as antagonist forces to competitiveness [74]. In another case [75] in Indonesia, it was demonstrated that some green issues need to be improved in line with lean and green criteria, namely the guidelines for "ISO 14000 and OHSAS Certificates", "Collaboration with Suppliers and Customers in Protecting the Environment", "Carrying out Industrial Waste Recycling", and "Product Design that can Reduce the Consumption of Energy and Raw Materials".

Finally, for the consolidation of the theoretical basis of this paper, and as presented in the research by Carvalho et al. (2014) [76], it is important to highlight that given the fact the lean and green paradigms can lead to opposite goals depending on the focus on each paradigm, an exploratory case study was conducted in the automotive supply chain context.

All companies belonging to the observed supply chain required higher implementation levels for all lean–green practices. Two separate sequences of capabilities were found, one for the automaker and another for the first-tier supplier. According to the authors, the first-tier supplier echelon should develop their "quality" first, then their "flexibility" and "delivery", and finally their "cost" and "environmental protection" aspects.

2. Materials and Methods

This article adopted the case study as a research strategy, since it is suitable when questions such as "why?", "what?", and "how?" are asked, which must be answered with a complete understanding of the nature and phenomenon studied and when the focus is on contemporary phenomena embedded in a real context [77]. Moreover, the main trend in all case studies is that they try to shed light on why a decision or set of decisions were made, how they were implemented, and what results were achieved [78]. This research sought to compare two cases of companies in the automotive chain, specifically two first-tier suppliers. In Figure 1, it is possible to see a summary of the research steps based on the case study method proposed by Yin (2017) [78].

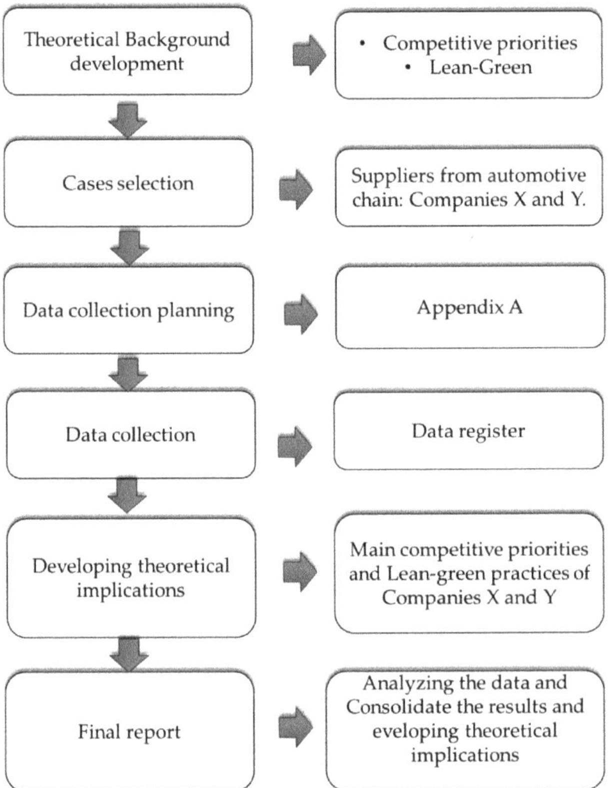

Figure 1. The followed research steps.

The first step of this research consisted of elaborating a theoretical background about the competitive priorities of the operations strategy and lean–green approach. Subsequently, the protocol for the case studies was elaborated and experts validated it. The next step was the pilot case study, followed by adjustments to the protocol and conducting the case studies.

The first step of the data collection consisted of a structured questionnaire that helped identify the main characteristics of the companies, the strategies adopted, the ranking priorities, and the practices adopted. The second part was a semi-structured interview to understand the major complementarities and conflicts generated between competitive priorities by the lean and green practices. The section about competitive priorities was based on the study by Ward et al. (1998) [79], the lean questions were based on the study by Shah and Ward (2003) [43], and the green section was based on practices presented by Thanki, Govindan, and Thakkar (2016) [80].

The criterion for selecting the companies was a search for two companies that acted as first-tier supplier companies of the automotive chain that had implemented lean and green practices in their operations. The study on these companies was conducted in May of 2021. Thus, the two selected companies will be referred to in this study as company X and company Y. They are in southern and southeastern Brazil. The interviewees were the production manager and the environmental manager of each company using the questionnaire in Appendix A. The questionnaire was sent to the interviewees in advance, and the interviews were recorded.

The reason for choosing companies from the automotive industry was attributed to the origins of lean manufacturing, as the lean management approach was created in this industry. Womack, Jones, and Roos (1991) [81] explained that the origin of lean manufacturing lies in the Toyota Production System. Moreover, as shown in the study by Caldera and Dawes (2017) [17], most of the lean and green models were developed in companies in the automotive industry. Additionally, this industrial segment is representative in terms of benchmarking for lean implementation [41,82–84]. Thus, companies in the automotive sector chain were the focus of this investigation, since it is a sector that can provide more empirical data on operations strategies regarding the adoption of lean and green approaches.

3. Results

3.1. Company Overview

Brazil produces passenger vehicles, trucks, and agricultural machinery. In 2020, 27 vehicle manufacturers and 446 auto parts companies were in operation in the country. In 2019, 2.94 million vehicles were produced [85]. Both of the studied companies are multinational automotive tier-one suppliers. Company X has over one thousand employees. Its operations are in southern Brazil, and it is headquartered in Germany. The products manufactured by company X are considered strategic components in the vehicle's final assembly, such as tires and mechanical belts. Company Y is also large and has over one thousand employees in southeastern Brazil. However, it is headquartered in the United States. Company Y produces components for assembly in its product portfolio, such as coatings, adhesives, and safety products. The products manufactured by Company Y are considered less strategic than those manufactured by company X.

3.2. Competitive Priorities of the Companies

Firstly, it is vital to highlight the aspects related to the companies' competitive strategies. They are mainly directed toward quality. In other words, quality is the main positioning factor for both companies in the market. According to the results, company X's main competitiveness factor is only quality. On the other hand, company Y also considers innovation capacity and flexibility. Regarding the main competitive advantages vis-à-vis competitors, company X competes on quality and price, while company Y competes only on quality. Figure 2 summarizes in a graphic main factors and their respective levels of importance for the competitiveness of these companies, according to the interviewees, with 1 being the least important and 5 being the most important.

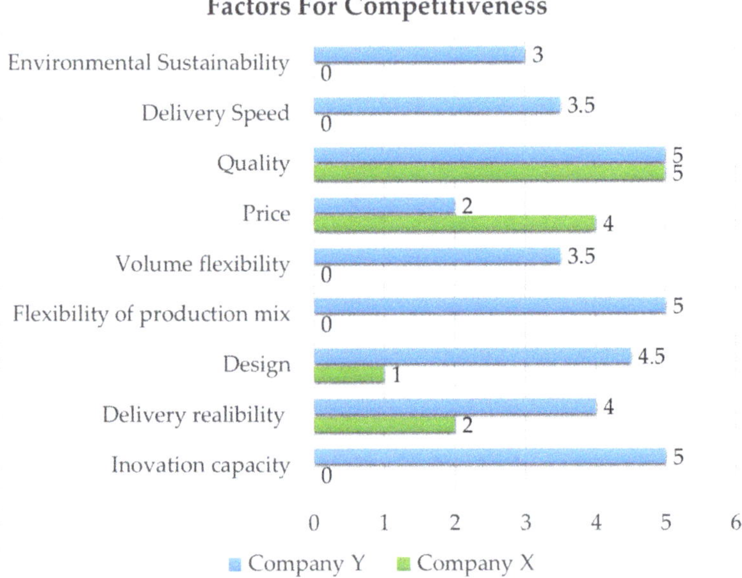

Figure 2. Factors for competitiveness.

In addition to quality, as already highlighted, it was observed that company X attributes a high degree of importance to price, and in a less critical manner to innovation, delivery, and flexibility, while considering the others not important at all. On the other hand, company Y considers other criteria important to ensure its competitiveness in the market. For example, flexibility in the production mix, product design process, and innovation capacity is considered very important, followed by the reliability and speed of delivery and finally the price and environmental sustainability.

Regarding the overall objective of the operations strategies, both companies consider cost reductions very important via defect reductions. In the case of defect reductions, these are convergent with the competitive strategy being driven by quality. Table 1 shows the order of competitive priorities as adopted by each company.

Table 1. Ranking of the companies' competitive priorities.

Company	1st Priority	2nd Priority	3rd Priority	4th Priority	5th Priority
X	Quality	Cost	Flexibility	Service and Delivery	Environment
Y	Quality	Cost and Delivery	Environment and Flexibility	Service	-

The results show that the quality priority is considered very important in all its aspects, being scored as the first priority, converging with corporate strategy, and demonstrating alignment between corporate strategic and operational objectives. On the other hand, cost comes in second place. Another observation point is the similarity in the ordering of priorities, differing in the prioritization of delivery and the environment. In the case of delivery, company X ranks it as the fourth most crucial priority, whereas company Y ranks it as the second, followed by cost. The environmental priority is considered the least essential priority by company X, while company Y considers it the third most important priority.

3.3. Lean–Green Practices

This topic highlights the main aspects of the lean–green practices in the studied companies. Tables 2 and 3 present the implemented practices and the stages of implementation.

Table 2. The companies' lean practices.

Lean Practices	Implementation Level	
	Company X	Company Y
Visual Management	Fully deployed and tracked	Partially deployed
Just-in-Time (JIT)	Nothing done	Partially deployed
Kaizen	Fully deployed and tracked	Fully deployed and tracked
Kanban (Pull Production)	Currently at "project" level but not yet implemented	Partially deployed
Cellular Manufacturing	Partially deployed	Fully deployed and tracked
Total Productive Maintenance (TPM)	Partially deployed	Fully deployed and tracked
Value Stream Mapping (VSM)	Fully deployed and tracked	Partially deployed
Poka-Yoke (mistake-proofing system)	Fully deployed and tracked	Partially deployed
Standardized work	Partially deployed	Fully deployed and tracked
5 whys/Ishikawa diagram (fishbone)	Fully deployed and tracked	Fully deployed and tracked
5s (five S's)	Fully deployed and tracked	Fully deployed and tracked
Other Lean practices		Tier Meetings—daily leadership briefings.

Table 3. The companies' green practices.

Green Practice	Implementation Level	
	Company X	Company Y
Environmental accreditation of suppliers	Early Implementation	Partially deployed
Product life cycle analysis	Early Implementation	Partially deployed
Reverse logistics	Early implementation	Incipiently implemented
Environmental Management Plan/System	Fully implemented and controlled	Fully deployed and controlled
Waste management policy	Fully implemented and controlled	Fully deployed and controlled
Water consumption reduction program	Early implementation	Partially implemented
Energy conservation program	Incipient implementation	Partially implemented
Recycling program	Fully implemented and controlled	Incipient implementation
Environmental education programs for the community	Incipient implementation	Incipiently implemented
Program to reduce material consumption	Fully implemented and controlled	Currently at "project" level but not yet implemented
Cross-process resource sharing programs	Early implementation	Currently at "project" level but not yet implemented
Cleaner production program	Incipient implementation	Nothing done
Publication of reports with environmental information	Fully deployed and controlled	Fully implemented and controlled
Effluent treatment	Fully implemented and controlled	Fully deployed and controlled

Company X has implemented lean processes since 2006, but only in 2010 did the program became part of the company's management system. As for their green practices, they have been implemented since 2002, since these practices were adopted to meet the environmental laws required at the time. According to the production manager, their lean approach focuses on quality improvements, and in as much as their green practices are concerned, according to the environmental manager, these are focused on meeting the current standards and legislation.

Currently, company X integrates lean and green issues through their strategy, whereby the company needs to comply with international corporate targets for reducing carbon emissions. To this effect, in addition to the practices taken to comply with environmental regulations, environmental indicators are being deployed utilizing the Hoshin–Kanri approach, a practice widely used in lean manufacturing to deploy the strategy for operations. Still, in the interviewees' view, this integration is very incipient, but it was emphasized that lean manufacturing reduces waste and that this leads to improvements of some environmental aspects, especially in terms of energy and material consumption. Furthermore, lean manufacturing is a facilitator for the inclusion of green practices and environmental improvements, since the environmental management system is based on its philosophy and practices.

Regarding company Y, the person responsible for managing their operations pointed out that the program made its first attempt to implement lean practices in 2013, but only in 2015 did the program become part of the management system to improve quality and achieve process standardization. Regarding green manufacturing, the company first implemented their practices in the 1990s, with their specific cleaner production initiatives being some of the forerunners in implementing this program. However, despite being a pioneer, the company operates these practices as an isolated program that seeks to encourage isolated projects to achieve environmental impact reductions for products or processes. The integration of green processes into lean procresses, as at company X, is still very nascent. Tables 2 and 3 present, respectively, the lean and green practices and their implementation levels.

It was possible to observe that only the safety indicator is addressed in the daily management of lean practices, and occasionally projects to improve environmental performance occur within the lean system. Additionally, according to the interviewees from company X, it is believed that the organizational structure and lean culture can contribute to an improvement of the environmental priority. Similarly, they exemplify lean projects in which reductions in energy and material consumption were achieved. However, despite the green gains being measured in some lean projects, there are still no environmental performance indicators in the lean management system.

What can be observed is that the companies are very close when it comes to the level of implementation of their practices, both lean and green. Regarding lean practices, company Y uses two practices, just-in-time and pulled production, which are in the preliminary stages. Still, it was possible to identify a similar implementation level or with a small difference between being partially implemented and totally implemented in the remaining lean practices. In the same way, company X presents a slightly lower level of green practices. However, it uses the practice of environmental accreditation of suppliers that is not implemented at company Y. On the other hand, company Y uses cleaner production and life cycle analysis practices at advanced levels.

A point of convergence between all interviewees is that lean practices, which help reduce waste, indirectly lead to the improvement of environmental aspects, especially concerning reductions in material and energy consumption. Furthermore, both companies mentioned that there is still a trade-off between cost and the environment. Projects that seek to reduce their environmental impact beyond what is required by law must bring some financial return to the company—regardless of the positive environmental impact. At the same time, cost reduction projects that generate environmental effects within the limits of the legislation and carbon reduction targets are not usually implemented.

4. Discussion

The main point to be discussed is the difference regarding the ordering of the competitiveness priorities. The environmental priority is positioned fifth and third. This may be attributed to the fact that only company Y considers environmental sustainability as a factor in competitiveness. In addition to practices for the compliance of environmental aspects, it also has the cleaner production program, which is a proactive strategy for the eco-efficiency of its processes, although it is still an isolated initiative from the lean program. Similarly, it is essential to note that despite the order of prioritization, this confirms what is shown in the literature, namely that environmental priority becomes included as a competitive priority in both companies [62].

Another point to be observed is that although the environmental priority is not considered the most important priority and is not a focus priority in lean manufacturing, the interviewees agreed with the literature [62]. They stated that the reduction in lean waste also impacts the environmental priority. Furthermore, the results confirm that lean manufacturing assists in achieving the main quality priority, as suggested by Ball (2015) and Suifan, Alazab, and Alhyari (2019) [7,44]. Moreover, the interviewees' account illustrates what is proposed in [29,57,58] when reporting that lean manufacturing can be a complementary facilitator for implementing green practices in management systems.

These results converge with the cumulative capabilities theory of the sandal cone approach presented by Ferdows and De Meyer (1990) [37] regarding competitive priorities when there are improvements in lean waste reductions; consequently, the companies achieve better environmental performance. Lastly, the results show that the lean and green level of company Y is higher than in company X. At the same time, the trade-offs between the environment and another competitive priority in company Y are less frequent. The environment is considered more important in company Y than in company X.

In the two cases, it was also observed that the investments in green initiatives depended on the investment level. The results showed that the trade-offs lens could be viewed from the perspective of Skinner (1969) [1] when companies do not adopt costly green initiatives. In addition, the trade-off between the cost and environment was exposed, as punctuated by Longoni and Cagliano (2015) and Vargas- Berrones, Sarmiento, and Whelan (2020) [4,39]. However, the question related to legislation pointed out by Mathiyazhagan et al. (2021) [74] was not clear, regarding how ecological aspects are seen as antagonist forces to competitiveness.

Given this background, based on the theory of cumulative capabilities, it is clear that by developing good practices that seek to improve the environmental priority, one also has the ability to reduce costs. Thus, integrated practices that make the environment a priority become essential. Additionally, compared with a study done in 2014 [76], it is possible to observe that environmental protection will become more important for first-tier suppliers. Finally, it is important to emphasize that previous studies [72,73,86] from the automotive industry also show that lean manufacturing can support the environmental priority.

5. Conclusions

This study indicated that the environmental priority has become a factor of competitiveness, whether incipient or behind other priorities, from the corporate strategy to the operations strategy. However, the priorities of quality, cost, and delivery in the cases presented here are still considered more important.

In the cases of the companies analyzed here, it was demonstrated that the integration between lean and green practices could facilitate the inclusion of the environmental priority in the operations strategy and in the management systems, as presented in the literature. Furthermore, as demonstrated in company X, this integration can help in unfolding long-term environmental and operational goals. In company Y, lean manufacturing is a facilitating factor, by means of the organizational structure and culture of this approach, for the implementation of green practices.

Additionally, it is essential to reinforce that the results show what the lean–green literature proposes, and most of the results are from cases in the same industry. Furthermore, the results show that the environment has become more important. They also show that legislation can be an influencing factor for strategic decisions regarding the adoption of greener practices.

Like all research, our study is not without its limitations. Consequently, these limitations can serve as the basis for further investigations. Although the results presented here can highlight relevant aspects regarding the integration of lean and green practices and their role in operations strategies, this is still a preliminary discussion. Only two cases were analyzed, and the choice of the sample was intentional due to the limited number of companies that already operate lean–green practices in the Brazilian context. This limitation prevents generalizations. Thus, the development of further research on this theme is necessary, especially in other relevant markets that have not yet been investigated. As a future research direction, empirical studies in other industries for comparative purposes are also recommended, as well as quantitative studies in larger samples, seeking to analyze in depth how much each practice contributes to each competitive priority when implementing lean and green practices. Finally, standardization in data collection can enable cross-country and cross-temporal analyses.

Author Contributions: Conceptualization, G.A.Q. and A.G.A.F.; methodology, G.A.Q.; validation, G.A.Q.; formal analysis, G.A.Q.; investigation, G.A.Q.; resources, A.G.A.F.; data curation, G.A.Q.; writing—original draft preparation, G.A.Q.; writing—review and editing, I.C.M.; visualization, G.A.Q.; supervision, A.G.A.F. and I.C.M.; project administration, G.A.Q. and I.C.M.; funding acquisition, I.C.M. All authors have read and agreed to the published version of the manuscript.

Funding: This research received no external funding.

Data Availability Statement: Not applicable.

Conflicts of Interest: The authors declare no conflict of interest.

Appendix A. Case Study Questionnaire

A. Company description

1. *Number of employees*
   Less than 50
   50 to 100 employees
   100 to 500 employees
   500 to 1000 employees
2. *Main products manufactured*

   ____

3. *Position in the supply chain (Nominal level ranging from Original Equipment Manufacturer to n-tier supplier)*
   First tier
   Second tier
   Third Tier
   Fourth Tier
   Fifth tier or more

B. Operations strategy

1. *Who is the main clientele of your products?*
   Final user
   Original equipment manufacturers
   Replacement market (aftermarket)
2. *Which of the options below is your organization's main competitive advantage?*
   Price
   Delivery

Customization
Less environmental impact
Service Level
Another:

3. *Concerning your competitive strategy, please rank in order of importance the 5 main factors for your company's competitiveness (5 being the most important, 1 being the less important).*
    5   4   3   2   1
    Price
    Design (product characteristics, technology)
    Quality
    Volume flexibility
    Flexibility of production mix (product variety)
    Delivery reliability
    Delivery speed
    Environmental sustainability
    Innovation capacity
    Another (Example: Location, aggregate services, etc.)

4. *Competitive Priorities (Please use this scale to signal the degree of importance for each competitive priority for your operations).*
    5   4   3   2   1
    Cost of production (total cost of products sold)
    Direct costs (labor and material)
    Overhead costs (administration, maintenance)
    Design quality (projected performance of main product characteristics)
    Conformance (a product manufactured according to design specifications)
    Reliability (probability of the product not failing)
    Product Flexibility (ability to adapt products to customer's needs)
    Volume Flexibility (ability to respond to variations in required quantities)
    Process Flexibility (includes production mix flexibility, sequencing flexibility, and routing flexibility)
    Reliability (probability of delivering the right product in the right quantity and on time)
    Speed of service (time elapsed between order and delivery of the product to the customer)
    Customer problem solving
    Supplier support (in-product development, process planning, and component production)
    Actions to reduce material waste, energy consumption, water consumption, and emissions.
    3R—Remanufacturing, reuse, and recycling.

C. Lean Manufacturing

1. *Indicate the year bracket in which Lean Manufacturing practices were implemented in your organization.*
    Before 1990
    Between 1990 and 2000
    Between 2001 and 2005
    Between 2006 and 2010
    Between 2011 and 2015
    From 2016 onwards

2. *To your knowledge, which of the following factors motivated the implementation of Lean Manufacturing practices in your organization? (You can choose several options if needed.)*
    Cost reduction
    Quality improvement
    Customer's requirement

Market competition
Corporate strategy
Another:

3. *Lean Manufacturing Practices. Please use this scale to signal the level of implementation of Lean Manufacturing in your current operations.*

> Nothing has been done
> Currently at the "project" level but not yet implemented
> Incipient implementation
> Partially deployed
> Fully deployed and tracked

Lean Manufacturing Practices:

> Kanban (pull production)
> Just-in-Time (JIT)
> Just-in-Sequence (JIS)
> Total Productive Maintenance (TPM)
> 5S (five S)
> Value Stream Mapping (VSM)
> Poka-Yoke (error-proofing system)
> Cellular Manufacturing
> Visual Management
> 5 why/Ishikawa (fishbone) diagram
> Kaizen
> Standardized work

D. Green Manufacturing

1. *Indicate the year bracket in which Green Manufacturing practices were implemented in your organization.*
Before 1990
Between 1990 and 2000
Between 2001 and 2005
Between 2006 and 2010
Between 2011 and 2015
From 2016 onwards

2. *To your knowledge, what factors motivated the implementation of Green Manufacturing practices in your organization? You can choose several options if needed.*
Cost reduction
Quality improvement
Customer's requirement
Market competition
Corporate strategy
Legislation
Another:

3. *Green Manufacturing Practices. Please use this scale to signal the level of implementation of Green Manufacturing in your current operations.*

> Nothing has been done
> Currently at the "project" level but not yet implemented
> Incipient implementation
> Partially deployed
> Fully deployed and tracked

Green Manufacturing Practices:

> Environmental Management Plan
> Waste Management Policy
> Effluent Treatment

Water consumption reduction program
Energy conservation program
Recycling program
Program to reduce material consumption
Publication of reports with environmental information
Product life cycle analysis
Environmental accreditation of suppliers
Environmental education programs for the community
Inter-process resource-sharing programs
Cleaner production program
Reverse logistics
Another (please, specify)

E. Understanding Lean–Green and Competitive priorities

1. *What is your opinion of the implementation of Lean and Green manufacturing practices in relation to the competitive priorities of your current operations?*
2. *In your opinion, what is the contribution of Lean manufacturing to Green manufacturing practices?*
3. *Do you consider that Lean manufacturing practices contribute to environmental performance? If so, in what way?*

References

1. Skinner, W. Manufacturing—Missing Link in the Corporate Strategy. *Harv. Bus. Rev.* **1969**, *47*, 136–145.
2. Gandhi, N.S.; Thanki, S.J.; Thakkar, J.J. Ranking of Drivers for Integrated Lean-Green Manufacturing for Indian Manufacturing SMEs. *J. Clean. Prod.* **2018**, *171*, 675–689. [CrossRef]
3. Garza-Reyes, J.A. Lean and Green-a Systematic Review of the State of the Art Literature. *J. Clean. Prod.* **2015**, *102*, 18–29. [CrossRef]
4. Longoni, A.; Cagliano, R. Environmental and Social Sustainability Prioritiesn: Their Integration in Operations Strategies. *Int. J. Oper. Prod. Manag.* **2015**, *35*, 216–345. [CrossRef]
5. Asadabadi, M.R.; Ahmadi, H.B.; Gupta, H.; Liou, J.J.H. Supplier Selection to Support Environmental Sustainability: The Stratified BWM TOPSIS Method. *Ann. Oper. Res.* **2022**, 1–24. [CrossRef]
6. Gavronski, I. Resources and Capabilities for Sustainable Operations Strategy. *J. Oper. Supply Chain Manag.* **2012**, *1*, 1–20. [CrossRef]
7. Suifan, T.; Alazab, M.; Alhyari, S. Trade-Off among Lean, Agile, Resilient and Green Paradigms: An Empirical Study on Pharmaceutical Industry in Jordan Using a TOPSIS-Entropy Method. *Int. J. Adv. Oper. Manag.* **2019**, *11*, 69–101. [CrossRef]
8. Thanki, S.; Thakkar, J.J. An Investigation on Lean–Green Performance of Indian Manufacturing SMEs. *Int. J. Product. Perform. Manag.* **2020**, *69*, 489–517. [CrossRef]
9. Leong, W.D.; Teng, S.Y.; How, B.S.; Ngan, S.L.; Rahman, A.A.; Tan, C.P.; Ponnambalam, S.G.; Lam, H.L. Enhancing the Adaptability: Lean and Green Strategy towards the Industry Revolution 4.0. *J. Clean. Prod.* **2020**, *273*, 122870. [CrossRef]
10. Cherrafi, A.; Elfezazi, S.; Govindan, K.; Garza-Reyes, J.A.; Benhida, K.; Mokhlis, A. A Framework for the Integration of Green and Lean Six Sigma for Superior Sustainability Performance. *Int. J. Prod. Res.* **2017**, *55*, 4481–4515. [CrossRef]
11. Florida, R. Lean and Green: The Move to Environmentally Conscious Manufacturing. *Calif. Manage. Rev.* **1996**, *39*, 80–105. [CrossRef]
12. Dües, C.M.; Tan, K.H.; Lim, M. Green as the New Lean: How to Use Lean Practices as a Catalyst to Greening Your Supply Chain. *J. Clean. Prod.* **2013**, *40*, 93–100. [CrossRef]
13. Jamali, G.; Asl, E.K.; Zolfani, S.H.; Šaparauskas, J. Analysing LARG Supply Chain Management Competitive Strategies in Iranian Cement Industries. *Econ. Manag.* **2017**, *20*, 70–83. [CrossRef]
14. Bouhannana, F.; Elkorchi, A. Trade-Offs among Lean, Green and Agile Concepts in Supply Chain Management: Literature Review. In Proceedings of the 2020 IEEE 13th International Colloquium of Logistics and Supply Chain Management (LOGISTIQUA), Fez, Morocco, 2–4 December 2020; IEEE: Piscataway, NJ, USA, 2020; pp. 2–4. [CrossRef]
15. Yin, S.; Zhang, N.; Ullah, K.; Gao, S. Enhancing Digital Innovation for the Sustainable Transformation of Manufacturing Industry: A Pressure-State-Response System Framework to Perceptions of Digital Green Innovation and Its Performance for Green and Intelligent Manufacturing. *Systems* **2022**, *10*, 72. [CrossRef]
16. Queiroz, G.A.; Junior, P.N.A.; Melo, I.C. Digitalization as an Enabler to SMEs Implementing Lean-Green? A Systematic Review through the Topic Modelling Approach. *Sustainability* **2022**, *14*, 14089. [CrossRef]
17. Caldera, H.T.S.; Desha, C.; Dawes, L. Exploring the Role of Lean Thinking in Sustainable Business Practice: A Systematic Literature Review. *J. Clean. Prod.* **2017**, *167*, 1546–1565. [CrossRef]
18. Skinner, W. Manufacturing Strategy: The Story of Its Evolution. *J. Oper. Manag.* **2007**, *25*, 328–335. [CrossRef]
19. Voss, C.A. Paradigms of Manufacturing Strategy Re-Visited. *Int. J. Oper. Prod. Manag.* **2005**, *25*, 1223–1227. [CrossRef]

20. Wright, S.C.W. Manufacturing Strategy: Defining the Missing Link. *Strateg. Manag. J.* **1984**, *5*, 77–91. [CrossRef]
21. Dangayach, G.S.; Deshmukh, S.G. Manufacturing Strategy Literature Review and Some Issues. *Int. J. Oper. Prod. Manag.* **2001**, *21*, 884–932. [CrossRef]
22. Filho, A.G.A.; Nogueira, E.; Bento, P.E.G. Operations Strategies of Engine Assembly Plants in the Brazilian Automotive Industry. *Int. J. Oper. Prod. Manag.* **2015**, *35*, 817–838. [CrossRef]
23. Kim, J.S.; Arnold, P. Operationalizing Manufacturing Strategy: An Exploratory Study of Constructs and Linkage. *Int. J. Oper. Prod. Manag.* **1996**, *16*, 45–73. [CrossRef]
24. Hayes, R.; Pisano, G.; Upton, D.; Wheelwright, S. *Produção, Estratégia e Tecnologia: Em Busca Da Vantagem Competitiva*; Bookman: Porto Alegre, Brazil, 2007; ISBN 8-57-780108-X.
25. Slack, N.; Lewis, M. *Operations Strategy*, 3rd ed.; Pearson Education Limited: London, UK, 2011; ISBN 978-0-27374-044-5.
26. Filho, A.G.A.; Pires, S.R.I.; Vanalle, R.M. On manufacturing competitive priorities: Trade-offs and implementation sequences. *Gestão Produção* **1995**, *2*, 173–180. [CrossRef]
27. Garvin, D.A. Manufacturing Strategic Planning. *Calif. Manag. Rev.* **1993**, *35*, 85–106. [CrossRef]
28. Alves, J.R.X.; Alves, J.M. Production Management Model Integrating the Principles of Lean Manufacturing and Sustainability Supported by the Cultural Transformation of a Company. *Int. J. Prod. Res.* **2015**, *53*, 5320–5333. [CrossRef]
29. Ng, R.; Low, J.S.C.; Song, B. Integrating and Implementing Lean and Green Practices Based on Proposition of Carbon-Value Efficiency Metric. *J. Clean. Prod.* **2015**, *95*, 242–255. [CrossRef]
30. Ben Ruben, R.; Vinodh, S.; Asokan, P. Implementation of Lean Six Sigma Framework with Environmental Considerations in an Indian Automotive Component Manufacturing Firm: A Case Study. *Prod. Plan. Control* **2017**, *28*, 1193–1211. [CrossRef]
31. Souza, J.P.E.; Alves, J.M. Lean-Integrated Management System: A Model for Sustainability Improvement. *J. Clean. Prod.* **2018**, *172*, 2667–2682. [CrossRef]
32. Azzone, G.; Noci, G. Identifying Effective PMSs for the Deployment of "Green" Manufacturing Strategies. *Int. J. Oper. Prod. Manag.* **1998**, *18*, 308–335. [CrossRef]
33. Johansson, G.; Winroth, M. Introducing Environmental Concern in Manufacturing Strategies: Implications for the Decision Criteria. *Manag. Res. Rev.* **2010**, *33*, 877–899. [CrossRef]
34. Skinner, W. The Focused Factory. *Harv. Bus. Rev.* **1974**, *52*, 113–121.
35. Sarmiento, R.; Thurer, M.; Whelan, G. Rethinking Skinner's Model: Strategic Trade-Offs in Products and Services. *Manag. Res. Rev.* **2016**, *39*, 1199–1213. [CrossRef]
36. Boyer, K.K.; Lewis, M.W. Competitive Priorities: Investigating the Need for Trade-Offs in Operations Strategy. *Prod. Oper. Manag.* **2002**, *11*, 9–20. [CrossRef]
37. Ferdows, K.; De Meyer, A. Lasting Improvements in Manufacturing Performance: In Search of a New Theory. *J. Oper. Manag.* **1990**, *9*, 168–184. [CrossRef]
38. Flynn, B.B.; Flynn, E.J. An Exploratory Study of the Nature of Cumulative Capabilities. *J. Oper. Manag.* **2004**, *22*, 439–457. [CrossRef]
39. Vargas-Berrones, K.X.; Sarmiento, R.; Whelan, G. Can You Have Your Cake and Eat It? Investigating Trade-Offs in the Implementation of Green Initiatives. *Prod. Plan. Control* **2020**, *31*, 845–860. [CrossRef]
40. Porter, M.E.; Linde, C. Van Der Green and Competitive: Ending the Stalemate Green and Competitive. *Harv. Bus. Rev.* **1995**, *73*, 120–134.
41. James, P.W.; Jones, D.T. *Lean Thinking: Banish Waste and Create Wealth*; Free Press: New York, NY, USA, 2003; ISBN 0-74-324927-5.
42. Kunkera, Z.; Tošanović, N.; Štefanić, N. Improving the Shipbuilding Sales Process by Selected Lean Management Tool. *Machines* **2022**, *10*, 766. [CrossRef]
43. Shah, R.; Ward, P.T. Lean Manufacturing: Context, Practice Bundles, and Performance. *J. Oper. Manag.* **2003**, *21*, 129–149. [CrossRef]
44. Ball, P. Low Energy Production Impact on Lean Flow. *J. Manuf. Technol. Manag.* **2015**, *26*, 412–428. [CrossRef]
45. Losonci, D.; Demeter, K. Lean Production and Business Performance: International Empirical Results. *Compet. Rev.* **2013**, *23*, 218–233. [CrossRef]
46. Pathak, P.; Singh, M.P. Sustainable Manufacturing Concepts: A Literature Review. *Int. J. Eng. Technol. Manag. Res.* **2017**, *4*, 1–13. [CrossRef]
47. Pampanelli, A.B.; Found, P.; Bernardes, A.M. A Lean & Green Model for a Production Cell. *J. Clean. Prod.* **2014**, *85*, 19–30. [CrossRef]
48. Bhattacharya, A.; Nand, A.; Castka, P. Lean-Green Integration and Its Impact on Sustainability Performance: A Critical Review. *J. Clean. Prod.* **2019**, *236*, 117697. [CrossRef]
49. Cobra, R.L.R.B.; Guardia, M.; Queiroz, G.A.; Oliveira, J.A.; Ometto, A.R.; Esposto, K.F. "Waste" as the Common "Gene" Connecting Cleaner Production and Lean Manufacturing: A Proposition of a Hybrid Definition. *Environ. Qual. Manag.* **2015**, *25*, 25–40. [CrossRef]
50. Vais, A.; Miron, V.; Pedersen, M.; Folke, J. "Lean and Green" at a Romanian Secondary Tissue Paper and Board Mill—Putting Theory into Practice. *Resour. Conserv. Recycl.* **2006**, *46*, 44–74. [CrossRef]
51. Vinodh, S.; Arvind, K.R.; Somanaathan, M. Application of Value Stream Mapping in an Indian Camshaft Manufacturing Organisation. *J. Manuf. Technol. Manag.* **2010**, *21*, 888–900. [CrossRef]

52. Teixeira, P.; Coelho, A.; Fontoura, P.; Sá, J.C.; Silva, F.J.G.; Santos, G.; Ferreira, L.P. Combining Lean and Green Practices to Achieve a Superior Performance: The Contribution for a Sustainable Development and Competitiveness—An Empirical Study on the Portuguese Context. *Corp. Soc. Responsib. Environ. Manag.* **2022**, *29*, 887–903. [CrossRef]
53. Chakravorty, S.S. An Implementation Model for Lean Programmes. *Eur. J. Ind. Eng.* **2010**, *4*, 228–248. [CrossRef]
54. Chakravorty, S.S.; Shah, A.D. Lean Six Sigma (LSS): An Implementation Experience. *Eur. J. Ind. Eng.* **2012**, *6*, 118–137. [CrossRef]
55. Cherrafi, A.; Elfezazi, S.; Chiarini, A.; Mokhlis, A.; Benhida, K. The Integration of Lean Manufacturing, Six Sigma and Sustainability: A Literature Review and Future Research Directions for Developing a Specific Model. *J. Clean. Prod.* **2016**, *139*, 828–846. [CrossRef]
56. Dieste, M.; Panizzolo, R.; Garza-Reyes, J.A. Evaluating the Impact of Lean Practices on Environmental Performance: Evidences from Five Manufacturing Companies. *Prod. Plan. Control* **2020**, *31*, 739–756. [CrossRef]
57. Ghobakhloo, M.; Azar, A.; Fathi, M. Lean-Green Manufacturing: The Enabling Role of Information Technology Resource. *Kybernetes* **2018**, *47*, 1752–1777. [CrossRef]
58. Salvador, R.; Piekarski, C.M.; De Francisco, A.C. Approach of the Two-Way Influence Between Lean and Green Manufacturing and Its Connection to Related Organisational Areas. *Int. J. Prod. Manag. Eng.* **2017**, *5*, 73. [CrossRef]
59. Carlos, S.; Reis, M.; Dinis-carvalho, J.; Silva, F.J.G.; Santos, G.; Ferreira, L.P.; Lima, V. The Development of an Excellence Model Integrating the Shingo Model and Sustainability. *Sustainability* **2022**, *14*, 9472.
60. Jasti, N.V.K.; Kodali, R. Lean Manufacturing Frameworks: Review and a Proposed Framework. *Eur. J. Ind. Eng.* **2016**, *10*, 547–573. [CrossRef]
61. US EPA. *The Lean and Environment Toolkit*; US EPA: Washington, DC, USA, 2007; p. 96.
62. Belhadi, A.; Touriki, F.E.; El Fezazi, S. Benefits of Adopting Lean Production on Green Performance of SMEs: A Case Study. *Prod. Plan. Control* **2018**, *29*, 873–894. [CrossRef]
63. Duarte, S.; Cruz Machado, V. Green and Lean Implementation: An Assessment in the Automotive Industry. *Int. J. Lean Six Sigma* **2017**, *8*, 65–88. [CrossRef]
64. Mollenkopf, D.; Stolze, H.; Tate, W.L.; Ueltschy, M. Green, Lean, and Global Supply Chains. *Int. J. Phys. Distrib. Logist. Manag.* **2010**, *40*, 14–41. [CrossRef]
65. Campos, L.M.S.; Vazquez-Brust, D.A. Lean and Green Synergies in Supply Chain Management. *Supply Chain Manag.* **2016**, *21*, 627–641. [CrossRef]
66. Rothenberg, S.; Pil, F.K.; Maxwell, J. Lean, Green, and the Quest for Superior Environmental Performance. *Prod. Oper. Manag.* **2001**, *10*, 228–243. [CrossRef]
67. Sawhney, R.; Teparakul, P.; Bagchi, A.; Li, X. En-Lean: A Framework to Align Lean and Green Manufacturing in the Metal Cutting Supply Chain. *Int. J. Enterp. Netw. Manag.* **2007**, *1*, 238–260. [CrossRef]
68. de Carvalho, A.C.V.; Granja, A.D.; da Silva, V.G. A Systematic Literature Review on Integrative Lean and Sustainability Synergies over a Building's Lifecycle. *Sustainability* **2017**, *9*, 1156. [CrossRef]
69. Konys, A. Green Supplier Selection Criteria: From a Literature Review to a Comprehensive Knowledge Base. *Sustainability* **2019**, *11*, 4208. [CrossRef]
70. Siegel, R.; Antony, J.; Garza-Reyes, J.A.; Cherrafi, A.; Lameijer, B. Integrated Green Lean Approach and Sustainability for SMEs: From Literature Review to a Conceptual Framework. *J. Clean. Prod.* **2019**, *240*, 118205. [CrossRef]
71. Garza-Reyes, J.A.; Kumar, V.; Chaikittisilp, S.; Tan, K.H. The Effect of Lean Methods and Tools on the Environmental Performance of Manufacturing Organisations. *Int. J. Prod. Econ.* **2018**, *200*, 170–180. [CrossRef]
72. Ojani, S.; Darestani, S.A. Green-Lean Operations Evaluation Framework: A Case from Iranian Automotive Industry. *Int. J. Product. Qual. Manag.* **2022**, *37*, 36. [CrossRef]
73. Motlagh, F.O.; Darestani, S.A. Evaluating the Impact of Lean Tools and MUDA on Organisation's Green Performance. *Int. J. Green Econ.* **2021**, *15*, 378. [CrossRef]
74. Mathiyazhagan, K.; Agarwal, V.; Appolloni, A.; Saikouk, T.; Gnanavelbabu, A. Integrating Lean and Agile Practices for Achieving Global Sustainability Goals in Indian Manufacturing Industries. *Technol. Forecast. Soc. Change* **2021**, *171*, 120982. [CrossRef]
75. Aisyah, S.; Purba, H.H.; Jaqin, C.; Amelia, Z.R.; Adiyatna, H. Identification of Implementation Lean, Agile, Resilient and Green (LARG) Approach in Indonesia Automotive Industry. *J. Eur. Systèmes Autom.* **2021**, *54*, 317–324. [CrossRef]
76. Carvalho, H.; Azevedo, S.; Cruz-Machado, V. Trade-Offs among Lean, Agile, Resilient and Green Paradigms in Supply Chain Management: A Case Study Approach. In Proceedings of the Seventh International Conference on Management Science and Engineering Management, Philadelphia, PA, USA, 7–9 November 2013; Springer: Berlin/Heidelberg, Germany, 2014; pp. 953–968.
77. Voss, C.; Tsikriktsis, N.; Frohlich, M. Case Research in Operations Management. *Int. J. Oper. Prod. Manag.* **2002**, *22*, 195–219. [CrossRef]
78. Yin, R.K. *Case Study Research and Applications: Design and Methods*, 6th ed.; Sage Publications: New York, NY, USA, 2017; Volume 1, ISBN 978-1-45224-256-9.
79. Ward, P.T.; McCreery, J.K.; Ritzman, L.P.; Sharma, D. Competitive Priorities in Operations Management. *Decis. Sci.* **1998**, *29*, 1035–1046. [CrossRef]
80. Thanki, S.; Govindan, K.; Thakkar, J. An Investigation on Lean-Green Implementation Practices in Indian SMEs Using Analytical Hierarchy Process (AHP) Approach. *J. Clean. Prod.* **2016**, *135*, 284–298. [CrossRef]

81. Womack, J.; Jones, D.; Roos, D. *The Machine That Changed the World: The Story of Lean Production*; HarperPerennial: New York, NY, USA, 1991.
82. Ohno, T. *Toyota Production System: Beyond Large Scale Production*; Productivity Press: New York, NY, USA, 1988; ISBN 0-91-529914-3.
83. Krafcik, J.F. Triumph of the Lean Production System. *Sloan Manag. Rev.* **1988**, *30*, 41.
84. Holweg, M. The Genealogy of Lean Production. *J. Oper. Manag.* **2007**, *25*, 420–437. [CrossRef]
85. ANFAVEA Associação Nacional Dos Fabricantes de Veículos Automotores. Available online: https://anfavea.com.br/site/ (accessed on 9 August 2022).
86. Sahu, A.K.; Sharma, M.; Raut, R.D.; Sahu, A.K.; Sahu, N.K.; Antony, J.; Tortorella, G.L. Decision-Making Framework for Supplier Selection Using an Integrated MCDM Approach in a Lean-Agile-Resilient-Green Environment: Evidence from Indian Automotive Sector. *TQM J.* **2022**. *ahead-of-print*. [CrossRef]

Article

Improving the Shipbuilding Sales Process by Selected Lean Management Tool

Zoran Kunkera [1,*], Nataša Tošanović [2] and Nedeljko Štefanić [2]

[1] Leaera Ltd., for Business Consulting, 10000 Zagreb, Croatia
[2] Department of Industrial Engineering, Faculty of Mechanical Engineering and Naval Architecture, University of Zagreb, 10000 Zagreb, Croatia
* Correspondence: zoran@leaera.hr

Abstract: Market positioning, i.e., the competitiveness of European shipyards, depends a lot on the measures of continuously improving the business processes, therefore meeting the criteria of environmental protection and sustainable energy. Lean management enables ongoing improvements of all system processes by recognizing and removing the unnecessary costs of the same, i.e., those activities which do not contribute to the added value for the customer. In this paper, the authors research the magnitude of improvements in the shipbuilding sales process achieved by applying the Lean tool "Value Stream Mapping" (VSM). The example of analysing the informational stream of the studied European shipyard's existing sales process, performed by implementing the VSM, has defined the measures to decrease the losses in the process, with an emphasis on waiting time in internal and external communication. Upon VSM of the future state, measuring improvements realised by applying key performance indicators began. Significant cost savings in the sales process and the simultaneous increase of productivity of the employees participating in those process activities have been noted, as well as the substantial growth in sales and the shipyard's income.

Keywords: shipbuilding; sales; lean management; value stream mapping; improvement

Citation: Kunkera, Z.; Tošanović, N.; Štefanić, N. Improving the Shipbuilding Sales Process by Selected Lean Management Tool. *Machines* **2022**, *10*, 766. https://doi.org/10.3390/machines10090766

Academic Editors: Luís Pinto Ferreira, Paulo Ávila, João Bastos, Francisco J. G. Silva, José Carlos Sá and Marlene Brito

Received: 25 July 2022
Accepted: 29 August 2022
Published: 2 September 2022

Publisher's Note: MDPI stays neutral with regard to jurisdictional claims in published maps and institutional affiliations.

1. Introduction

The European shipbuilding industry is increasingly facing Asian competition in the construction of cruise ships, passenger ships, ferries, and even vessels of special purpose, and for decades it has almost entirely lost its market position in niches of cargo ship containers, bulk cargo, and oil. According to [1], at the beginning of the 21st century, Asian countries dominate in large shipbuilding, with a greater world share in shipbuilding than European countries had 100 years ago, whose strong decline in large shipbuilding volume started in 1955. The first reason for that was the great rise of Japan in delivered gross registered tonnage, and then the remaining Asian countries—South Korea, then China.

The strengthening of Asian industrial conglomerates was mostly supported by indirect or even, in the case of China and South Korea, direct government subsidies. The Asian conglomerates also benefited from lower labour costs compared to Europe. European shipyards, with strong trade union activity and set fees in accordance with the extremely cyclical changes in freight rates on the shipping market [2], could not compete on price [1], thus leaving Asian countries with complete primacy in cargo shipbuilding. As quoted in [2], in 2019, the main Asian countries accounted for as much as 95.5% of the deadweight tonnage in the global newbuilding orderbook, of which the share of China was 45.4%, South Korea 28.1%, and Japan 22%, while Europe accounted for only 1.9% of the world's total ordered deadweight tonnage. Due to the lack of competitiveness in cargo shipbuilding in the last two decades, the European shipbuilding sector has been oriented primarily in building technologically sophisticated ships—cruise ships, vessels for oil and gas industry, suppliers for offshore wind farms, and others.

However, due to a decline in their share of the global shipbuilding orderbook in favour of China, Japan [2,3] and South Korea [4] enlarged their portfolio of newbuilds to capture a market niche of big cruise ships at the beginning of the 21st century. The aforementioned strategies were additionally strengthened by the consequences of the financial crisis from 2008 to 2012, which saw a decline in cargo vessel [3]. The ultimate and greatest challenge to European shipbuilding is the increasingly strong inclusion of China in the aforementioned market niche in the last decade, driven by the government-led policy of low selling prices [3], with the aim of adopting the necessary technologies as quickly as possible.

Therefore, European shipbuilding faces a great challenge to continuously improve business processes to at least maintain but also increase its competitiveness while meeting high environmental protection and energy efficiency criteria. The current market demands for highly complex and technologically advanced vessels affects all stages of shipbuilding—design, construction, operation, and maintenance—creates new challenges for the R&D sector, and enhances the growth and development of various support industries, such as IT, design, specialised outfit production, and similar. The importance of such a chain of added value, originated and led by the shipbuilding industry and having a major economic impact on the overall economy, has to be acknowledged by European countries so that shipbuilding will be considered as long-term economic strategies are instituted. Development and research activities must be the backbone of modern European shipbuilding so that the industry can be a leader of current trends in energy-efficient design and environmentally friendly vessels, so-called "smart ships".

Improving business processes to maintain or increase competitiveness is possibly approached from at least three intertwined aspects:

- Using funds from the European Union to encourage development activities, research, and innovation, as well as investment in digital technologies, infrastructure, and production resources;
- Implementing Industry 4.0 trends and technologies;
- Organizational changes in terms of transforming the shipyard into a "smart" (digital transformation) and "green factory", respectively.

Through "A New Industrial Strategy for Europe" adopted in 2020 [5], which is based on the "European Green Deal" strategy adopted a year earlier [6] on reaching the climate neutrality of Europe by 2050, the European Commission emphasises the primary significance of digital, i.e., the green transformation of European industry (dual transition), as a contributor to the process of achieving a climate-neutral European economy, defining the grant schemes to SMEs focused on development and research projects. Most of the leading countries in the maritime sector, not only in Europe but also globally, have recognised the importance of applying digital technologies to achieve an increase in the competitiveness of their industries in the sphere of sustainable growth and development [7,8]. They have also implemented new operational processes in shipbuilding [9] (e.g., Germany issued "Industrie 4.0", Spain adopted an agenda for Industrial Competitiveness, South Korea issued "Manufacturing Innovation 3.0 Strategy", China issued strategic declaration "Made in China 2025", etc.). Nevertheless, research work on applying digital technologies, i.e., the results achieved by implementing Industry 4.0 doctrines and trends in the shipbuilding sector, are still in the early stages.

The analysed research evaluates the importance of certain technological achievements of Industry 4.0 to the transformation of traditional business systems into smart ones differently. Therefore, some of the research [10] recognises Cloud, Big Data Analytics, Internet of Things, and Artificial Intelligence as technologies of primary importance in improving shipbuilding processes followed by other digital trends and technologies such as Digitisation (of all workshops' machinery), Additive Manufacturing (3D printing), Cybersecurity, Autonomous Guided Vehicles, Virtual and Augmented Reality, Collaborative Robotics, Blockchain, and Internet of Services. Other research [11–13] emphasises the importance of the Cyber–Physical System (CPS) platform in creating Shipyard 4.0, whereas Digital Twin has been defined as the leading technology for realising the CPS platform [14]. Moreover, a case study of Shipyard 4.0 [11] assumes all other digital technologies as secondary and is built upon Digital Twin technology.

Digital Twin represents a virtual replica of a physical entity (product, machine, process, and the like) and is related to it by selected smart sensors and reference software applications that monitor the parameters in real-time by analysing the collected data [14,15]. As such, Digital Twin can be utilised as the basis for a smart manufacturing system (SMS) configuration, recognising and preventing misconceptions in the early stage of the SMS design process [16]. Digital Twin technology, on the other hand, implies the comingling of three entities—the physical object, its virtual replica, and their mutual digital link/IT component [17,18]. Considering the specifics of the European shipyard studied in this paper related to constructing prototype vessels according to the client's preferences, the shipyard has recognised Digital Twin as the core technology for the transformation into Shipyard 4.0. In relation to the already performed research [19] regarding the importance of applying Digital Twin to sales and marketing activities, the studied shipyard recognises the advantages of applying 3D modelling with software tools in the earliest, pre-contractual, and sales phases. The ship's development in its virtual equivalent enables the client and the shipyard to improve the ship's parameters (e.g., the shape and the stability of the ship) which results in lower costs, i.e., better project profitability for the client. Furthermore, design activities in the digital twin ship enable the client, shipyard, and key suppliers to define the optimal technical and technological solutions with the aim of simplifying and thereby reducing the cost, resulting in an additional increase in added value for the client.

The complexity of financing the shipbuilding process [20,21] includes a large number of stakeholders [22], and after the financing is approved, an entire supply chain is involved in the transaction processes. With the aim of maintaining but also improving competitiveness, the studied European shipyard gives priority, along with Digital Twin technology, to the implementation of the digital achievements of Industry 4.0, to blockchain technology, and its software application of smart contracts [23–26]. It also takes into account the high potential of its applicability in the shipbuilding industry [27–29]. Blockchain technology represents the distributed database [28] which enables the execution of direct transactions between networked users without the need for verification. Once adopted by consensus of the users, transactions remain unchangeable, which ensures transparency and trust. Adopting blockchain technology by eliminating the need for third-party mediation not only allows for a significant reduction in the duration of transaction approval procedures, but also eliminates the high financing costs, thus increasing the competitiveness of the shipyard.

By implementing the applicable trends and technologies of Industry 4.0, the main directions of the integration of shipyard business processes are realised [10], namely:

- Vertical, by integrating the business processes from workshops, through departments and up to management, by transmitting data between virtual reality and sensors, i.e., actuators installed on work equipment (automatic machines, production lines, etc.) in real-time (CPS environment), thus achieving significant savings in processes;
- Horizontal, by integrating all the supply chain stakeholders in order to create possible new networks among the same, adjusted to the potential new projects;
- End-to-End engineering, by connecting the supply chain stakeholders with the process functions (design, supply, production, maintenance) to monitor the product lifecycle by analysing data assigned to the product, i.e., real data from construction and use of the product to optimise the shipbuilding process and improve the efficiency of product maintenance.

However, a prerequisite for access to a digital shipyard transformation is the Lean transformation of business processes, necessary from the aspect of achieving business competitiveness by eliminating or reducing unnecessary costs, while increasing company profits, productivity, and employee motivation, ultimately increasing customer satisfaction.

In the past decades, Lean management has been popular because of its potential to improve not just discrete manufacturing processes, where it had emerged, but also in industrial processes, as well as service processes, such as healthcare, banks, and universities. The major point of interest of Lean management in the shipbuilding industry, as the literature reveals, was the manufacturing process itself. Since premanufacturing processes play a huge role and have a huge influence on the actual manufacturing process, the motivation of this research was to study the possibility of improving the premanufacturing phase, which in turn would improve the manufacturing phase.

This paper demonstrates the importance of improving the sales process of the studied European shipyard by using Value Stream Mapping (VSM) as the chosen Lean management tool. After introducing and reviewing the literature in the first two sections, the third section highlights the features of the Lean management method and describes Value Stream Mapping as the selected Lean tool for improving the sales process. The fourth section describes the applied research methodology for this work. In the fifth section, the current phases and subphases of the sales process are determined, upon which in the sixth section a map of the current situation will be described and a map of the future state of the sales process is presented and its implementation is planned. The sales process improvement theoretical results are presented in the seventh section. In the eight section accomplishments of the experimental application of improvement measures are first presented and recommendations for further research are discussed. Finally, the ninth section gives a conclusion and argues the significance of the achieved sales process improvement results in the shipyard's business.

2. Literature Review

The following paragraphs review the available literature and scientific research on Lean management and manufacturing, respectively, in shipbuilding.

Unlike scientific research on the development and application of Lean principles in other manufacturing industries, there are not so many publications in the field of Lean shipbuilding. The authors Sullivan et al., as well as Suresh et al., claim that the reason is the novelty and restrictions of such an approach in this type of industry [30,31]. However, in their study, Suresh and other authors give an overview of lean shipbuilding in countries such as Japan, the US, and Norway and show that it is possible to achieve significant improvements in productivity with some of the Lean tools such as multi-skill training and five 'S' Value Stream Mapping and standardized work [31]. Sullivan et al. claim that in lean transformation in maritime design, it is necessary for three aspects to be considered: process, people, and technology, and their study is focused on the design process of vessels and improving that process with Lean principles [30]. Applying lean manufacturing principles in US shipyards was described in the article [32]. The authors claimed that there is the potential for significant productivity improvements in shipyards by implementing Lean principles. The development of the Lean shipbuilding MES system is described in an article by Song and Zhou [33]. They studied the case of a shipbuilding company's production process that had problems with a long manufacturing cycle, a low utilization rate of personnel, and an unbalanced production line which were addressed using the Lean shipbuilding MES system. Two studies dealing with the panel line assembly process in shipbuilding show significant improvements by introducing Lean principles in this part of the shipbuilding process. In the first of these two mentioned papers, only by analysing the time chart and current Value Stream Mapping are possibilities for improvements detected in the manner of the sequence modification of the manufacturing flow. The Value Stream Mapping of the manufacturing process was a key element in analysing and improving production efficiency [34]. The second mentioned paper also applied Value Stream Mapping to analyse the production process. The results were significant regarding man-hour and lead time savings of about 60 and 80%, respectively [35]. Another work by Kolich et al. describes applying Value Stream Mapping in shipyard panel assembly lines. They suggested that by implementing Kanban and pull principles, as well as group technology and production configurations, significant improvements in terms of man-hours and production time can be seen [36]. The case study of applying Lean principles in the European shipbuilding industry by Value Stream Mapping is presented in the paper [37]. The authors described the case study of applying VSM and simulation of pylon production in the shipbuilding process, which have resulted in a future state map of the process. They showed possibilities for reducing lead time and local inventory levels compared to the current state. Another interesting study was given by Niansheng et al. [38]. The authors give a systematic approach for the Lean supply chain in the shipbuilding industry by analysing specific characteristics of the shipbuilding processes to improve productivity and lower costs. The approach is established with procedures on three basic levels, and these are: foundation, execution, and index level. The first level incorporates the foundation of the management principles, the second level includes the operation directions, and the index level represents the performance measurement. Another study of Value Stream Mapping application in the shipbuilding industry, specifically in micro panel assembly, was given by Kolich et al. [39]. The authors describe the case study of flow improvement in this specific part of the production process.

The ship production process from the point of purchasing to the point of delivering the final product can be as long as two years. This is a very long and complex process that consists of many complex and interdependent steps, and focusing only on the production process itself in terms of applying Lean principles and improving the process can lead to improvements. However, the significant results in reducing the lead time and increasing productivity and production efficiency could only lead if the whole process, including the contracting phase, is taken into account for improvements. Implementing Lean principles in the sales process, specifically in the contracting phase, must be one of the key strategic decisions of the shipyard's management, given the effect achieved by improving the process in terms of revenue growth or production capacity employment. The reviewed literature showed the gap in this field, and this study aimed to research the possibility of Lean implementation in this specific area of the shipbuilding process, thereby detecting the losses in the sales process and accordingly defining the improvement measures to enhance the shipbuilding sales performance and competitiveness of the shipyard, respectively. Thus, the research questions aimed to be addressed in this case study were:

- Q1: What are the wastes in the sales process in shipbuilding according to the Lean principles?
- Q2: Is the Value Stream Mapping tool applicable in the sales department processes in the shipbuilding industry?

3. Lean Methodology

Since it was used to describe the specificity of Toyota Motor Company's production system [40], comparing it with other automotive industries, particularly Western ones, Lean management has become one of the widely used methodologies to improve, at first, production processes, but later also service companies as well hospitals and universities. As the method emerged, it was recognized as an improvement philosophy that can be applied in many industries.

The main goal of Lean manufacturing is to continuously improve the processes by eliminating all waste from the process, which Taiichi Ohno has classified into six categories: overproduction, transport, waiting, inventory, over-processing, unnecessary movement, and rework [41]. In the Lean approach, waste is everything that does not add value to the customer [42]. In addition, one of the definitions of Lean manufacturing is that Lean manufacturing, when implemented, reduces the time between the customer order and delivery of the final product by eliminating non-value-added activities. The goal is to achieve a short and agile process that can easily respond to customer demands [43].

In their book *Lean Thinking*, Womack and Jones define five basic principles of Lean manufacturing: value, value chain, flow, pull, and perfection [42]. The principle of value means that the process has to be organized from the perspective of the customer and the premise of what brings value to the customer. Value chain principles summarized the approach that the process is the sum of all activities necessary for bringing that value to the customer. The flow principles define the idea that the process has to flow smoothly, without interactions and interruptions, and the ideal state is to have a one-piece flow. The pull principle is the most important principle of Lean management and is very often associated with Kanban, a signal card system that is developed and used in Toyota for pulling products to the customer instead of pushing products to the customer. Basically, every activity in the process is an internal customer, and the previous activity does not produce until the upstream activity does not demand, and that way, the product is pulled towards the end customer. The principle of perfection represents the philosophy of continuous improvement, which never ends.

The principles of value, value-chain, and the flow, through one of the tools of Lean management, which is Value Stream Mapping (VSM), are the subject of this research.

Value Stream Mapping is the tool for identifying and analysing the value chain, thus analysing all activities which bring the product to the customer. It utilizes some common graphical icons and symbols to represent the process and, in that way, becomes a common language about processes [44]. As supposed from Western production thinking, which was concentrated mainly on the flow of the material, in the Toyota production system, many decades ago, the different value flows were identified, and those are: the flow of material, the flow of information, which is as important as material flow and even more significant because it dictates how the material will move through the production process. Thus, the flow of information directly influences the flow of material. The third flow recognized in Toyota is the flow of people.

Value Stream Mapping allows one to identify all activities, all flows, both the flow of material and information and the flow of people, with the primary purpose of analysing and eliminating all wastes from the process. Previously, researchers have developed many techniques for optimizing individual operations, but most of these techniques and tools cannot comprehend and visualize the whole picture of the entire production process. That way, the operations are analysed separately, which can bring isolated improvements which do not necessarily lead to improving the whole production process. Value Stream Mapping is the primary tool used to identify the opportunities for Lean tools to improve the entire process [45–48]. The Value Stream Mapping process is shown in Figure 1.

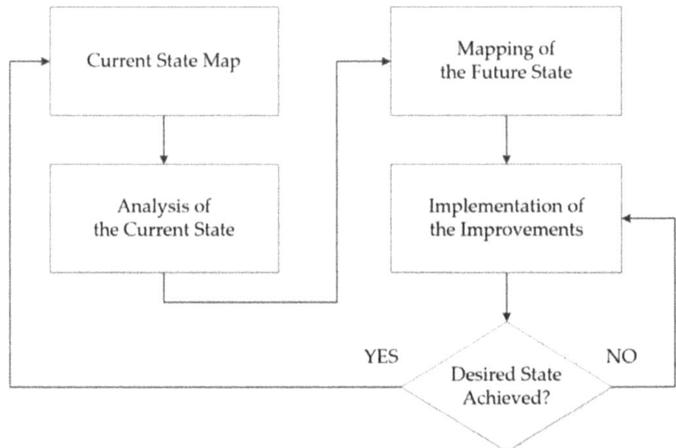

Figure 1. Value Stream Mapping process.

The first step in Value Stream Mapping is case selection and deciding for which product, product family, or service the Value Stream Map is going to be made. The next step is making the current Value Stream Map, and it is recommended to create a map of the current state in the field to collect reliable data. The current map is then analysed, and ideas for improvements are generated which are then presented in the future state map. According to the future state, an action plan for implementation is made. These steps were implemented in this case study, and details are given in section four.

4. Research Methodology

The purpose of this study was to research the adoption of Value Stream Mapping for analysing and improving the sales process in the shipbuilding industry, an event over which research does not have control. As Leonardo et al. suggest, in this situation the methodology applicable is a case study [49]. In addition, Rashid et al. suggest that for in-depth phenomena research, a qualitative case study methodology is a good choice [50].

In his work, Yin explains that there are three types of case studies, and those are exploratory, descriptive, and explanatory. Since the explanatory study is applicable for studies of the processes of the companies, this type of case study is chosen in this research. In addition, according to Yin, a case study can be a single-case case study or a multiple-case case study. Since in this research there were no other available cases for replication, a single-case case study was adopted [51]. Before conducting the case study, good preparation was necessary and as Rashid et al. in their work point out, issues such as clarity, selection, and operationalization of the case study can lead to wasting time, confusion, and wrong decisions. Thus, they suggest four phases of the case study and those are: step 1—the foundation phase, step 2—the pre-field phase, step 3—the field phase, and step 4—the reporting phase. The steps of this case study are given in Figure 2.

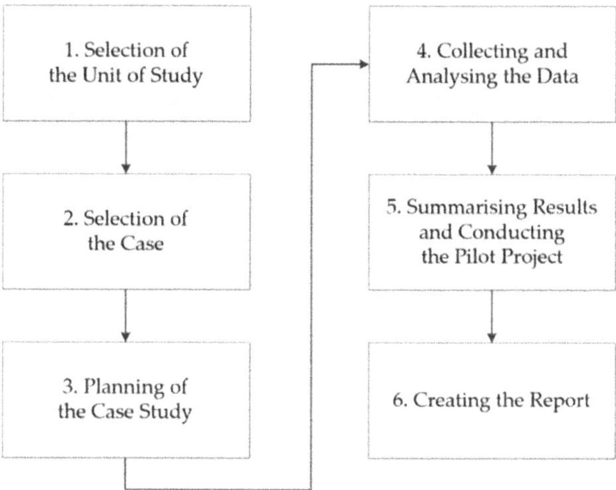

Figure 2. Case study steps.

The foundation phase begins with a selection of the unit study and the case. The unit selected was a European shipbuilding factory whose strategic investment plans were oriented toward implementing Industry 4.0 doctrines in technological and production processes and business systems using the available digital technology tools to transform the shipyard into a Smart factory.

According to the size of its infrastructure and number of employees, it belongs to the group of large European shipyard plants, organizationally including all business functions—administrative (sales, marketing, project management, purchasing, finances, and human resources), as well as all those technical and technological (planning, technology, design, engineering) and production (maintenance and transport included), with the possibility of project realisation exclusively or primarily by its own human resources.

The shipyard provides the services of developing/initial designing, designing, and constructing the most complex ships, with demanding special infrastructure constructions (e.g., bridges), offshore objects for the oil and gas industry, as well as components for offshore wind farms. It is present in the entire international market.

The shipyard currently employs a couple of thousand workers, of which more than 100 are designers who develop and optimise different types of the most technologically advanced ships. By using the most modern software tools, they virtually model and design a "digital prototype" (Eng. "digital twin" or "ship 0") [52,53] to define the optimal technical and technological solutions before the object/vessel construction launches (begins).

Since the Sales and Marketing Sector and the Project Management Office are the backbones of the company, the selected case was the company's sales sector. Implementing the Lean system in the sales process was one of the key strategic decisions of the shipyard's management to improve the process in terms of revenue growth, i.e., employment of production capacities.

The pre-field phase included planning the case study. Interviewing managers and designated personnel from the Sales and Marketing Sector, Initial Design Department, Planning and Technology Sector, Project Management Office, Key Material Sourcing Department, Legal Department and Financial Sector, was planned, all together in average 12 employees involved, within the period of about 3 months, mostly as a part of the regular (weekly) operational meetings. Additionally, the field walk with personnel to gather all the necessary data for the current Value Stream Map was planned.

The field phase, consisting of steps four and five of the case study, included the field walk, gathering the historical and present data, and constructing the current Value Stream Map. Data were collected from the archive (IT, server) of each and every involved sector and department detailing the time their employees' had engaged in the sales process during the past 2 years. Brainstorming meetings attended by at least three representatives from the Sales and Marketing Sector representatives along with directors, managers, and personnel, in various compositions were held on bi-weekly basis through approximately a nine-month period The purpose of these meetings was to analyse the current state and gain solutions for the future state map. The future state map was the result of this phase of the case study, and the pilot project was carried according to the future VSM.

The reporting phase included creating and structuring the results achieved with the VSM tool.

In the next section, the field and reporting phase will be thoroughly described.

5. Shipbuilding Sales Process in the Selected Unit

The Sales and Marketing Sector, with the associated Project Management Office, is the backbone of the shipbuilding system organisation; the sector is divided into the shipbuilding department, represented by three employees, and the off-shipbuilding department with up to eight employees, including technical and technological support.

The Initial Design Department also makes up part of this sector (the development/initial design of the ship), which is actively involved in the data processing of inquiries (for newbuilds), i.e., in offer preparation and project contracting.

The shipbuilding department of the Sales and Marketing Sector receives inquiries directly from the customers or gains them via agents ("brokers") based on the following:

- The shipyard's references, i.e., its market position (word of mouth, marketing activities—fairs, interviews, and advertising in specialised publications/editions);
- Direct approach to potential customers by the sales representatives regarding long-term business strategy (opening new market niches such as ship repairs and refits, constructing offshore objects, conversions and constructing megayachts);
- The ongoing projects and previous deliveries—re-ordering of "sister" or similar new vessels by the former and actual clients.

The value created by the shipbuilding sales process is the shipbuilding contract for the vessel optimised so that the ship parameters are improved during the development and design phase. This is accomplished by implementing the shipyard's available software tools to reach the best possible indicators of the feasibility study of the customer's project.

Processing inquiries for new construction, i.e., preparing the offer for the initial design of the ship, and designing and constructing the ship up to contracting the specific project, evolves in four phases: the inquiry selection phase, the Preliminary offer preparation phase, the Letter of Intent defining phase, and the contracting phase.

5.1. Inquiry Selection Phase

The selection of inquiries sent by the customer or agent is performed according to the following criteria:

1. Conformity with the market niches which are the shipyard's strategical mid-term and long-term orientation (passenger ships, ferries, cruise ships, special purpose vessels—oil industry, offshore wind farms), as well as the acceptable volume of the ships to be constructed (preferred gross tonnage depending on the shipyard's profile). Smaller ships are acceptable if they are from the mega-yacht or navy program niche, if they are contracted as a large serial of vessels by the basic shipbuilding contract and not as optional vessels, and if the customer is recognized as a long-term strategic partner in the shipbuilding or shipping sector;
2. Shipbuilding capacities regarding the available and short-term planned shipyard infrastructure (lifting capacity of cranes, the shipyard's hall size, slipway size, maximal depths of the shipyard's port, and the like);
3. The project realization capacity within the planned delivery date according to the shipyard's standard and usual delivery deadlines for of design and building a ship of similar parameters and, if required by the inquiry, according to the obligated class notations and requirements for its seaworthiness.

Inquiry analysis is performed by the Sales and Marketing Sector according to the selection criteria mentioned above and with the possible assistance of the Planning and Technology Sector relating to project realisation capacities. The inquiry selection phase usually lasts for 3 to 4 days (in average some 84 h), mostly conditioned by communication with the customer or its agent to clarify certain inquiry details. Here, the mentioned duration of this phase includes internal communication with the Planning and Technology Department.

5.2. Preliminary Offer Preparation Phase

The Preliminary offer preparation phase consists of two subphases: the Non-Disclosure Agreement defining the subphase and the Preliminary offer budgeting subphase. Preparation of the data framework for initial calculation is based on input documents sent by the potential client/agent, whereas the shipyard defines the minimum requirement for the potential client's documentation—conceptual design/preliminary general arrangement plan and the ship's outline specification. Before submitting the relevant documents, the potential client delivers the Non-Disclosure Agreement (NDA) proposal no more than three days from when the shipyard Sales Sector has confirmed its interest in processing the inquiry. The Sales and Marketing Sector contacts the company's Legal Department to prepare the Non-Disclosure Agreement proposal for each future revision after the potential client sends the answer regarding the company's comments. To review the initial Non-Disclosure Agreement proposal, the Legal Department takes 4 days. Every other future revision is generally exchanged within 3 days, which usually includes four revisions from the initial document. Once the Non-Disclosure Agreement is approved, it is delivered to the Management Board Office for signature. After that, and usually within 3 days, it is re-sent to the Sales Sector and electronically to the potential client.

Preparing a preliminary data framework for initial calculation is performed by the Sales representative based on the ship's technical parameters (gross tonnage, deadweight, number of passengers, number of crew members, vessel dimensions, and other relevant information), requirements for the ship's seaworthiness, requests according to required class notations (regulations of the ship classification societies), applicable international rules and regulations, and comparing it with ships of reference (assigning the coefficients regarding the complexity of the ship's requirements). The Sales and Marketing Sector contacts the Initial Design Department to analyse the applicable launching technology and the Planning and Technology Sector regarding the preliminary analysis of the possible shipbuilding technological methods, all to estimate project realisation possibilities regarding the available infrastructure and the shipyard's current technological achievements. Furthermore, depending on the inquiry, the Sales representative asks the Planning and Technology Sector to estimate the working hours for the ship's hull construction and outfitting, and from the (related) engineering company, the cost estimation for the anti-corrosion protection of the ship. In addition, the Planning and Technology Sector delivers the Sales Sector a detailed estimation of the project realisation period (elaborated in phases: effectiveness of the shipbuilding contract, the start of steel cutting, keel laying, launching, and delivery), all under the assumption that the relevant project is contracted within up to 8 months from receiving the inquiry. The average time to prepare the preliminary data framework for calculation is 7 to 10 working days, depending on the intensity of the involvement of the pre-listed departments, i.e., the related company. The preliminary data framework for the initial calculation, outlined in the table form, is delivered to the Management Board Office for the approximate estimation of the ship's cost. Within 5 days, the Management Board member gives information on the initial sales price with comments related to the ship's estimated construction/delivery deadline along with the suggested dynamics for paying based on which the Sales Sector prepares the Preliminary offer and delivers the same to the client within 1 day. The client usually takes 14 days to comment on the Preliminary offer. In total, the Preliminary offer preparation phase on average takes 44.5 days (1068 h).

5.3. Letter of Intent Defining Phase

If the potential client should find the Preliminary offer sufficient for continuing negotiations, the Financial Sector gets involved in the sales process by submitting the insurance instruments available for monitoring the relevant project within 3 days. Upon receiving the relevant information, the potential client usually asks the shipyard for a statement on the capacities of the financial support of the project, which means that within approximately 10 days, the Financial Sector, accompanying banks of the company and insurance companies, analyses the financial reports of the potential client, its business plan, and feasibility indicators of the project, and the statement on conclusions regarding the possibility of the company's participation in financing the project is delivered to the potential client. If all parties—the potential client, shipbuilder, and financial institutions—agree on options related to financing and payment dynamics, defining the Letter of Intent can begin. The period of mutual consent, starting from the shipyard's statement on the possibilities of financial participation in the project to the eventual agreement with the potential client about the acceptable financial arrangements to follow the newbuild's construction (or its series), usually takes 10 days. A Letter of Intent proposal is generally prepared by the company's Legal Department according to commercial instructions delivered by the Sales Sector within 2 days after agreeing with the potential client that the Letter of Intent will be prepared. Regarding the commercial–legal specificities of the project, preparing the Letter of Intent proposal usually lasts 4 days, after which it is delivered to the potential client for analysis. Generally, until the Letter of Intent is adopted, contracting parties exchange four revisions of the relevant documents, with an average period of three days to prepare comments between revisions. The agreed Letter of Intent is submitted by the Legal Department to the Management Board Office with a summary of the legal interpretation of the document and upon adoption and signing, is electronically sent to the Sales Sector, usually within 3 days. Hence, the average duration of the Letter of Intent defining phase is around 41 days (984 h).

5.4. Contracting Phase

The contracting phase comprises two simultaneously running subphases: the contractual ship price calculation subphase and the shipbuilding contract review subphase. Upon signing the Letter of Intent (LOI), the Initial Design Department begins with the development, i.e., the ship's initial design, and with defining/elaborating the technical specifications of the same to prepare contractual technical documentation within 120 days. The ship's initial design process implies the much-needed constant cooperation with the client's technical–commercial team on the shipbuilder's site to optimise the ship according to the client's wishes, so the best solutions are achieved with minimal expenses. This ensures that the ship's top quality and the fastest return on the client's investment are met. At the same time and during the first 45 days, the Initial Design Department prepares the ship's technical basis and submits them to the Key Material Sourcing Department to prepare the final calculation and the final contractual ship price, respectively. The ship's technical basis is delivered simultaneously by each profession (propulsion systems, electro components, navigation and communication equipment, and others) to relevant subdivisions of the Key Material Sourcing Department. Within 15 days from the last received technical basis, the Key Material Sourcing Department, based on market processing results and the best pre-contracted supply prices for the defined equipment and systems, delivers the detailed cost sheet to the Management Board Office, elaborated according to the functional division of the ship, for analysis and defining the ship's final cost.

Usually within 7 days, the Management Board sets the contractual ship price, which is further negotiated with the client by the Sales Sector. Approaching the final contractual ship price agreement is completed within 30 days. Then, the Legal Department intensifies the proposal review for the Shipbuilding contract—the same is delivered to the client by the Legal Department immediately after signing the LOI. After approximately 150 days (3600 h) of total average duration of the contracting phase and from signing the LOI, respectively, the final version of the Shipbuilding contract, along with the contractual technical documentation previously agreed upon and verified by the client, is given to each contracting party to be signed. To finalize the Shipbuilding contract and fulfil all financial prerequisites (issuing the advance payment guarantees, advance payment executed) takes around 90 days. Upon signing the Shipbuilding contract, the Project Management Office names the project manager of the relevant project.

6. Value Stream Mapping of the Shipbuilding Sales Process

6.1. Lean Metrics

The Management Board has to be provided with the metrics for measuring the performance and effectiveness of the shipbuilding system processes over a defined period of time, among which the Sales and Marketing Sector secures the measures monitoring sales and revenue growth, respectively, sales process cycle average duration, inquiry-to-shipowner conversion effectiveness, and average sales process cost.

The sales growth performance indicator measures the accomplishments of sales goals considering the company's turnover growth over a determined period, comparing the actual revenue with the forecasted one [54]. The average sales cycle length performance indicator, initially aligned with the duration of the actual state of the process, measures success in optimisation of the sales cycle: the shorter the cycle is, the more sales will be achieved [54]. The inquiry-to-shipowner ratio measures success in converting inquiries into shipbuilding contracts; a higher ratio indicates higher effectiveness of employees engaged in the sales process, particularly in the later phases [54,55]. The cost per client acquisition (CCA) performance indicator refers to all sales process costs generated in reaching the shipbuilding contract, including but not limited to marketing, travel and lodging, and research and development costs [54,55]. The CCA compares and analyses accrued costs per client acquisition, thereby enabling determination of sales priorities; furthermore, it helps achieve cost savings by identifying waste in the process that needs to be eliminated.

Key performance indicators to be applied to measure the sales process improvement by applying Value Stream Mapping as the chosen Lean tool, refer to:

1. Monitoring the transformation of the contacts into the clients, on annual average with the indicators as follows:

    - Conversion rate (C_1) of the Contact (CO) of the Potential Client (PC), as the result of accepting the Inquiry from the Customer (inquiry selection phase);
    - Conversion rate (C_2) of the Potential Client (PC) into the Apparent Client (AC), completed once the Preliminary offer is accepted by the Potential Client;
    - Conversion rate (C_3) of the Apparent Client (AC) into the Client (CL), completed by signing the LOI;
    - Conversion rate (C_4) of the Client (CL) into the Shipowner (SO)—the ultimate conversion completed by signing the Shipbuilding contract,

2. Removal, i.e., decreasing losses in the sales process, and realising the sales results by applying the following Lean metrics:

    - Average length of the sales process by each realised shipbuilding contract;
    - Costs of the sales process, annual average;
    - Number of the contracted man-hours in production, annual average.

6.2. Value Stream Mapping of the Current State

The average duration length of the shipbuilding sales process is defined by:

1. The average number of the "shipyard–customer" phase interactions (as presented in Appendix A, Table A1, based on the average historical data of the Sales and Marketing Sector archive, observed within the last two-year period);
2. the average phase times of:

    - Sales process activities that make added value;
    - Activities necessary for the sales process but do not contribute any value to the customer.

The average durations of the inquiry selection phase activities are shown in Table A2, the Preliminary offer preparation phase in Table A3, the LOI defining phase in Table A4, and the contracting phase in Table A5; considering value added and non-value added sales process activities durations, whereby non-value-added activities relate at most to the waiting time in internal and external communication, day time unit expressed in Tables A2–A5 spreads to the 24 h time span.

The Value Stream Map of the current state of the shipbuilding sales process is shown in Figure 3.

Value-added time of, for example, the Letter of Intent defining phase, equals the number of entries (32 Apparent Clients, AC, as shown in Table A1) into the subject phase, multiplied by the total average time (expressed by day time unit) of value-added activities of the subject phase, as presented in Table A4.

By analysing the current Value Stream Map of the sales process, it is noted that a large number of inquiries is transferred into Phase 2 of the sales process, and accordingly, a large quantity of (simultaneous) interactions with customers even into the next phases, which makes it drastically harder to recognize the buyers who ask for a stronger involvement in the offer-making process, to understand their requests better and to communicate with them constantly during the value creation. Due to processing several simultaneous inquiries, i.e., making the offer, legal analysis, and defining the contractual and technical documentation within the minimum or negligibly partial phase shifts, the average length of the individual sales process from receiving the inquiry to realising the shipbuilding contract usually takes 8 months, whereas activities which do not contribute to the added value of the sales process last an average of 6 months, which is 75% of the overall process duration time. Insufficient involvement with the customers regarding the analysis of their potential as clients, the common impossibility of not recognizing the customer's key representatives with decision power, and insufficiently monitored communication with the customer during the higher phases of the sales process, all due to the overload of the stakeholders in the sales process by the (over)sized quantity of inquiries "downloaded" into the sales process, eventually results in the opposite of what was expected, i.e., fewer realised shipbuilding contracts (average is 2.33 per year) versus the planned (average of 2.7 annually). Consequently and due to the available production capacity average of the shipyard (that matches the construction of about 2.5 newbuilds), for part of its capacities (around 120,000 production man-hours annually), the shipyard has to subcontract (most commonly other shipyards) or apply other labour law instruments (rescheduling the working hours, etc.).

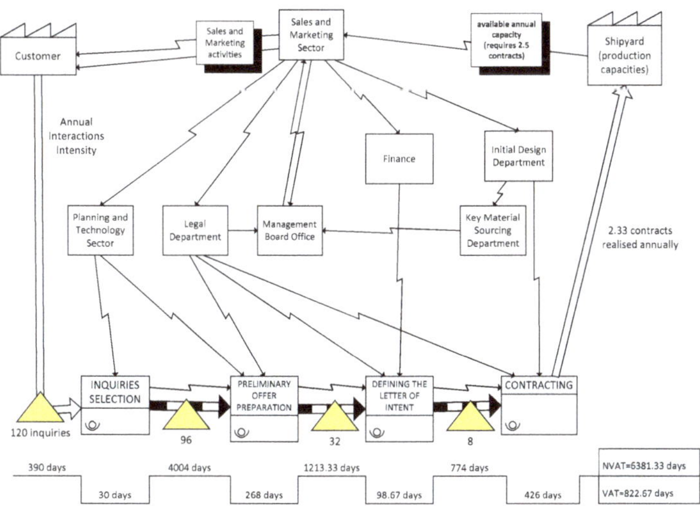

Figure 3. Current state Value Stream Map.

To perform the shipbuilding sales process according to the current state, 37 employees from the seven organisational units of the shipyard required, as shown in Table 1.

Table 1. The necessary number of employees in the sales process—annual average—existing state.

	Average Number of the Phase Interactions, Yearly	Sales and Marketing Sector (h)	Initial Design Dept. (h)	Key Material Sourcing Dept. (h)	PMO (h)	Mgmt. Board Office (h)	Planning and Technology Sector (h)	Legal Dept. (h)	Financial Sector (h)
					Number of working hours, average				
1st phase/Inquiries (Contacts, CO)	120	720	0	0	0	0	360	0	0
2nd phase/Potential Clients, PC	96	3072	4608	0	0	384	3072	1152	0
3rd phase/Apparent Clients, AC	32	960	0	0	0	128	0	1088	1088
4th phase/Clients, CL	8	1536	41,200	5760	0	416	384	1728	304
Total number of working hours, per year		6288	45,808	5760	0	928	3816	3968	1392
Number of employees engaged per dept./sector, yearly		3.4	24.79	3.12	0	0.5	2.06	2.15	0.75

To conclude, the large number of phase interactions with the customers as a consequence of the high percentage of accepted inquiries compared to the received ones (80% of the "approved" inquiries), results in a 2.7% share of the average annual expense of the shipbuilding sales process (EUR 3,112,000, as shown in Table 2,) in the shipyard's overall annual revenue (nearly EUR 116,500,000), which is approximately 34% more than the expenditure plan of the planned annual cost of the sales process (nearly EUR 2,330,000, which is 2% of the average annual revenue, excluding the marketing costs).

Table 2. Average annual sales process cost—current state.

Phases/Cost	Current State		
	PI [1]	Phase Cost, Average (€)	Annual Phase Cost, (€)
Inquiries (Contacts, CO)	120	360	43,200
Potential Clients, PC	96	6520	625,920
Apparent Clients, AC	32	8580	274,560
Clients, CL	8	271,040	2,168,320
Total cost of the sales process, annual average (€)			3,112,000

[1] Average number of phase interactions, annually.

6.3. Value Stream Mapping of the Future State

Value Stream Mapping of the future state of the shipbuilding sales process is anticipated by defining the measures which could realise changes in the sales process.

Improvements in the sales process will be achieved firstly by strict selection of the inquiries received, i.e., by detailed analysis and consequently with a lower number of phase interactions with the customers and therefore more effective engagement of the shipyard workers when communicating during the higher phases of the sales process, eventually resulting in high added value for the customer.

Decreasing the number of phase interactions with the customers will be achieved by improving the sales process according to the following:

1. The Sales Sector will engage a renowned, international consulting service agency specialising in the maritime sector to complete the study analysis of the actual, mid-term, and long-term expected competitiveness of the shipyard within specific market niches (passenger ships, ferries, special purpose vessels, cruise ships, cargo ships, yachts, naval vessels) and their mid-term and long-term prospective growth potential, risks, new markets, etc. The analysis will contain the review of current and forecast trends in the shipbuilding and shipping industry. The study analysis will be ordered in six month intervals and used by the Sales Sector as the tool during the inquiry selection phase. It will be the base according to which, with additional time engagement of the Sales Sector employees, the inquiries will be analysed in further detail, i.e., to be selected more rigorously. If the inquiry should define specific requests for maritime features of the ship and the scope of required class notations, the Sales Sector will include the Initial Design Department in Phase 1 of the sales process to receive an opinion about the complexity of the criteria and parameters set by the inquiry, i.e., their implication to the aimed delivery deadline, all from the beginning due to the more rigorous choice of the inquiry to transfer it to the next phase of the sales process. Furthermore, because of the same reason and regarding the same criteria, i.e., the possibility of realising the project within the aimed delivery deadline, the Planning and Technology Sector will be included in the inquiry selection phase, with its additional engagement power, for a more detailed analysis of the project's realisation and actual mid-term planning, i.e., the shipyard's capacities which, besides

the already contracted projects, also includes projects from the 3rd and 4th phase of the sales process.

2. The Sales and Marketing Sector will join the legal specialist as the permanent employee of the sector for faster internal communication with the shipyard's employees, therefore inciting the customers to more agile interaction, with an emphasis on reviewing the pre-contractual and contractual documentation (Non-Disclosure Agreement, LOI, and Shipbuilding contract). When defining the Non-Disclosure Agreement, to make a better choice amongst potential clients to continue with the negotiations, the Sales Sector, i.e., its legal specialist, will ask for the legal/organisational structure of the customer, and based on the same, along with available network data, will perform a preliminary legal due diligence in a reduced and narrow scope. Furthermore, most commonly, the customer's request to pay for an indefinite indemnity charge if the NDA provisions would be breached will not be accepted, but the shipyard will insist on limiting the damages to a certain amount; only if the potential client was recognised as a possible strategic or a long-term partner, and/or the shipyard is a major source of the disclosed information, can the shipyard accept the initial, common request of the customer. When the NDA is signed, and as the additional criterion for selecting potential clients, the Financial Sector will conduct one "mini" financial due diligence of the customer, based on data available on the Internet and possible financial statements delivered by the customer. The Preliminary offer will include two additional "conditions" for the customer:

    - Approximate deadlines to realise further sales process phases—accepting the Preliminary offer, signing the LOI, defining and accepting the contractual ship price as well as signing the Shipbuilding contract—all as the prerequisites for handing over the ship within the deadline desired by the client, or within the first next possible deadline and according to the production plan, i.e., the availability of the shipyard's capacities;

    - The preliminary price of the Initial design, i.e., contractual technical documentation (defining the general arrangement plan and technical specification of the ship, developed from the ship's initial design activities), along with the payment conditions of the same: in installments during the contracting phase or at once if the shipyard and the customer do not sign the shipbuilding contract or the same has not become effective. Accepting the aforementioned will be considered the main prerequisite for signing the LOI. Only if the client is recognized as a strategic one, with the possibility of building more ships in a serial, and whose models of financing depend on developing and contracting the (first) ship, will the shipyard estimate the possibilities for acknowledging to take over 50% maximum of the Initial design expenses. To conclude, the validity deadline of the Preliminary offer will be shortened, i.e., defined as not longer than 30 days, depending on the characteristics of the project and current plans of the shipyard.

1. The Finance Sector will additionally ask the customer for evidence of their existing funds or to reserve funds to finance the construction or pay for the ship upon its delivery. It will be performed in the form of deposited funds or with the confirmation that the customer's financial institution guarantees that the payment is ready to be issued. Depending on the customer's compliance with these additional criteria, the shipyard will estimate if the negotiations with the customer should continue and should the LOI, previously agreed upon, be signed. These criteria will not be applied in the event that the customer should be a Public Administration Authority (ministries and the like).

2. Upon signing the LOI, the Project Management Office will nominate the manager of the relevant project, who will, as a coordinator of the Client's commercial–technical teams and the shipbuilder, will participate in the initial design of the ship, i.e., defining and elaborating the technical specification of the same, for the extra evaluation of the project requirements, to contribute to the optimal implementation of the same

during the ship's development, and amongst all due to managing the dynamics of the communication between the teams pursuant to the Preliminary offer and "conditioned" deadlines for defining the ship's price and agreeing on the Shipbuilding contract. The Project Manager will actively be involved in the work of legal teams during the revision process of the Shipbuilding contract and will initialise the final, executed contract version.

A Value Stream Map of the future state of the shipbuilding sales process is shown in Figure 4.

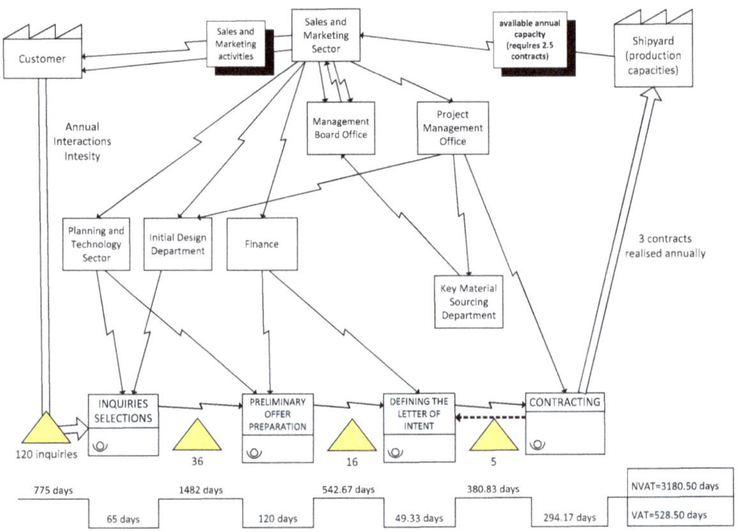

Figure 4. Future state Value Stream Map.

6.4. Implementing the Improvements—Theoretical Model

Applying the aforementioned improvement measures does not necessarily impact the change in the average phase time—except for doubling the duration length of the inquiry selection phase activities (Table A6), but it changes the relationship of the average time of the activities. Namely, within the Preliminary offer preparation phase, and especially in the contracting phase, it decreases the average duration time of the activities, which does not create added value (Table A7, i.e., Table A9). When defining the LOI for the average length of the activity, which does not create added value for the customer, the time is shortened, while the duration of the activities which develop the added value remains unchanged (Table A8). The stated decreases of losses in the process, with an emphasis on those referring to the waiting time to receive the answer from the customer, are predominantly realised by introducing the conditions for defining the approximate realisation deadlines of the higher phases of the sales process, starting from the deadline for accepting the Preliminary offer, which "assigns" the expected dynamics of the communication with the shipyard (defining the approximate reaction time of the customer to technical–commercial and legal questions of the shipyard's teams—the customer's "response time" within the process) to achieve the possible project realisation at least close to the client's desired deadline for the ship's delivery. If the customer should significantly step out of the approximately defined dynamics of interacting with the shipyard's teams, the new dates to realise higher sales process phases will depend on the engagement possibilities of the shipyard's employees in the involved sectors/departments on the relevant project and according to their availability concerning other actual projects, i.e., customers.

A big contribution to the reduction in losses and shortening the duration of the activities which do not create added value in the contracting phase is performed by naming the manager of the relevant project, knowing that they are in charge of monitoring the compatibility of the activities performance dynamic with approximately defined dates to realise the key points of the phase (process "follow up") by managing the cross-teams communication in a way that the same is kept within the assigned time intervals (optimal "response time" of the shipyard as well as the customer to their mutual questions regarding, for example, the initial design and optimisation of the ship, defining the ship's price, Shipbuilding contract revisions, etc.).

In addition to the Inquiry selection phase, introducing the improvements has a neutral effect on the average costs of the sales process phases.

A significant decrease in the average number of phase interactions regarding the "shipyard–customer" (Table A10) is primarily achieved by rigorous inquiry selection, and in later phases by all other process improvement measures, with an emphasis on efficacy selection of the Potential Clients by introducing the additional requirements for the customer in the Preliminary offer. The reduction mentioned above in the average number of phase interactions with the customers, significant reduction of the losses is achieved in the sales process and concerning working hours spent, and the representation and travel expenditures for the inquiries, i.e., customers of the lower and low potential of conversion to the next, higher process phases, which in the end results in a decrease in the average annual cost of the sales process according to the facts presented in the next section. Implementing the improvement measures for the sales process results in a nearly 38% decrease in the average number of employees necessary annually to perform the sales activities. The average number of employees in departments/sectors engaged in the sales process after improving the same is shown in Table 3.

Table 3. The necessary number of employees in the sales process—annual average—future state.

	Average Number of the Phase Interactions, Yearly	Sales and Marketing Sector (h)	Initial Design Dept. (h)	Key Material Sourcing Dept. (h)	PMO (h)	Mgmt. Board Office (h)	Planning and Technology Sector (h)	Legal Dept. (h)	Financial Sector (h)
				Number of working hours, average					
1st phase/Inquiries (Contacts, CO)	120	1140	720	0	0	0	840	0	0
2nd phase/Potential Clients, PC	36	1728	1728	0	0	144	1152	0	324
3rd phase/Apparent Clients, AC	16	896	0	0	0	64	0	0	640
4th phase/Clients, CL	5	1915	25,750	3600	880	260	240	0	190
Total number of working hours, per year		5679	28,198	3600	880	468	2232	0	1154
Number of employees engaged per dept./sector, yearly		3.07	15.26	1.95	0.48	0.25	1.21	0	0.62

7. Theoretical Results

Introducing improvement measures within the sales process, and with the primarily realised decrease in the average number of phase interactions, has relieved the employees engaged in the sales process of the most common comparative analysis of too many customers by, therefore, providing them with the possibility of a quality "follow-up" of high-prospect ones, with an emphasis on the faster recognition of the customers' representatives who have the decision-making power and an on-time "response time" in communicating with the same, which has resulted in a far better, i.e., higher rate of contact conversion into clients and eventually into shipowners. As shown in Table 4, besides the conversion rate of the Contacts into the Potential Client, which, as the consequence of the rigorous inquiry selection, has decreased by 62% compared to the sales process state before introducing improvement measures, all other conversion indicators have improved significantly, i.e., the conversion rate of the Potential Client into the Apparent Client has increased by 33%, of the Apparent Client into the Client by 25%, and eventually the Customer into the Shipowner by a high 106%.

Table 4. Key performance indicators—conversion rates.

Average Number of the Phase Interactions, Yearly			
	Current state	Future state	
Inquiries (Contacts, CO)	120	120	
Potential Clients, PC	96	36	
Apparent Clients, AC	32	16	
Clients, CL	8	5	
Shipbuilding contracts (Shipowners, SO)	2.33	3	
Key Performance Indicators—Conversion Rates			
	Current state	Future state	Change	Change (%)
Conversion rate C_1, C_1 = PC/CO	0.8	0.3	0.38	−62
Conversion rate C_2, C_2 = AC/PC	0.33	0.44	1.33	33
Conversion rate C_3, C_3 = CL/AC	0.25	0.31	1.25	25
Conversion rate C_4, C_4 = SO/CL	0.29	0.6	2.06	106

Better identifying those customers that should be focused on and communicated with was achieved by better understanding the customer's requests in the project, more efficient team cooperation when defining the contractual technical documentation, and generally improving the parameters of the ship in its initial design/development phase, which at the end of the process results in the high added value for the customer, i.e., with a project/ship of optimal feasibility. At the same time, by decreasing the average number of phase interactions with customers, the sales process is shortened, i.e., the average duration of each sales process is decreased by 15 days. In such an improved state, the individual process, from receiving the inquiry to realising the shipbuilding contract, is performed during an average of 7.5 months, while during one realised shipbuilding contract, the average sales process duration after introducing improvement measures decreases by 60%. The achievements in decreasing the average time of the sales process, i.e., the activities which do not create the added value and those which do after improvements are implemented, are shown in Table 5, whereby the number of phase interactions per realised shipbuilding contract (PI_S) equals the average yearly number of phase interactions divided by the average number of realised shipbuilding contracts (2.33 in the existing state and 3 in the future one, annually).

Furthermore, improving the shipbuilding sales process using Value Stream Mapping as the chosen Lean tool has created a decrease in the average annual sales process cost by 38%, with the simultaneous higher number of realised shipbuilding contracts (average of 3 per year) in comparison to the one assigned by the plan (2.7 per year). Comparing the average annual cost of the sales process before and after implementing improvements is shown in Table 6.

Table 5. Key performance indicator—average sales process duration.

| | Current State | | | | | Future State | | | | | Change (%) | |
| | | Phase Times Duration, Total | | By Each Realised Shipbuilding Contract | | | Phase Times Duration, Total | | By Each Realised Shipbuilding Contract | | | |
Phases/Duration	PI$_S$ [1]	VAT [2]	NVAT [3]	VAT	NVAT	PI$_S$	VAT	NVAT	VAT	NVAT	VAT	NVAT
Inquiries (Contacts, CO)	51.5	6	78	309	4017.2	40	13	155	520	6200	68.3	54.3
Potential Clients, PC	41.2	67	1001	2760.5	41,242.9	12	80	988	960	11,856	−65.2	−71.3
Apparent Clients, AC	13.7	74	910	1016.3	12,497.9	5.3	74	814	394.7	4341.3	−61.2	−65.3
Clients, CL	3.4	1278	2322	4388	7972.5	1.7	1412	1828	2353.3	3046.7	−46.4	−61.8
Total per sales process (h)		1425	4311				1579	3785			10.8	−12.2
Total per sales process (days)		59.4	179.6				65.8	157.7			10.8	−12.2
Total per shipbuilding contract (h)				8473.8	65,730.5				4228	25,444	−50.1	−61.3
Total per shipbuilding contract (days)				353.1	2738.8				176.2	1060.2	−50.1	−61.3

[1] Average number of phase interactions, annually per realised shipbuilding contract; [2] value added activities average duration (h); [3] non-value-added activities average duration (h).

Table 6. Key performance indicator—average annual sales process cost.

| | Current State | | | Future State | | | |
Phases/Cost	PI [1]	Phase Cost, Average (€)	Annual Phase Cost, (€)	PI	Phase Cost, Average (€)	Annual Phase Cost, (€)	Change	Change (%)
Inquiries (Contacts, CO)	120	360	43,200	120	1300	156,000	3.61	261.11
Potential Clients, PC	96	6520	625,920	36	7040	253,440	0.4	−59.51
Apparent Clients, AC	32	8580	274,560	16	8500	136,000	0.49	−50.47
Clients, CL	8	271,040	2,168,320	5	277,080	1,385,400	0.64	−36.11
Total cost of the sales process, annual average (€)			3,112,000			1,930,840	0.62	−37.96

[1] Average number of phase interactions, annually.

To emphasise, by improving the sales process, the average annual cost of the same was 17% lower than the one assigned by the expenditure plan, defined for the observed financial year—the one during which the improvements of the sales process were introduced.

In conclusion, the most noted result of process improvement is seen in the average number of contracted production man-hours per year, which stands for almost 30% growth in realising sales in comparison to the average number of production hours contracted before implementing the improvement measures within the sales process. Knowing that work has an almost 30% share within the ship's price, better sales realisation (average of 3 contracts per year, in comparison to the previous 2.33 before implementing the improvement measures into the sales process) has raised the annual revenue of the shipyard by 20%, i.e., nearly EUR 140,000,000 in comparison to EUR 116,500,000 realised before the improvement measures were introduced. Consequently, the average annual cost of the sales process has significantly decreased (EUR 1,930,840, as shown in Table 6,) in the total annual revenue of the shipyard (EUR 140,000,000, in conditions of the improved state of the process) from 2.7% to only 1.4%.

The shortage in the annual production capacity (for the difference between available capacity sufficient for construction of around 2.5 newbuilds, and required one for realisation of 3 newbuilds on average, contracted in the improved state of the sales process), the shipyard may substitute by employing or subcontracting additional workers of the required specialties.

8. Pilot Project and Discussion

An experimental application of the measures for improving the shipbuilding sales process begins in the Preliminary offer preparation phase by indicating the preliminary price of the Initial design and the payment method of the same. By accepting a Preliminary offer supplemented in this way by the Potential Client, which is the main prerequisite for signing the LOI, the conversion rate C_2 has increased by 103% compared to the existing sales process in the observed six-month period, which is three times more than expected from the theoretical model of implementing the improvement measures. Namely, in the observed six-month period, the relevant improvement measure was implemented to all 12 Potential Clients (2. sales process phase—Preliminary offer preparation phase) from the market niches and according to the following:

- 4 mega-yachts, average project value above EUR 100 mil;
- 2 series of naval vessel newbuilds, an average project value above EUR 100 mil;
- 4 cruise ships, average project value above EUR 300 mil;
- 2 ferries, average project value above EUR 50 mil

Seven of the abovementioned Potential Clients have accepted the obligation to pay the cost of preparing the Initial design, i.e., contractual technical documentation, as a new prerequisite for signing the LOI and paying the same during or after the last phase of the sales process—the contracting phase. Before introducing the relevant improvement measure, the costs of providing services by the Initial Design Department, as an integral part of the Sales and Marketing Sector, have been calculated in full as the expenditure within the planned (annual) cost of the shipbuilding sales process. As the eighth Apparent Client, one of the buyers of the cruise ships has been accepted, recognised as the strategic one, with the potential of constructing one more ship from the serial.

Furthermore, within the previous six months, the measure of entitling the project manager is implemented for four Clients (all from the mega-yachts niche). The Project Management Office has nominated one project manager for two projects. In three projects, during their contracting phase, a significant reduction in the duration of activities that do not create additional value was noticed thanks to the continuous monitoring of how the communication is led, especially the communication among the commercial–technical and legal teams of the Clients and shipyard, i.e., thanks to maintaining the dynamics according to the planned time for realising the key points of the relevant phase. By reducing the waiting time within the internal communication, and especially within the communication with the Clients' teams, the duration of the contracting phase time is reduced by 10 to 30 days in relation to the shipbuilding sales process state before implementing relevant improvement measure. Yet, the duration time of the contracting phase related to the fourth mega-yacht project is prolonged significantly due to the great increase of waiting time in external communication, exclusively due to the inadequate "response-time" by the commercial-technical Client's teams, and by nearly 50 days in relation to the theoretical model of the future process state.

This research needs to be continued with the periodical analyses of realised improvements in the sales process according to further dynamics and the schedule of implementing suggested measures of the improvements into the process, along with the verification of the theoretical results of improvement. By continuously comparing the shipbuilding sales process improvements resulting from the experiment with the value stream of the theoretical model of the future state, the reflection possibilities on new measures for the process improvement are enlarged, with the continuous increase of added value for the buyer accordingly.

Results achieved, as well as the future improvement results, can additionally be improved by introducing the Kaizen Lean tool in a way that the shipyard employees should be encouraged, especially those involved in the sales process, including them in "workshops" to make them continuously think about further measures of improvement to result in the additional cost reduction in the process.

Regarding the extreme contribution of the Initial Design Department's activities in creating the added value and optimisation of the ship, it is suggested that the future research topic analyses the justification of the business strategy of integrating initial design activities (Initial or Conceptual design) into the business system of the shipyard in relation to the approach of engaging the "outsourced" design offices to perform the same. Indeed the analysis calls for a distinction of a prototype shipyard or shipyards for smaller serials, such as European ones, from those oriented to larger series of new constructions (such as Asian shipyards), regarding the differences in the requirements for the Initial design.

Furthermore, it is necessary to presume that the conversion rates, as part of the Lean metrics defined by this paper, are not key performance indicators applicable for the shipyards of larger series of new constructions. Therefore, all stated here has to be analysed additionally. If the presumptions appear to be correct, research must be performed on the possibility of adjusting the relevant performance indicators to such systems or the need to define the alternative ones.

9. Conclusions

Under conditions of the actual customers' requirements in the international shipbuilding market for energy-efficient and ecologically acceptable vessels, shipbuilders need to implement the digital transformation from traditional into "smart" and eventually into "green", At the same time, according to the European Commission strategies adopted ("European Green Deal", then attached to it "A New Industrial Strategy for Europe", and others) on achieving climate neutrality by 2050 at the latest, European shipyards aim to adopt the applicable Industry 4.0 doctrines and technologies also from that aspect, therefore realising the prerequisites for sustainable growth and development of their business systems. Implementing the Lean method for managing business processes is generally the basic prerequisite to approach the shipyard's digital transformation to maintain, i.e., to enlarge the competitiveness, and grow its value.

The authors of this paper recognise the (sub)process of sales and its flow of value, i.e., information in the process of realising shipbuilding projects in the studied European shipyard as the basis for structuring the assumptions of the overall improvement of business results, from the growth of employment and profitability to increasing client satisfaction and employee motivation.

Therefore, Lean transformation of the studied shipyard starts from the shipyard's Sales and Marketing Sector, as the leading functional units of the sales process, whereas Value Stream Mapping (VMS) has been chosen as the Lean tool for identifying and decreasing/removing the losses from the process. After analysing the current sales process, the value it creates was confirmed and the applicable key performance indicators were defined. The shipbuilding contract for the vessel additionally improved in parameters compared to the initial requirements of the customer.

After mapping the existing state of the sales process and analysing the value stream through its four phases—from the selection of the inquiry, preparation of the Preliminary offer, then defining the Letter of Intent, and finally, the contracting phase—the key shortcomings were identified as (i) an extremely lengthy sales process, (ii) smaller number of realised shipbuilding contracts compared to the one defined by the business plan, (iii) higher average yearly expenditures of the shipbuilding sales process than planned, and (iv) all as the consequence of a large number of simultaneous interactions with customers as they were not properly selected during the sales process interphases. In accordance with what was seen, the improvement measures were defined, primarily in order to decrease the number of phase interactions between shipyard employees involved in the sales process with customers. Improvements in the sales processes were planned to be achieved, among other things, by (i) hiring a renowned international consulting firm specialised in the maritime sector in order to better understand the situation of certain market niches, (ii) adding a legal specialist in the Sales and Marketing Sector as permanently employed staff, (iii) requiring customers to follow the deadlines for realising the sales process phases as well as for covering the expenses of the initial design, all as the prerequisite for signing the Letter of Intent, and (iv) appointing a project manager by the shipyard already in the contracting phase. According to the same, a diagram was created of the future state of the theoretical model of shipbuilding sales process. The value stream of the improved state resulted in a more than significant sales process improvement, measured as follows:

- Shortening the average duration of the sales process of one realised shipbuilding contract by 60%;
- Decreasing the average annual cost of the sales process by 38%;
- Increasing annual sales by almost 30%, followed by an increase in the productivity of employees involved in the sales process by a high 93%.

Implementing the improvement measures significantly decreases the average number of sales employees needed—from the current 37 to 23 annually. Eventually, the improved state of the sales process results in a 20% increase in the average annual turnover of the shipyard compared to its realisation in the current state.

The authors of this paper foresee further opportunities to improve the shipbuilding sales process, not only by continuously monitoring it by implementing the VSM tool, but also by presenting the Kaizen Lean tool to shipyard employees, i.e., the philosophy of its contribution to achieving (further) savings in the process.

Author Contributions: Conceptualization, Z.K.; methodology, N.T.; validation, Z.K., N.T. and N.Š.; formal analysis, Z.K.; investigation, Z.K.; resources, Z.K.; data curation, Z.K.; writing—original draft preparation, Z.K. and N.T.; writing—review and editing, Z.K. and N.T.; visualization, Z.K.; supervision, N.Š. All authors have read and agreed to the published version of the manuscript.

Funding: The Article Processing Charges (APCs) is funded by the European Regional Development Fund, grant number KK.01.1.1.07.0052.

Data Availability Statement: Not applicable.

Conflicts of Interest: The authors declare no conflict of interest.

Appendix A

Table A1. The average number of the "shipyard—customer" interactions by conversion phases—the current state.

Phases	Monthly	Quarterly	Yearly	Three-Yearly	SC [1]
Inquiries (Contacts, CO)	10	30	120	360	51.5
Potential Clients, PC	8	24	96	288	41.2
Apparent Clients, AC	2.67	8	32	96	13.73
Clients, CL	0.67	2	8	24	3.43
Shipbuilding contracts (Shipowners, SO)	0.19	0.58	2.33	7	1

[1] Average number of phase interactions, annually per realised shipbuilding contract.

Table A2. The average durations of the inquiry selection phase activities—the current state.

	Inquiry Selection Phase, Total				
	(Days)	(h)	VAT [1]	NVAT [2]	Total (h)
	3.5	84			
Customer				24	24
Sales and Marketing Sector			4.5	32	36.5
Planning and Technology Sector			1.5	22	23.5
Total			6	78	84

[1] Value added activities average duration (h); [2] non-value-added activities average duration (h).

Table A3. The average durations of the Preliminary offer preparation phase activities—the current state.

	Non-Disclosure Agreement Defining Subphase				Preliminary Offer Budgeting Subphase				Phase Activities Duration, Total		
	Total				Total						
	(Days)	(h)	VAT [1]	NVAT [2]	(Days)	(h)	VAT	NVAT	VAT	NVAT	(h)
	16	384			28.5	684					
Customer				144				336		480	480
Sales and Marketing Sector			4	40			20	28	24	68	92
Initial Design Dept.							14	36	14	36	50
Mgmt. Board Office			1	71			3	117	4	188	192
Planning and Technology Sector							18	112	18	112	130
Legal Department			7	117					7	117	124
Total			12	372			55	629	67	1001	1068

[1] Value added activities average duration (h); [2] non-value-added activities average duration (h).

Table A4. The average durations of the Letter of Intent defining phase activities—the current state.

	Letter of Intent Defining Phase, Total				
	(Days)	(h)	VAT [1]	NVAT [2]	Total (h)
	41	984			
Customer				384	384
Sales and Marketing Sector			24	56	80
Management Board Office			4	68	72
Legal Department			22	142	164
Financial Sector			24	260	284
Total			74	910	984

[1] Value added activities average duration (h); [2] non-value-added activities average duration (h).

Table A5. The average durations of the contracting phase activities—the current state.

| | Contractual Ship Price Calculation Subphase | | | | Shipbuilding Contract Review Subphase | | | | Phase Activities Duration, Total | | |
| | Total | | | | Total | | | | | | |
	(Days)	(h)	VAT [1]	NVAT [2]	(Days)	(h)	VAT	NVAT	VAT	NVAT	(h)
	97	2328			150	3600					
Customer								744		744	744
Sales and Marketing Sector			84				92	192	176	192	368
Initial Design Dept.			480				104	360	584	360	944
Key Material Sourcing Dept.			160				80	240	240	240	480
Mgmt. Board Office			18				24	264	42	264	306
Planning and Technology Sector			24				8	72	32	72	104
Legal Department							176	330	176	330	506
Financial Sector							28	120	28	120	148
Total			766				512	2322	1278	2322	3600

[1] Value added activities average duration (h); [2] non-value-added activities average duration (h).

Table A6. The average durations of the inquiry selection phase activities—the future state.

| | Inquiry Selection Phase, Total | | | | |
	(Days)	(h)	VAT [1]	NVAT [2]	Total (h)
	7	168			
Customer				24	24
Sales and Marketing Sector			6	54.5	60.5
Initial Design Department			3	33	36
Planning and Technology Sector			4	43.5	47.5
Total			13	155	168

[1] Value added activities average duration (h); [2] non-value-added activities average duration (h).

Table A7. The average durations of the Preliminary offer preparation phase activities—the future state.

| | Non-Disclosure Agreement Defining Subphase | | | | Preliminary Offer Budgeting Subphase | | | | Phase Activities Duration, Total | | |
| | Total | | | | Total | | | | | | |
	(Days)	(h)	VAT [1]	NVAT [2]	(Days)	(h)	VAT	NVAT	VAT	NVAT	(h)
	14	336			30.5	732					
Customer				168				336		504	504
Sales and Marketing Sector			17	79			20	28	37	107	144
Initial Design Dept.							14	36	14	36	50
Mgmt. Board Office			1	71			3	117	4	188	192
Planning and Technology Sector							18	112	18	112	130
Financial Sector							7	41	7	41	48
Total			18	318			62	670	80	988	1068

[1] Value added activities average duration (h); [2] non-value-added activities average duration (h).

Table A8. The average durations of the Letter of Intent defining phase activities—the future state.

	Letter of Intent Defining Phase, Total				
	(Days)	(h)	VAT [1]	NVAT [2]	Total (h)
	37	888			
Customer				360	**360**
Sales and Marketing Sector			42	106	**148**
Management Board Office			4	68	**72**
Financial Sector			28	280	**308**
Total			**74**	**814**	**888**

[1] Value added activities average duration (h). [2] Non-value-added activities average duration (h).

Table A9. The average durations of the contracting phase activities—the future state.

	Contractual Ship Price Calculation Subphase				Shipbuilding Contract Review Subphase [3]				Phase Activities Duration, Total		
	Total				Total						
	(Days)	(h)	VAT [1]	NVAT [2]	(Days)	(h)	VAT	NVAT	VAT	NVAT	(h)
	100	2400			135	3240					
Customer								520		520	520
Sales and Marketing Sector			84				248	340	**332**	340	672
Initial Design Dept.			480				104	330	**584**	330	914
Key Material Sourcing Dept.			160				80	190	**240**	190	430
Project Mgmt. Office			98				56	48	**154**	48	202
Mgmt. Board Office			18				24	230	**42**	230	272
Planning and Technology Sector			24				8	60	**32**	60	92
Financial Sector							28	110	**28**	110	138
Total			**864**				**548**	**1828**	**1412**	**1828**	**3240**

[1] Value added activities average duration (h). [2] Non-value-added activities average duration (h). [3] Subphases are running simultaneously, therefore only NVAT of the subphase that starts and ends same as the subject phase, applies.

Table A10. The average number of the "shipyard—customer" interactions by conversion phases—the future state.

Phases	Monthly	Quarterly	Yearly	Three-Yearly	SC [1]
Inquiries (Contacts, CO)	10	30	120	360	40
Potential Clients, PC	3	9	36	108	12
Apparent Clients, AC	1.33	4	16	48	5.33
Clients, CL	0.42	1.25	5	15	1.67
Shipbuilding contracts (Shipowners, SO)	0.25	0.75	3	9	1

[1] Average number of phase interactions, annually per realised shipbuilding contract.

References

1. Tenold, S. The Declining Role of Western Europe in Shipping and Shipbuilding, 1900–2000. In *Shipping and Globalization in the Post-War Era—Contexts, Companies, Connections*; Petersson, N.P., White, N.J., Tenold, S., Eds.; Springer Nature Switzerland AG: Cham, Switzerland, 2019; pp. 9–36.
2. Hess, M.; Pavić, I.F.; Kos, S.; Brčić, D. Global shipbuilding activities in the modern maritime market environment. *Sci. J. Marit. Res.* **2020**, *34*, 270–281. [CrossRef]
3. Kamola-Cieślik, M. Changes in the Global Shipbuilding Industry on the Example of Selected States Worldwide in the 21st Century. *Eur. Res. Stud. J.* **2021**, *XXIV*, 98–112. [CrossRef]
4. Parc, J.; Normand, M. Enhancing the Competitiveness of the European Shipbuilding Industry: A Critical Review of its Industrial Policies. *Asia-Pac. J. EU Stud.* **2016**, *14*, 73–92.
5. Communication from the Commission to the European Parliament, the European Council, the Council, the European Economic and Social Committee and the Committee of the Regions. Available online: https://ec.europa.eu/info/sites/default/files/communication-eu-industrial-strategy-march-2020_en.pdf (accessed on 1 May 2022).
6. Communication from the Commission to the European Parliament, the European Council, the Council, the European Economic and Social Committee and the Committee of the Regions. Available online: https://ec.europa.eu/info/sites/default/files/european-green-deal-communication_en.pdf (accessed on 1 May 2022).
7. Zhong, R.J.; Xu, X.; Klotz, E.; Newman, S.T. Intelligent Manufacturing in the Context of Industry 4.0: A Review. *Engineering* **2017**, *3*, 616–630. [CrossRef]
8. Leng, J.; Ruan, G.; Jiang, P.; Xu, K.; Liu, Q.; Zhou, X.; Liu, C. Blockchain-empowered sustainable manufacturing and product lifecycle management in industry 4.0: A survey. *Renew. Sustain. Energy Rev.* **2020**, *132*, 110112. [CrossRef]
9. Kunkera, Z. *Lean Application in the Sales and Marketing Sector of the Observed Shipyard*; Seminar paper from the "Production and Project Management" University course, Postgraduate Doctoral Studies; University of Zagreb, Faculty of Mechanical Engineering and Naval Architecture: Zagreb, Croatia, 2020.
10. Sanchez-Sotano, A.; Cerezo-Narváez, A.; Abad-Fraga, F.; Salguero-Gómez, J. Trends of Digital Transformation in the Shipbuilding Sector. In *New Trends in the Use of Artificial Intelligence for the Industry 4.0*; IntechOpen: London, UK, 2020.
11. Recamán Rivas, A. Navantia's Shipyard 4.0 model overview. *Ship Sci. Technol.* **2018**, *11*, 77–85. [CrossRef]
12. Pang, T.Y.D.; Pelaez Restrepo, J.; Cheng, C.-T.; Yasin, A.; Lim, H.; Miletic, M. Developing a Digital Twin and Digital Thread Framework for an 'Industry 4.0' Shipyard. *Appl. Sci.* **2021**, *11*, 1097. [CrossRef]
13. Jagusch, K.; Sender, J.; Jericho, D.; Fluegge, W. Digital thread in shipbuilding as a prerequisite for the digital twin. *Procedia CIRP* **2021**, *104*, 318–323. [CrossRef]
14. Wang, Z. *Digital Twin Technology*; IntechOpen: London, UK, 2020; Volume 80974, pp. 95–114.
15. Vidal-Balea, A.; Blanco-Novoa, O.; Fraga-Lamas, P.; Vilar-Montesinos, M.M.; Fernández-Caramés, T. Creating Collaborative Augmented Reality Experiences for Industry 4.0 Training and Assistance Applications: Performance Evaluation in the Shipyard of the Future. *Appl. Sci.* **2020**, *10*, 9073. [CrossRef]
16. Leng, J.; Wang, D.; Shen, W.; Li, X.; Liu, Q.; Chen, X. Digital twins-based smart manufacturing system design in Industry 4.0: A review. *J. Manuf. Syst.* **2021**, *60*, 119–137. [CrossRef]
17. Valk, H.; Strobel, G.; Winkelmann, S.; Hunker, J.; Tomczyk, M. Supply Chains in the Era of Digital Twins—A Review. *Procedia Comput. Sci.* **2021**, *00*.
18. Rabah, S.; Assila, A.; Khouri, E.; Maier, F.; Ababsa, F.; Bourny, V.; Maier, P.; Merienne, F. Towards improving the future of manufacturing through digital twin and augmented reality technologies. *Procedia Manuf.* **2018**, *17*, 460–467. [CrossRef]
19. Lukas, U. *Virtual and Augmented Reality in the Maritime Industry*; Publishing House of Poznan University of Technology: Poznan, Poland, 2006; pp. 193–201.
20. Laurent, D.; Cenk, Y. Ship Finance Practices in Major Shipbuilding Economies. In *OECD Science, Technology and Industry Policy Papers, No. 75*; OECD Secretariat: Paris, France, 2019.
21. Gostomski, E.; Nowosielski, T. Financing the construction and purchase of sea ships. *Sci. J. Marit. Iniversity Szczec.* **2020**, *64*, 61–69.
22. Lee, M.-H.; Han, J.-J. Study on the Optimal Number of Creditors in Shipping Finance Syndication. *J. Int. Logist. Trade* **2016**, *14*, 200–210. [CrossRef]
23. Kõlvart, M.; Poola, M.; Rull, A. Smart Contracts. In *The Future of Law and eTechnologies*; Kerikmäe, T., Rull, A., Eds.; Springer International Publishing: Cham, Switzerland, 2016; pp. 133–147.
24. Allam, Z. On Smart Contracts and Organisational Performance: A Review of Smart Contracts through The Blockchain Technology. *Rev. Econ. Bus. Stud.* **2018**, *11*, 137–156. [CrossRef]
25. Leng, J.; Sha, W.; Lin, Z.; Jing, J.; Liu, Q.; Chen, X. Blockchained smart contract pyramid-driven multi-agent autonomous process control for resilient individualised manufacturing towards Industry 5.0. *Int. J. Prod. Res.* **2022**, *60*, 1–20. [CrossRef]
26. Leng, J.; Chen, Z.; Sha, W.; Ye, S.; Liu, Q.; Chen, X. Cloud-edge orchestration-based bi-level autonomous process control for mass individualization of rapid printed circuit boards prototyping services. *J. Manuf. Syst.* **2022**, *63*, 143–161. [CrossRef]
27. Czachorowski, K.; Solesvik, M.; Kondratenko, Y. The Application of Blockchain Technology in the Maritime Industry. In *Green IT Engineering: Social, Business and Industrial Applications*; Kharchenko, V., Kondratenko, Y., Kacprzyk, J., Eds.; Springer Nature Switzerland AG: Cham, Switzerland, 2019; pp. 561–577.
28. Peronja, I.; Lenac, K.; Glavinović, R. Blockchain technology in maritime industry. *Sci. J. Marit. Res.* **2020**, *34*, 178–184. [CrossRef]

29. Zhu, J.; Wu, M.; Liu, C. Research on the Application Mode of Blockchain Technology in the Field of Shipbuilding. In Proceedings of the IEEE International Conference on Artificial Intelligence and Computer Applications (ICAICA), Dalian, China, 27–29 June 2020.
30. Sullivan, B.P.; Rossi, M.; Terzi, S. A Customizable Lean Design Methodology for Maritime. In Proceedings of the 15th IFIP International Conference on Product Lifecycle Management (PLM), Turin, Italy, 2–4 July 2018; pp. 508–519.
31. Sharma, S.; Gandhi, P.J. Production concept of lean shipbuilding which is combination of lean manufacturing and lean construction. *Procedia Eng.* **2017**, *194*, 232–240. [CrossRef]
32. Lang, S.; Dutta, N.; Hellesoy, A.; Daniels, T.; Liess, D.; Chew, S.; Canhetti, A. Shipbuilding and Lean Manufacturing—A Case Study. In Proceedings of the Ship Production Symposium, Ypsilanti, MI, USA, 13–15 June 2001.
33. Song, T.; Zhou, J. Research and Implementation of Lean Production Mode in Shipbuilding. *Processes* **2021**, *9*, 2071. [CrossRef]
34. Oliveira, A.; Gordo, J.M. Lean tools applied to a shipbuilding panel line assembling process. *Shipbuilding* **2018**, *69*, 53. [CrossRef]
35. Kolich, D.; Storch, R.L.; Fafandje, N. Lean Built-Up Panel Assembly in a Newbuilding Shipyard. *J. Ship Prod. Des.* **2017**, *00*, 1–9.
36. Kolich, D.; Storch, R.L.; Fafandjel, N. Optimizing shipyard interim product assembly using a value stream mapping methodology. In Proceedings of the World Maritime Technology Conference, Providence, RI, USA, 3–7 November 2015.
37. Gjeldum, N.; Veža, I.; Bilić, B. Simulation of Production Process reorganized with Value Stream Mapping. *Tech. Gaz.* **2011**, *18*, 341–347.
38. Niansheng, C.; Xiang, N.; Jiang, X.; Kunlin, L. A systematic approach of lean supply chain management in shipbuilding. *SN Appl. Sci.* **2021**, *3*, 572. [CrossRef]
39. Kolich, D.; Yao, Y.L.; Fafandjel, N.; Hadjina, M. Value stream mapping micropanel assembly with clustering to improve flow in a shipyard. In Proceedings of the International Conference on Innovative Technologies, Leiria, Portugal, 10–12 September 2014; pp. 85–88.
40. Womack, J.P.; Jones, D.T.; Roos, D. *The Machine That Changed the World: The Story of Lean Production*; Simon & Schuster UK Ltd.: London, UK, 2007.
41. Monden, Y. *Toyota Production System: An Integrated Approach to Just-In-Time*, 4th ed.; Taylor & Francis Group: Abingdon, UK, 2012.
42. Womack, J.P.; Jones, D.T. *Lean Thinking*; Free Press: New York, NY, USA, 2003.
43. Liker, J.K. *The Toyota Way 14 Management Principles from the World's Greatest Manufacturer*; McGraw-Hill: New York, NY, USA, 2004.
44. Rother, M.; Shook, J. *Learning to See*; The Lean Enterprise Institute: Boston, MA, USA, 2003.
45. Shararah, M.A. A value stream map in motion. *Ind. Eng.* **2013**, *45*, 46–50.
46. Nash, M.A.; Shela, R. *Mapping the Total Value Stream: A Comprehensive Guide for Production and Transactional Processes*; Productivity Press, Taylor & Francis Group: Abingdon, UK, 2008.
47. Martin, K.; Osterling, M. *Value Stream Mapping: How to Visualize Work and Align Leadership for Organizational Transformation*, 1st ed.; McGraw-Hill Education: New York, NY, USA, 2013.
48. Fawaz, A.A.; Rajgopal, J. Analyzing the benefits of lean manufacturing and value stream mapping via simulation: A process sector case study. *Int. J. Prod. Econ.* **2007**, *107*, 223–236. [CrossRef]
49. Leonardo, D.G.; Sereno, B.; Sant, D.; da Silva, A.; Sampaio, M.; Massote, A.A. Implementation of hybrid Kanban-CONWIP system: A case study. *J. Manuf. Technol. Manag.* **2017**, *28*, 714–736. [CrossRef]
50. Rashid, Y.; Rashid, A.; Warraich, M.A.; Sabir, S.S.; Waseem, A. Case Study Method: A Step-by-Step Guide for Business Researchers. *Int. J. Qual. Methods* **2019**, *18*, 1–13. [CrossRef]
51. Yin, R.K. *Case Study Research: Design and Methods*; Sage Publications: London, UK, 2013.
52. Gierin, J.; Dyck, A. Maritime Digital Twin architecture. *at–Automatisierungstechnik* **2021**, *69*, 1081–1095. [CrossRef]
53. Stachowski, T.-H.; Kjeilen, H. Holistic Ship Design—How to Utilise a Digital Twin in Concept Design through Basic and Detailed Design. In Proceedings of the International Conference on Computer Applications in Shipbuilding, Singapore, 26–28 September 2017.
54. Sales KPIs & Metrics—Explore The Best Sales KPI Examples. Available online: https://www.datapine.com/kpi-examples-and-templates/sales (accessed on 27 June 2022).
55. 8 KPIs to Improve Your Marketing and Sales | Formstack. Available online: https://www.formstack.com/resources/blog-sales-marketing-kpis-to-track (accessed on 27 June 2022).

Article

Developing and Implementing a Lean Performance Indicator: Overall Process Effectiveness to Measure the Effectiveness in an Operation Process

Lisbeth del Carmen Ng Corrales [1,2,*], **María Pilar Lambán** [1], **Paula Morella** [1], **Jesús Royo** [1], **Juan Carlos Sánchez Catalán** [3] **and Mario Enrique Hernandez Korner** [1,2]

[1] Design and Manufacturing Engineering Department, Universidad de Zaragoza, 50018 Zaragoza, Spain; plamban@unizar.es (M.P.L.); pmorella@unizar.es (P.M.); jaroyo@unizar.es (J.R.); mario.hernandez2@utp.ac.pa (M.E.H.K.)

[2] Department of Industrial Engineering, Universidad Tecnológica de Panamá, Ciudad de Panama 0819-07289, Panama

[3] Smart Systems, Tecnalia, Basque Research and Technology Alliance (BRTA), 20009 Donostia-San Sebastian, Spain; jcarlos.sanchez@tecnalia.com

* Correspondence: lisbeth.ng@utp.ac.pa

Citation: Ng Corrales, L.d.C.; Lambán, M.P.; Morella, P.; Royo, J.; Sánchez Catalán, J.C.; Hernandez Korner, M.E. Developing and Implementing a Lean Performance Indicator: Overall Process Effectiveness to Measure the Effectiveness in an Operation Process. *Machines* **2022**, *10*, 133. https://doi.org/10.3390/machines10020133

Academic Editors: Luís Pinto Ferreira, Paulo Ávila, João Bastos, Francisco J. G. Silva, José Carlos Sá and Marlene Brito

Received: 29 December 2021
Accepted: 9 February 2022
Published: 12 February 2022

Publisher's Note: MDPI stays neutral with regard to jurisdictional claims in published maps and institutional affiliations.

Abstract: The purpose of this paper is to build up and implement a framework of a lean performance indicator with collaborative participation. A new indicator derived from OEE is presented, overall process effectiveness (OPE), which measures the effectiveness of an operation process. The action research (AR) methodology was used; collaborative work was done between researchers and management team participation. The framework was developed with the researchers' and practitioners' experiences, and the data was collected and analyzed; some improvements were applied and finally, a critical reflection of the process was done. This new metric contributes to measuring the unloading process, identifying losses, and generating continuous improvement plans tailored to organizational needs, increasing their market competitiveness and reducing the non-value-add activities. The OEE framework is implemented in a new domain, opening a new line of research applied to logistic process performance. This framework contributes to recording and measuring the data of one unloading area and could be extrapolated to other domains for lean performance. It was possible to generate and validate knowledge applied in the field. This study makes collaborative participation providing an effectiveness indicator that helps the managerial team to make better decisions through AR methodology.

Keywords: OEE; KPIs; action research; lean; operation process; performance

1. Introduction

To remain competitive, companies must constantly review the performance of their operations and production processes. To improve the operation process, it is not always necessary for investments in technology, equipment, and facilities, but instead, the increase of the efficiency could be made with small improvements associated with process sustainability [1]. It is useful for the company to find the appropriate tools to measure the efficiency and productivity of a transport process to eliminate or reduce non-value-added activities [2]. Previous research in transport value has established through interviews with practitioners that there are two types of value-added activities, transportation, and unloading/loading activities [3]. This study's novelty is that it develops and implements an adaptation of the Overall Equipment Effectiveness (OEE) indicator to measure the effectiveness of a loading or unloading process. In this research, this indicator is adapted and implemented in a logistic process to evaluate and measure every step of the selected process. The OEE has been used as a lean metric to identify and eliminate process losses [4,5].

OEE indicator is used to measure the effectiveness level, identifying losses and allowing with a later analysis to identify the non-value-adding activities. OEE is an indicator of effectiveness, although it is often confused with efficiency. However, effectiveness indicates the objectives achieved and efficiency relates to the use of resources and the achievements attained [6]. A considerable amount of research has been published in recent years [7] concerning the OEE, such as case studies, implementations, original OEE slightly modified, and new adaptations to other areas. The OEE has been adapted to various domains depending on industry need, such as line manufacturing [8], transport [9], mining equipment [10], and assembly tasks [11]. Authors proposed areas in which OEE could be applicable, in the service sector and logistics processes, such as goods reception or performing selection in a warehouse [1,12,13]. This indicator is appreciated by managers because it is an overall metric, simple and clear, which identifies areas of potential improvement.

The purpose of this paper is to build up and implement a framework of a lean performance indicator with collaborative participation. The implementation of this framework makes it possible to establish a procedure to record, measure, and study the process of lateral unloading trucks. An OEE framework adapted to the logistic field will be implemented and the identification of the losses that reduce the effectiveness of the process is intended. To achieve this objective, the OEE indicator traditionally used in production will be adapted to the logistic area to measure the effectiveness of the unloading truck process. A new metric, named Overall Process Effectiveness (OPE), is developed to quantify the effectiveness of the logistic process, identifying the hidden losses that most affect the unloading truck process.

To fulfill the goal of this study, action research (AR) methodology was selected. The AR used in this study seeks to solve an organizational issue while building up scientific knowledge. AR is a collaborative work of research with participation in the field. The term was used in works of the German psychologist Kurt Lewis [14] that considers the impact linked between research and action to create change in the organizational process. The AR must be researched in action, participative, a sequence of events, concurrent with action, and provide an approach to problem-solving [15].

The paper is organized as follows: in Section 2 described and explained the performance measurement systems, the origin of the OEE concept, and the six big losses. Section 3 the proposed methodology and case description are displayed. The framework to be implemented is developed in Section 4 and in the next section, the analysis of the results is presented. Finally, Sections 6 and 7 show the discussion and conclusion of the research, respectively.

2. Theoretical Underpinning

2.1. Performance Measurement System

Companies have a great interest in performance measurement systems as they provide insight into the performance of a specific activity. Proper implementation and use of the performance measurement systems provide a clearer picture of the performance of the company's operational processes.

A performance measurement system can guide and influence the decision-making process. Since the end of the 1980s, performance measurement systems have become a relevant issue for scholars and practitioners [16]. Over the last years, companies increase their competitiveness by implementing performance measurement systems as platforms of continuous improvements in organizational processes [17]. The performance measurement systems can be defined as the process of quantifying the efficiency and effectiveness of a certain activity with a set of metrics. It can be examined at three different levels: individual performance measures, set of performance measures, and environment within which the performances measurement system operates [18].

Over the years, companies have implemented performance measurement systems in different areas and sectors such as finance, process, human resources, and service. These systems facilitate monitoring the efficiency of time and resources, helping to set realistic goals, objectives, and improvements. A good performance measurement system provides the company feedback on job performance, a guide for defining new and clear goals, and generating increases in the profit margins. Caplice and Sheffi in 1995 [19] state that a good logistics performance measurement system should be comprehensive, causally oriented, vertically integrated, horizontally integrated, internally comparable, and useful. From a strategic point of view, performance measurement is not only the manager's responsibility, but field operators must also be involved in the implementation. A logistics performance measurement system must include multiple goals to be viewed within the organizational performance. Caplice and Sheffi [20] argued that must be considered in a logistics performance indicator, validity, robustness, usefulness, integration, economy, compatibility, level of detail, and behavior soundness.

According to the literature [21,22], the three steps to develop a performance measurement system are:

a. Design: this phase involves planning what and how to measure. It also identifies the key objectives to be achieved and the areas involved in the data collection.
b. Implementation: the systems and procedures are put into action. In this phase, the output of the results is verified, and the system is refined if necessary.
c. Use: the results are reviewed. The process or resource understudy is checked to see if it is being effective and under the company's goals.

To quantify organizational and individual performance, Key Performance Indicator (KPI) can enable an objective measurement, based on organizational goals. KPIs monitor critical business activities and provide information to increase organizational performance [23]. Industry 4.0 has encouraged digital transformation in the organization. Digital transformation allows having reliable, truthful, and real-time information to feed the management information system to make better decisions [7,24], though digital data is possible to enhance a data-driven and more intelligent approach toward the continuous improvement process [25]. OEE is a KPI widely accepted as a tool well implemented in lean manufacturing to monitor the performance of a process.

Traditionally in general practice, individual performance was measured from utilization, productivity, and effectiveness metrics [20]. For this research, a multi-dimensional perspective is considered for a new metric [26]. The proposed indicator is the OEE, a key performance indicator used to measure the availability, performance, and quality in a process. Moreover, the indicator allowed us to identify productivity losses during the process for continuous improvements.

2.2. Overall Equipment Effectiveness

Initially, the OEE was a quantitative metric of the total productive maintenance launched by Nakajima in 1988. Muchiri and Pintelon in 2008 [6] defined OEE as a tool of performance that measures any losses in production and identifies the areas of process improvement. This metric measures the availability, performance, and quality of individual equipment showing the productivity in the manufacturing operations. OEE is the results in a percentage of the multiplication of these three parameters:

$$\text{OEE (\%)} = \text{Availability} \times \text{Performance} \times \text{Quality}$$

Availability measures the productive time of the machine or equipment when scheduled or available. This parameter is affected when there are losses due to equipment failure or setup and adjustment. Some activities are not included in the available time like the planned downtime such as planned maintenance, labor meeting, among others. If the planned downtime is considered in the production time, the result would be lower, but the true availability would be shown [27].

Performance refers to the real production versus the productive capacity of the machine. It can be also measured by the actual time cycle time against the ideal cycle time. Performance is affected by speed losses due to idling and minor stoppage or reduced speed.

Quality focuses on the quantity of product that meets the standards versus the total production. Quality may be affected by-products that do not meet established standards, damaged products, or scrap.

The OEE range has been the subject of discussion over the years by different authors. Ref. [28] considers that there is no optimal number of the indicator since different criteria depend on the industry where it is applied. Ref. [29] proposes 85% as the ideal OEE value. Worldwide studies indicate that the average OEE rate in manufacturing plants is 60% [30]. Other authors differ in the indicator range between 30 to 80%, with being the 50% the more realistic value [27].

Several authors define the OEE depending on the application and situation. Table 1 presents a comparison between [29,31] variables definition. Nakajima defined the variables depending on equipment performance on the other hand Braglia established the variable based on the labor performance.

Table 1. Variable definition OEE from Nakajima and OEE adaptation from Braglia.

Variable	[29]	Variable	[31]
Quality (Q)	$\dfrac{input - volume\ of\ quality\ defects}{input}$	Overall workplace effectiveness (OWE)	$\dfrac{value\ time}{operating\ time}$
Performance (P)	$\dfrac{ideal\ cycle\ time * output}{operating\ time}$	Management effectiveness (ME)	$\dfrac{operating\ time}{effective\ available\ time}$
Availability (A)	$\dfrac{loading\ time - downtime}{loading\ time}$	Worker availability (WA)	$\dfrac{effective\ available\ time}{net\ available\ time}$
OEE	$A \times P \times Q$	ROLE	$OWE \times ME \times WA$

Authors like [32] consider that the gains in OEE, while important and ongoing, are insufficient because no machine is isolated. OEE is widely accepted as a quantitative tool, but it is limited to the productivity behavior of individual equipment [33]. Due to the needs presented in the industry, the OEE has led to modification and enlargement of the original OEE tool to fit a broader perspective [6]. OEE is an effective management metric that has adapted and gained importance in different industries becoming one of the most important measures in the modern industry [34,35].

OEE is traditionally used to monitor production performances, but it can also be used as a metric for process improvement activities in other contexts [36]. In the production context, OEE has been used to measure the productivity and performance of a line manufacturing system by detecting and quantifying the line's critical points [8,37,38]. It has also been possible to measure losses of resources associated with human, material, and machine factors [39–41]. In the mining sector, the OEE has been modified to measure the effectiveness of mining equipment such as draglines, shovels, and trucks [10,42]. Garcia-Arca et al. [1] improved efficiency in road transport and Ref. [9] optimized the effectiveness of urban freight transportation by adapting OEE. Concerned about the environment and raising awareness to contribute to a more sustainable world, Refs. [4,43] proposed modifications to the OEE, including measurement of sustainability and performance in the production system.

2.3. Six Big Losses

The OEE indicator is designed to identify losses that reduce the effectiveness, Ref. [29] define six large losses that affect the availability (downtime losses), performance (speed losses), and quality (quality losses) factors:

Losses that affect the availability:

- Equipment failure or breakdown of these losses are caused by major shutdowns producing losses time and quality losses.
- Setup and adjustment of these losses of time are caused by the change of requirements in production at the end of one item to another item.

Losses that affect the performance:

- Idling and minor stoppage losses when production is interrupted by a temporary malfunction or when a machine is idling
- Reduced speed losses refer to the difference between the equipment design speed and the actual operating speed.

Losses that affect the quality

- Reduced yield occurring from the start-up to stabilization
- Quality defects and reworks losses

Losses or disturbances could also be classified depending on how often they occur as chronic or sporadic [27,44]. Chronic disturbances are usually small, hidden, and complicated to identify because they appear as normal, whereas the sporadic are obvious, occur quickly, irregularly, and with dramatic effects.

3. Methodology and Unloading Process Description

For this study, AR methodology was followed to encourage the study in a logistics process. One of the aspects that have motivated the application of this methodology is to contribute with qualitative studies in the logistics sector [45]. Another advantage of using AR is that it allows addressing a gap in theory with an explorative approach [46]. AR is one form of a case of study in which a sequence of activities and an approach is used to solve a problem. These practices involve data gathering, reflection on the action as it is presented through the data, generating evidence from the data, and making claims to knowledge based on conclusions drawn from validated evidence [47].

This paper attempts to solve an organizational issue and validate the application of OEE in a logistic process by applying AR to enrich the field's body of knowledge. We based our study on the AR cycle, and the phases of planning, acting, observing, and reflecting were included. Figure 1 presents the action research model and the phases applied in this study.

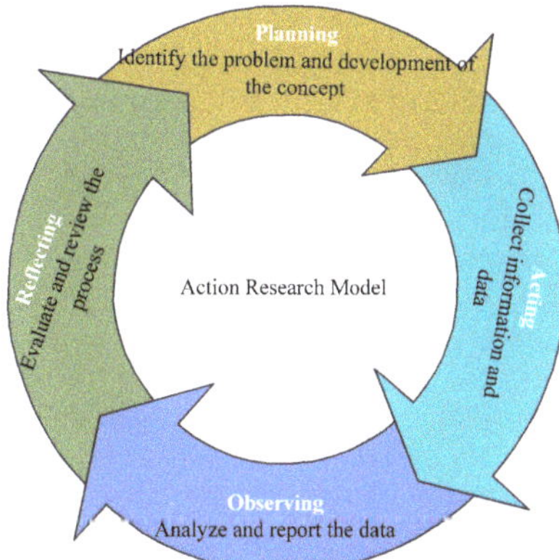

Figure 1. Action Research Model Phases used in this study.

According to the literature, three major categories, design, data collection, and data analysis aspects [48], make up a rigorous and relevant research approach. In the first category, research question, contribution in the practice and the science, motivation of the research, and the case description are reported subsequently. Although all the aspects require accompaniment by the company, this first required the experience of the researchers, since it was necessary to define the research questions, to know and understand the needs of the company and how the theoretical aspects of the indicator were going to be adapted and implemented in the chosen process. In the second category, the description, method, and access for data collection, the role of the team in charge, and the triangulation of experiences were considered during the action process. In team meetings (researchers and practitioners) of reflection and discussion, the roles of each team member were defined and reliable systematic methods for data collection were implemented. Finally, the data were categorized and validated through the proposed indicator, the results were presented and after reflection, meetings with the whole team, the model was updated.

As established in the literature for the development of rigorous research, research questions are established, which we will answer using the AR methodology. This paper attended to answer these research questions:

RQ1. Can the new indicator measure the effectiveness of the lateral unloading truck process?

RQ2. Does the AR methodology provide solutions to the organizational issue?

RQ3. Has this study identified opportunities for improvements in the unloading process?

This methodology helps to create knowledge with collaborative participation, commonly in a specific and practical context to improve conditions and practices in the domain.

This research is carried out in one of the assembly plants of a multinational company in the automotive sector in Spain. The sub-department of material handling that belongs to the supply chain department, in search of improving the efficiency of its processes wants to implement a procedure to record, measure, and study the process of lateral unloading trucks. The project team is comprised of a researcher team, hereinafter researchers, and a management team. The researchers are the authors of this study specialized in the following areas:

- Experts in logistics processes;
- Experts in the OEE indicator applied in other engineering fields.

The management team is the practitioners been part of the company and consists of:

- Head of material handling department;
- Shifts supervisors;
- Shifts operators.

Through a reflection phase, with the professional experience and the theoretical knowledge of the researchers, the variables that influence the unloading process were defined and characterized. To broaden the theoretical knowledge of the performance measurement system, an indicator based on OEE was proposed for the logistics process.

This work was developed in four phases, where the researchers were presented and participated together with the company for the development and implementation of the indicator. Firstly, in the planning phase, the problem presented in the material handling sub-department was identified, a variation in unloading effectiveness was observed according to the work shift based on the number of trucks unloaded per shift. This phase had a duration of one month in which the researchers gained knowledge from the process at the company. During this phase, the researchers met with the management team to understand and define the scope and parameters of the research. Subsequently, to define an indicator to measure and improve the process of unloading material from different suppliers, an effectiveness indicator named OPE will be developed to implement it in the following phases.

In the second phase, the data is obtained from two sources of information: the routes registration system (LST) and the unloading operator. New data collection forms were created to obtain additional information from the unloading operator. The data were collected for five months. For data reliability, the company provided access to the LST system and proposed data collection from field operators. Meetings and training on the new measuring instruments were held with the operators, in this phase, the researchers actively participated.

Once the data is collected, in a third phase the indicator is calculated, and the results were analyzed. The results were analyzed through different perspectives for the support and implementation of improvements. Several meetings were held between the researchers and the management team, in which feedback on the results was given, changes on data acquisition were proposed, and process improvements were suggested. This research involved the collaborative participation of the company and the researchers to reach reflections from different points of view. Through these reflections we gain a wider scope, reducing bias, and by validating the analysis to increased scientific rigor, achieving the triangulation mentioned by some authors [48,49].

In the last phase, there is a further clarification of the initial situation, a meeting for self-reflecting takes place among researchers, and the management team considers defining the improvements in the process. Despite reflection being placed as the last phase of the RA, it was carried out in each of the phases of the development and implementation of the new indicator in the company.

Unloading Process Description

The area chosen for the study is the unloading area with the highest flow of trucks, where the material is received from approximately 150 suppliers, and in 1780 different references are handled. The company works 24 h a day divided into three shifts: morning shift (6:00–14:00), afternoon shift (14:00–22:00), and night shift (22:00–6:00). All shifts have four breaks (one 8-min break and three 9-min breaks) and an 18-min snack break. Depending on the circumstances and the number of work breaks that can be worked on, but by law, it is not possible to work during the snack break. The unloading area has a reception area (temporary storage) and two unloading tracks, allowing two trucks to be parked in the unloading position (Figure 2). The unloading process is performed by a single operator, which could increase the waiting time between trucks.

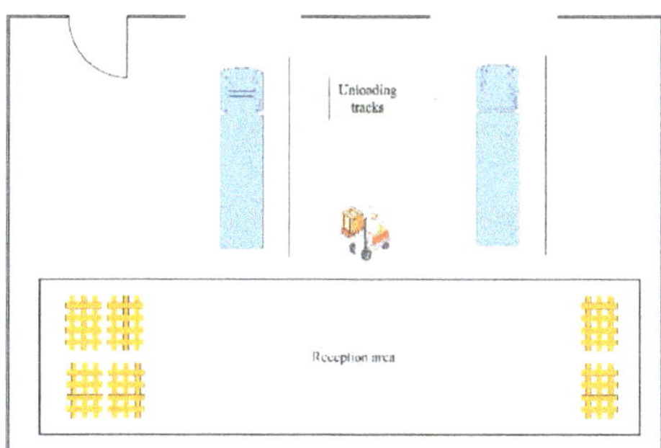

Figure 2. Physical distribution of the unloading area.

The process begins with the arrival of the truck at the external part of the unloading area, the driver gets off with the documentation corresponding to the transported material he is transporting. Then, the driver enters the warehouse, hands over the documents to the unloading operator who checks if the material belongs to that unloading section. If it does, he tells the driver to park the truck in the unloading track, otherwise, he is shown where to go. The operator records the following data in the document: route, time window, time of entry, time of exit, supplier, and registration.

Once the truck is on the unloading track, the driver gets off, puts on helmets and boots to start the process of preparing the truck to unload the material. The driver opens the side tarp of the truck, removes the side safety boards, and waits for the unloading operation. The operator begins the process by verifying that the material inside the truck is in good condition, if so, he proceeds to unload the containers. Otherwise, if the products contained are disordered in the truck, not loaded sideways, or in poor condition (wet, damaged, dumped, etc.), he calls the supervisor to verify the integrity of the container to proceed with the unloading process or asks the truck to leave. The operator uses a long-shovel forklift to unload the material and place it in the reception area, where it is temporarily stored and then placed in the warehouse. The material is stored in containers, the containers of the same type are placed together. It is also considered that all the material transported in the same truck must be placed in the reception area as close as possible.

Once the unloading process is finished, the operator signs the document confirming, the driver proceeds to place the lateral safety boards, close the tarp, and leave the premises, leaving the unloading track for the next truck. Before continuing with the next truck, the operator must perform a verification of the products in the containers. The list of materials detailed in the documentation provided by the driver upon arrival must coincide with the material that was unloaded from the truck. If there is a difference, the operator takes note of it and notifies the supervisor to report the discrepancy. Figure 3 presents a flowchart of the unloading process.

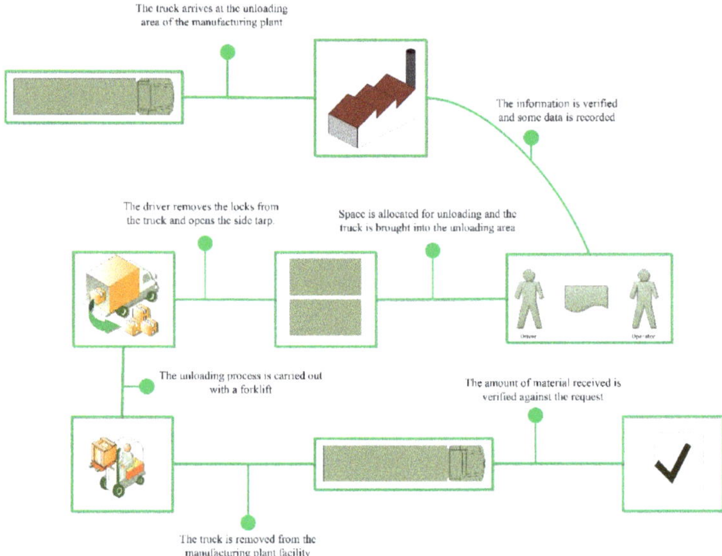

Figure 3. Unloading process flowchart.

The company has a delivery truck that travels within the manufacturing plant. This truck is used to transport the materials that were unloaded in another zone that does not correspond to it. The delivery truck has priority when it arrives at an unloading area and must be attended, once the operator has finished with the in-process truck unloading.

Once the researchers understood the unloading process, a framework was proposed that would meet the objective of this study. Through the framework, it was possible to find activities that do not add value and establish actions for continuous improvement. The OEE is an indicator widely used in manufacturing that looks forward to the effectiveness of the process [1]. The reason to choose this indicator is to adapt it to this research.

4. Developing the Framework

Through this framework, the company is oriented to reach a strategic objective to increase the overall efficiency in the production support activities. Based on the review of the existing OEE approaches, researchers proposed to the management team a new indicator OPE. The aim of this new indicator OPE is to measure the effectiveness of a logistic process. This framework considers that the effectiveness of the process is affected by internal and external factors. Internal factors are those situations that occur within the company and are managed and controlled internally in the organization. In contrast, external factors originated outside the company, depending on the participation or collaboration of external agents to manage and control them. Both factors influence and affect positively or negatively the performance of the company.

The factors that were analyzed for the development of the indicator are availability, performance, quality, and punctuality. The availability measures the real total time the system is operating without time losses due to the lack of resources that prevent the process from running smoothly. The resources that affect the availability are the lack or failure of the forklift, lack of operator, among others.

$$Availability\ (A) = \frac{operating\ time}{total\ time} \times 100\% \tag{1}$$

$$Operating\ time = total\ time - downtime$$

$$Total\ time - workday - planned\ downtime$$

The workday can be calculated for the entire day or the work shift.

The performance measures the ratio of the ideal speed of the unloading and actual unloading speed. The performance is affected by losses that slow down the process.

$$Performance\ (P) = \frac{Ideal\ time}{Real\ time} \times 100\% \tag{2}$$

The ideal time is the historical average of the unloading time multiplied by the number of downloads made in the workday.

The quality rate measures the number of trucks that arrive with the material in satisfactory order so the process of lateral unloading can be carried out without delay.

$$Quality\ (Q) = \frac{Number\ of\ trucks\ as\ requested}{Total\ number\ of\ trucks} \tag{3}$$

Besides, the availability, performance, and quality, a new factor is added, related to the timely arrival of the truck. Punctuality is a ratio that indicates the portion of trucks that arrive in the assigned time window.

$$Punctuality\ (U) = \frac{Number\ of\ trucks\ arrive\ on\ time}{Total\ number\ of\ trucks} \tag{4}$$

The quality and punctuality factors are affected by internal and external situations, whereas the availability and performance factors are affected by situations entirely internal to the company. Table 2 shows a detailed list of losses classified by availability, performance, quality, and punctuality.

Table 2. Breakdown of losses by availability, performance, quality, and punctuality.

Category		Losses
Availability	Internal	Damage to the access door Forklift discharge or breakdown Lack of forklift
Performance	Internal	Occupational accident The material drop during the unloading Documentation check Loading or unloading of the internal truck Temporary storage area full
Quality	Internal	Truck for another uploading point
	External	Disordered material in the truck Material not loaded sideways The material is in poor condition (wet, damaged, dumped, etc.)
Punctuality	External	The truck does not arrive at the assigned time window

This framework goes in pursuit to integrate important aspects of the process in a single measure of effectiveness (Figure 4). The perspectives to be measured with this framework are the availability of resources, the performance of resources, the quality of the process, and the added factor of punctuality. This last factor seeks to include a broader picture that affects the development of the process.

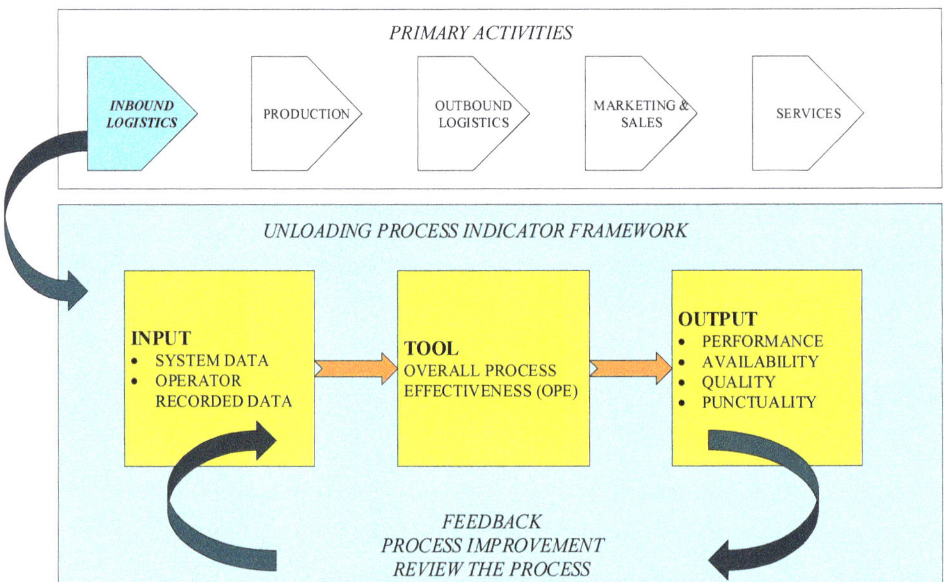

Figure 4. Unloading process indicator framework.

This new metric measures the effectiveness of the unloading process considering the four factors described above (see Equation (5)). It will be possible to identify and measure the losses that help to implement corrective actions to improve the effectiveness of the process.

$$OPE = A \times P \times Q \times U \tag{5}$$

The application of this integrated indicator can reveal through global OPE monitoring which of the four factors (availability, performance, quality, and punctuality) could be improved to increase the productivity of the process.

5. Results of the Framework Implementation

The results presented below are the product of several meetings between the researchers and the management team. The development of the research results is based on the acting and observing phases where the information is collected and analyzed. The proposed framework for measuring the performance of the unloading process through the OPE was presented to the management team. Meetings were held with the operators to indicate the process for template records to collect the data.

The information was worked through a spreadsheet that collects the data for later processing. The data is obtained by the operator and the LST system that provides information about the routes.

The operator is responsible to record the following information:

1. Route: route of a truck that comes from a supplier and periodically supplies the same material.
2. Entry time: the time the truck enters the unloading tracks.

3. Exit time: the time the truck leaves the unloading tracks.
4. Incidents: the record of any incident that involves a loss of time during the download process.

The LST system will provide the following information:

1. Time window: the planned time interval of one hour during which the truck must arrive.
2. Unloading points: different areas of the plant where the truck must be unloaded, depending on the material being transported.
3. Supplier: the company that supplies the material.

Once the data is recorded in the spreadsheet, the following information is calculated:

1. Gross time: total time that the unloading track has been occupied by a truck.
2. Net time: time that the unloading track has been occupied by a truck without considering the break time and the snack break.
3. Unloading time: actual time spent unloading the truck.
4. Punctuality: it is considered punctual if the truck enters the unloading track in the planned time window.

In addition to the data described above, the following information must be recorded:

1. Shift: For night shift TN, TA, and TB for morning and afternoon shifts, respectively, which are rotated.
2. Day: the day the download is made.
3. Week: number of the week according to the calendar.

Table 3 presents a sample of the information collected from the process. It shows an abstract of the three shifts, the day of the week, week number, route code, time window, unloading points, entry time, exit time, gross time, net time, unloading time, and punctuality. Some of the information was modified for company confidentiality.

Table 3. The collected information from the unloading process to calculate the indicator.

N°	86	87	88	107	108	109	117	118	119
Shift	TN	TN	TN	TA	TA	TA	TB	TB	TB
Date	2/13/2020	2/13/2020	2/13/2020	2/13/2020	2/13/2020	2/13/2020	2/13/2020	2/13/2020	2/13/2020
Week	7	7	7	7	7	7	7	7	7
Route	MB23X	PTC3X	PT63Y	MB14H	SK14M	PT24L	E624R	Delivery	E334N
Time window 1	21:00	21:00	22:00	10:00	10:00	10:00	15:00	-	14:00
Time window 2	22:00	22:00	23:00	11:00	11:00	11:00	16:00	-	15:00
Unloading points	D51E	D51E	D51E	RHE	D51E	D51E	D51E		D51E
Unloading points				D51E					
Entry time	22:50	23:15	23:40	10:40	11:05	11:30	15:25	16:00	16:15
Exit time	23:10	23:30	23:55	11:10	11:30	12:25	16:00	16:15	16:35
Gross time	0:20:00	0:15:00	0:15:00	0:30:00	0:25:00	0:55:00	0:35:00	0:15:00	0:20:00
Net time	0:20:00	0:15:00	0:15:00	0:30:00	0:25:00	0:55:00	0:35:00	0:15:00	0:20:00
Unloading time	0:20.00	0:15:00	0:15:00	0:15:00	0:20:00	0:55:00	0:35:00	0:15:00	0:20:00
Supplier	ABC	DEF	GHI	JKL	MNO	PQR	STU		WVX
Punctuality	NOK	NOK	NOK	OK	NOK	NOK	OK		NOK

Time spent in the data collection phase was five months. During this period, approximately 2135 trucks unloading were recorded. In some periods and shifts, no data were reflected because the company was closed due to the health crisis caused by COVID-19.

The OPE indicator was calculated daily by shift and each factor was studied separately. Each factor was calculated with the equations indicated above and considering the losses established for each one of them. The total time of the shift for the availability calculation was 7 h and 42 min, and only 18 min break for the snack was considered. The ideal time for the performance calculation was considered as the average unloading time multiplied by the number of trucks served during the shift. Figure 5 shows the average rate of the four factors per shift and week. The shifts are divided into morning (MO), afternoon (AF), and night (TN).

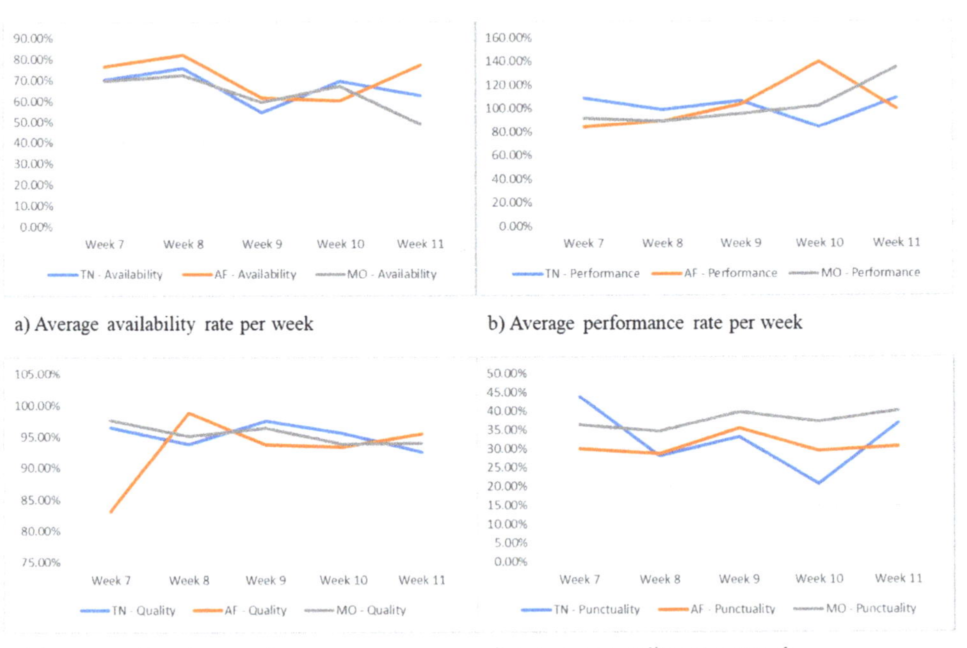

a) Average availability rate per week

b) Average performance rate per week

c) Average quality rate per week

d) Average punctuality rate per week

Figure 5. The average rate of the four factors per shift and week (**a**) availability; (**b**) performance; (**c**) quality; (**d**) punctuality.

The factors with the lowest averages that affect process performances are punctuality and availability. These two factors are the most influential in decreasing the final OPE result regardless of the shift. Availability is around 50 to 70%, being higher in the afternoon shift most of the time. The results indicated that the greatest loss of time is due to the lack of trucks to attend. The punctuality factor was added as an important part of the indicator, values between 25 and 35% were presented. Analyzing this result in-depth, the company detected that more than 50% of the trucks arrive outside their assigned time window. Most of the time, the delay is due to traffic jams, carrier outsourcing, and supplier material availability. The performance and quality factors are very close to the desired value. The performance on certain occasions provided values above 100%. After some meetings with the management team, it is notable that the average ideal time may be the cause of performance results, which will be evaluated in the future.

Figure 6a shows the OEE calculation (availability * performance * quality) and Figure 6b shows the OPE (availability * performance * quality * punctuality) during June 2020. In the OEE the morning and afternoon shifts maintain a similar behavior, the average is around 58%, whereas the night shift maintains a lower performance. The OPE presents great variations in its results attributed mainly to the punctuality factor. Results demonstrate that with the punctuality factor from morning and afternoon shifts, OPE varies on average around 23%. During this month, the night shift was irregular and was only worked from Monday to Friday, due to the restrictions of the health crisis for both cases. Figure 6c presents a comparison between the OEE and OPE of the different work shifts. It can be highlighted how the punctuality factor decreases the performance of the unloading process.

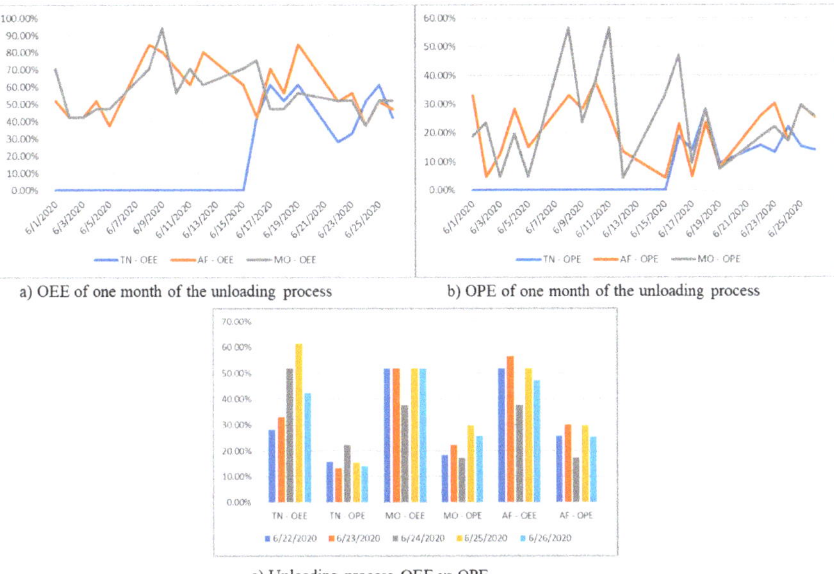

a) OEE of one month of the unloading process

b) OPE of one month of the unloading process

c) Unloading process OEE vs OPE

Figure 6. Graphic results (**a**) OEE of one month; (**b**) OPE of one month; (**c**) One week of OEE vs. OPE.

All these results were analyzed in multiple meetings with the management team, originating initiatives of improvements in the unloading process. The successful implementation of the OPE was due to a critical collaborative inquiry, where there was learning through action and reflection.

The proposed framework enables collaborative participation within the AR methodology used in this study, motivating a cultural change. Moreover, it was favorable for the company because it promotes participation from different organizational levels that conformed the management team in all the AR phases. The continuous improvement cycle that our framework focuses on is the result of the cohesion and synergy of the teamwork.

6. Discussion

This framework contributed to data digitization to encourage faster and more transparent results. Industry 4.0 and its pillars are tools that help the evolution of management systems [24]. This framework can lay the bases for other organizations to digitally transition their processes as part of a continuous improvement cycle [50]. The continuous improvement cycle is fed by KPIs information to eliminate activities that do not add value to a process. The constant evolution of production processes forces the organization to look for data management options to make data-driven decisions rather than intuition.

Through the framework proposed it can be possible to record and measure the process to obtain data for continuous improvement. Besides, the OPE allows the losses identification to take action and generate continuous improvement for lean performance.

The success of this collaborative research provides a practical contribution to the company that will improve the unloading process with the OPE implementation. Besides, it generates theoretical knowledge that can be applied to other departments or companies.

Once the AR cycle ended, we can provide answers to the three initial research questions.

1. OPE indicator results do measure the effectiveness of the lateral unloading process. The results indicated that the categorized variables with the project team provided information on the four factors of the OPE and their influence on the unloading process.
2. The AR methodology provided solutions related to the unloading process procedure and documentation. It was possible to record and organize the information

to calculate the performance indicator. Moreover, through the synergy between the researchers and the management team, it was possible to reduce bias in data collection and analysis of the results. Bias is reduced by the participation of external and internal personnel, improving the objectivity of the information analyzed. Through the collaborative participation of the researchers and the practitioners during the study, the new OEE adaptation could be measured and validated.

3. The study identified several opportunities for improvement, which are detailed in the theoretical and practical contributions.

6.1. Theoretical Contributions

Some authors consider the original OEE insufficient [28], as it was only used to measure the effectiveness of particular equipment. Many manufacturers have customized OEE to fit their particular industrial requirements [8]. This study makes two strong contributions. The first contribution is the adaptation of the OEE indicator to other domains, in this case to a logistic process. The modification contributes to the identification of the three OEE parameters (availability, performance, and quality) and punctuality as an external factor that is considered important in the process. It was possible to measure the actual operating time during the unloading process with the availability factor. The performance allowed us to calculate the ratio of the standard time vs. the real-time of the unloading process. The quality factor was measured based on the number of trucks that fulfill cargo requirements for unloading conditions among the total number of trucks unloaded in the shift. Finally, with the punctuality factor, we can identify the number of trucks that arrive within their time window. The punctuality factor measures the goods on-time delivery from the suppliers, one of the most important performance variables in a logistic process. [47].

The second contribution is to increase scientific knowledge in the logistics sector through the AR methodology [15]. Closing the gap between theory and practice [16] by the synergy between the company and the researchers in developing AR to solve an organizational issue. Through this AR, self-reflection of the team members was achieved to obtain improvements in the unloading process. The collaboration among members for problem-posing and answering questions for decision-making was crucial. The use of the AR methodology during the research contributed to a practical transformation within the company and the advancement of theoretical knowledge. Data gathering and analysis were accomplished by the active participation of the researchers during approximately seven months and the involvement of the management team in the implementation of the framework.

The collaboration between researchers and practitioners was valuable. It was possible to generate knowledge for the creation of the new indicator by the researchers, which was then validated by the management team of the company. This new metric allows to measure the unloading process and generate continuous improvement plans tailored to organizational needs, increasing their market competitiveness.

The ability to reach a high level of practical relevance of business research is one of the major advantages of this method [14]. The effectiveness of the proposed method has been substantiated here. To increase the scientific rigor of these results, different points of view with the triangulation were achieved, reducing the bias during the study, and through the collaboration with the management team, we had access to reliable information.

6.2. Practical Contributions

To the best of our knowledge, this is the first time that the OEE has been adapted to measure the effectiveness of an unloading process. We believe that this framework can be replicated in other companies that perform loading or unloading of goods. This new metric for measuring process effectiveness provided a breakdown of losses that occur during truck unloading. It was possible to identify the factors that affect the process, many of which were attributed to the internal situation, but others were noted to be external to the company, these external factors unsetting the planning and unloading time of a truck.

Through the AR methodology applied in this research, it was possible firstly, to strengthen the relationship between the practitioners and researchers providing scientific knowledge to the field and practical knowledge for an organizational issue. This practical knowledge allowed for the identification of a set of best practices to increase market competitiveness. Secondly, the OPE became one of the performance measurement indicators of the organization, and the reflection meetings that were held to validate it were adopted by the organization.

At the organizational level were identified several improvements starting with the developed framework that can guide the application in other processes. Moreover, data collection that was held with several templates and registered in specialized spreadsheets improves the way organizations analyze process information. The framework allows the company to build up a transition to a data-driven management approach.

With the OPE results, the company identified two improvements that help resource efficiency. First, it was noticed that more than 50% of trucks do not arrive in the planned time window, affecting the unloading schedule by shifts. Improvements are being made in the planning of truck arrivals, and meetings have also been held with suppliers to take joint action. Establishing a collaborative process with the supplier will strengthen the process flow between them [48]. The second improvement is the time redistribution of the operator to the unloading process. It was identified that the operator has a lot of leisure time with the availability parameter, so it was decided that the operator would attend two unloading areas.

Also, as a consequence of the self-reflections, a reevaluation of the calculation of the theoretical or ideal unloading time was being considered. This seeks to improve the presented results of the performance parameter above 100% on some occasions. A more accurate result could be obtained by calculating the average unloading time per route.

After a critical discussion with the management team, we defined the ideal OPE range for the company in the unloading process as between 40 and 50%. We want to highlight that this range considers that OPE results are influenced by external and internal factors. Some factors are not directly controlled by the company and cannot be compared with the OEE ranges obtained in production. Few authors proposed ideal ranges of OEE, establishing between 30 to 80% [29] and worldwide studies in manufacturing plants estimate an average of 60% [32]. The company continues evaluating improvement options with the OPE results. A permanent committee was established after the first OPE results to reflect on the necessary improvements to the unloading process, to achieve a higher percentage of the indicator.

This proposed framework is the first approximation to be developed for other logistics processes or equipment.

7. Conclusions

The OEE has been used as an indicator to control and monitor productivity and improve performance [49]. In this paper, a new indicator OPE has been introduced as a measure of a logistic process. Besides, the availability, performance, and quality factors considered in the original OEE, the punctuality factor is included. Punctuality was considered a determining factor to avoid bottlenecks in the process. The indicator helps to identify losses that occur during the process, intending to implement improvements that help the company's competitiveness.

This study was done with a collaboration between researchers and practitioners. AR methodology used considers the spiral cycle of plan, act, observe, reflect, and then reviews the process again.

First, the problem was established as the parameters and scope of the project were identified. Data from different sources were collected, and the indicator was calculated and analyzed.

Finally, theoretical and practical contributions could be evidenced in the process. OEE was implemented in a different domain than production. Losses identification that occurs during the lateral unloading truck process was imperative for the adapted OEE. Improvements in the process were identified to increase work efficiencies, such as action plans to regulate the timing of the truck's arrival and distribution of the operator availability in the unloading areas. The AR methodology is flexible, and it can be applied in various industry domains since it involves the participation of multidisciplinary teams and promotes organizational changes. With this collaboration in action between researchers and practitioners, it was possible to generate and validate theoretical knowledge applied in the field. The AR provides a practical transformation and contributed to the advancement of scientific knowledge. It also encourages active and collaborative participation where a systematic learning and reflective process is developed throughout the action research cycle.

The study is not devoid of limitations. Some of the limitations that emerged in the application of the OPE were the global health crisis, which affected the company's operations and data collection. Other limitations were the study of only one unloading area. The company has several unloading areas, but for this research, the largest one was taken, which has two unloading tracks. Finally, for further research, the authors consider this indicator can be applied to the loading process with slight modifications. Moreover, it could be applied to other types of organization and cargo.

Author Contributions: Conceptualization, M.P.L. and J.R.; Methodology, M.E.H.K. and L.d.C.N.C.; Validation, L.d.C.N.C.; Formal analysis, L.d.C.N.C.; Investigation, L.d.C.N.C.; Data curation, M.E.H.K. and L.d.C.N.C.; Writing—original draft preparation, L.d.C.N.C.; Writing—review and editing, M.P.L., M.E.H.K., P.M. and L.d.C.N.C.; Supervision, M.P.L. and J.R.; Project administration, J.R. and J.C.S.C. All authors have read and agreed to the published version of the manuscript.

Funding: This research received no external funding.

Institutional Review Board Statement: Not applicable.

Informed Consent Statement: Not applicable.

Data Availability Statement: The data presented in this study are available on request from the corresponding author. The data are not publicly available due to the privacy policy of the organization.

Acknowledgments: The authors would like to acknowledge the scholarship granted by the government of Panama under the IFARHU—SENACYT program for a Ph.D. candidate from 2018 to 2021.

Conflicts of Interest: The authors declare no conflict of interest.

References

1. García-Arca, J.; Prado-Prado, J.C.; Fernández-González, A.J. Integrating KPIs for Improving Efficiency in Road Transport. *Int. J. Phys. Distrib. Logist. Manag.* **2018**, *48*, 931–951. [CrossRef]
2. Burdzik, R.; Ciesla, M.; Sładkowski, A. Cargo Loading and Unloading Efficiency Analysis in Multimodal Transport. *Promet Traffic Transp.* **2014**, *26*, 323–331. [CrossRef]
3. Simons, D.; Mason, R.; Gardner, B. Overall Vehicle Effectiveness. *Int. J. Logist. Res. Appl.* **2004**, *7*, 119–135. [CrossRef]
4. Domingo, R.; Aguado, S. Overall Environmental Equipment Effectiveness as a Metric of a Lean and Green Manufacturing System. *Sustainability* **2015**, *7*, 9031–9047. [CrossRef]
5. Gibbons, P.M.; Burgess, S.C. Introducing OEE as a Measure of Lean Six Sigma Capability. *Int. J. Lean Six Sigma* **2010**, *1*, 134–156. [CrossRef]
6. Muchiri, P.; Pintelon, L. Performance Measurement Using Overall Equipment Effectiveness (OEE): Literature Review and Practical Application Discussion. *Int. J. Prod. Res.* **2008**, *46*, 3517–3535. [CrossRef]
7. del Carmen Ng Corrales, L.; Lambán, M.P.; Hernandez Korner, M.E.; Royo, J. Overall Equipment Effectiveness: Systematic Literature Review and Overview of Different Approaches. *Appl. Sci.* **2020**, *10*, 6469. [CrossRef]
8. Nachiappan, R.M.; Anantharaman, N. Evaluation of Overall Line Effectiveness (OLE) in a Continuous Product Line Manufacturing System. *J. Manuf. Technol. Manag.* **2006**, *17*, 987–1008. [CrossRef]
9. Muñoz-Villamizar, A.; Santos, J.; Montoya-Torres, J.; Jaca, C. Using OEE to Evaluate the Effectiveness of Urban Freight Transportation Systems: A Case Study. *Int. J. Prod. Econ.* **2018**, *197*, 232–242. [CrossRef]
10. Elevli, S.; Elevli, B. Performance Measurement of Mining Equipments by Utilizing OEE. *Acta Montan. Slovaca Ročník* **2010**, *15*, 95–101.
11. Braglia, M.; Gabbrielli, R.; Marrazzini, L. Overall Task Effectiveness: A New Lean Performance Indicator in Engineer-to-Order Environment. *Int. J. Product. Perform. Manag.* **2019**, *68*, 407–422. [CrossRef]

12. Sharma, R.; Singh, J.; Rastogi, V. The Impact of Total Productive Maintenance on Key Performance Indicators (PQCDSM): A Case Study of Automobile Manufacturing Sector. *Int. J. Product. Qual. Manag.* **2018**, *24*, 267. [CrossRef]
13. Supriyanto, H.; Mokh, S. Performance Evaluation Using Lean Six Sigma and Overall Equipment Effectiveness: An Analyzing Tool. *Int. J. Mech. Eng. Technol.* **2018**, *9*, 487–495.
14. Reason, P.; Bradbury, H. *Handbook of Action Research: Concise Paperback Edition*; SAGE Publications Inc.: London, UK, 2006.
15. Coughlan, P.; Coghlan, D. Action Research for Operations Management. *Int. J. Oper. Prod. Manag.* **2002**, *22*, 220–240. [CrossRef]
16. Gutierrez, D.M.; Scavarda, L.F.; Fiorencio, L.; Martins, R.A. Evolution of the Performance Measurement System in the Logistics Department of a Broadcasting Company: An Action Research. *Int. J. Prod. Econ.* **2015**, *160*, 1–12. [CrossRef]
17. Fawcett, S.E.; Bixby Cooper, M. Logistics Performance Measurement and Customer Success. *Ind. Mark. Manag.* **1998**, *27*, 341–357. [CrossRef]
18. Neely, A.; Gregory, M.; Platts, K. Performance Measurement System Design: A Literature Review and Research Agenda. *Int. J. Oper. Prod. Manag.* **2005**, *25*, 1228–1263. [CrossRef]
19. Caplice, C.; Sheffi, Y. A Review and Evaluation of Logistics Performance Measurement Systems. *Int. J. Logist. Manag.* **1995**, *6*, 61–74. [CrossRef]
20. Caplice, C.; Sheffi, Y. A Review and Evaluation of Logistics Metrics. *Int. J. Logist. Manag.* **1994**, *5*, 11–28. [CrossRef]
21. Lohman, C.; Fortuin, L.; Wouters, M. Designing a Performance Measurement System: A Case Study. *Eur. J. Oper. Res.* **2004**, *156*, 267–286. [CrossRef]
22. Nudurupati, S.S.; Bititci, U.S.; Kumar, V.; Chan, F.T.S. State of the Art Literature Review on Performance Measurement. *Comput. Ind. Eng.* **2011**, *60*, 279–290. [CrossRef]
23. Badawy, M.; El-Aziz, A.A.A.; Idress, A.M.; Hefny, H.; Hossam, S. A Survey on Exploring Key Performance Indicators. *Future Comput. Inform. J.* **2016**, *1*, 47–52. [CrossRef]
24. Martinez, F. Process Excellence the Key for Digitalisation. *Bus. Process Manag. J.* **2019**, *25*, 1716–1733. [CrossRef]
25. Peças, P.; Encarnação, J.; Gambôa, M.; Sampayo, M.; Jorge, D. PDCA 4.0: A New Conceptual Approach for Continuous Improvement in the Industry 4.0 Paradigm. *Appl. Sci.* **2021**, *11*, 7671. [CrossRef]
26. Chow, G.; Heaver, T.D.; Henriksson, L.E. Logistics Performance: Definition and Measurement. *Int. J. Phys. Distrib. Logist. Manag.* **1994**, *24*, 17–28. [CrossRef]
27. Jonsson, P.; Lesshammar, M. Evaluation and Improvement of Manufacturing Performance Measurement Systems—the Role of OEE. *Int. J. Oper. Prod. Manag.* **1999**, *19*, 55–78. [CrossRef]
28. Dal, B.; Tugwell, P.; Greatbanks, R. Overall Equipment Effectiveness as a Measure of Operational Improvement A Practical Analysis. *Int. J. Oper. Prod. Manag.* **2000**, *20*, 1488–1502. [CrossRef]
29. Nakajima, S. *Introduction to TMP*; Productivity Press: Cambridge, MA, USA, 1988.
30. Iannone, R.; Nenni, M.E. Managing OEE to Optimize Factory Performance. In *Operations Management*; IntechOpen: Rijeka, Croatia, 2013; pp. 31–50.
31. Braglia, M.; Castellano, D.; Frosolini, M.; Gallo, M.; Marrazzini, L. Revised Overall Labour Effectiveness. *Int. J. Product. Perform. Manag.* **2020**, *70*, 1317–1335. [CrossRef]
32. Scott, D.; Pisa, R. Can Overall Factory Effectiveness Prolong Moore's Law? *Solid State Technol.* **1998**, *41*, 10.
33. Huang, S.H.; Dismukes, J.P.; Shi, J.; Su, Q.; Wang, G.; Razzak, M.A.; Robinson, D.E. Manufacturing System Modeling for Productivity Improvement. *J. Manuf. Syst.* **2002**, *21*, 249–259. [CrossRef]
34. Gupta, P.; Vardhan, S.; al Haque, M.S. Study of Success Factors of TPM Implementation in Indian Industry towards Operational Excellence: An Overview. In Proceedings of the IEOM 2015—5th International Conference on Industrial Engineering and Operations Management, Proceeding, Dubai, United Arab Emirates, 23 April 2015; Institute of Electrical and Electronics Engineers Inc.: New York, NY, USA, 2015.
35. Tsarouhas, P.H. Overall Equipment Effectiveness (OEE) Evaluation for an Automated Ice Cream Production Line: A Case Study. *Int. J. Product. Perform. Manag.* **2019**, *69*, 1009–1032. [CrossRef]
36. Andersson, C.; Bellgran, M. On the Complexity of Using Performance Measures: Enhancing Sustained Production Improvement Capability by Combining OEE and Productivity. *J. Manuf. Syst.* **2015**, *35*, 144–154. [CrossRef]
37. Braglia, M.; Frosolini, M.; Zammori, F. Overall Equipment Effectiveness of a Manufacturing Line (OEEML). *J. Manuf. Technol. Manag.* **2008**, *20*, 8–29. [CrossRef]
38. Raja, P.N.; Kannan, S.M.; Jeyabalan, V. Overall Line Effectiveness—A Performance Evaluation Index of a Manufacturing System. *Int. J. Product. Qual. Manag.* **2010**, *5*, 38. [CrossRef]
39. Eswaramurthi, K.G.; Mohanram, P.V. Improvement of Manufacturing Performance Measurement System and Evaluation of Overall Resource Effectiveness. *Am. J. Appl. Sci.* **2013**, *10*, 131–138. [CrossRef]
40. Sheu, D.D. Overall Input Efficiency and Total Equipment Efficiency. *IEEE Trans. Semicond. Manuf.* **2006**, *19*, 496–501. [CrossRef]
41. Puvanasvaran, A.P.; Yoong, S.S.; Tay, C.C. Effect of Hidden Wastes in Overall Equipment Effectiveness Calculation. *ARPN J. Eng. Appl. Sci.* **2017**, *12*, 6443–6448.
42. Mohammadi, M.; Rai, P.; Gupta, S. Performance Evaluation of Bucket Based Excavating, Loading and Transport (BELT) Equipment—An OEE Approach. *Arch. Min. Sci.* **2017**, *62*, 105–120. [CrossRef]
43. Durán, O.; Capaldo, A.; Duran Acevedo, P. Sustainable Overall Throughputability Effectiveness (S.O.T.E.) as a Metric for Production Systems. *Sustainability* **2018**, *10*, 362. [CrossRef]

44. Ljungberg, Ö. Measurement of Overall Equipment Effectiveness as a Basis for TPM Activities. *Int. J. Oper. Prod. Manag.* **1998**, *18*, 495–507. [CrossRef]
45. Näslund, D. Logistics Needs Qualitative Research—Especially Action Research. *Int. J. Phys. Distrib. Logist. Manag.* **2002**, *32*, 321–338. [CrossRef]
46. Coughlan, P.; Draaijer, D.; Godsell, J.; Boer, H. Operations and Supply Chain Management: The Role of Academics and Practitioners in the Development of Research and Practice. *Int. J. Oper. Prod. Manag.* **2016**, *36*, 1673–1695. [CrossRef]
47. McNiff, J.; Whitehead, J. *Action Research: Principles and Practice*, 2nd ed.; Routledge Falmer: New York, NY, USA, 2002.
48. Näslund, D.; Kale, R.; Paulraj, A. Action Research In Supply Chain Management-A Framework For Relevant And Rigorous Research. *J. Bus. Logist.* **2010**, *31*, 331–355. [CrossRef]
49. Farooq, S.; O'Brien, C. An Action Research Methodology for Manufacturing Technology Selection: A Supply Chain Perspective. *Prod. Plan. Control.* **2015**, *26*, 467–488. [CrossRef]
50. Li, F. Leading Digital Transformation: Three Emerging Approaches for Managing the Transition. *Int. J. Oper. Prod. Manag.* **2020**, *40*, 809–817. [CrossRef]

machines

Analysis of the Implementation of the Single Minute Exchange of Die Methodology in an Agroindustry through Action Research

Murilo Augusto Silva Ribeiro [1], Ana Carolina Oliveira Santos [1,*], Gabriela da Fonseca de Amorim [2], Carlos Henrique de Oliveira [1], Rodrigo Aparecido da Silva Braga [3] and Roberto Silva Netto [4]

[1] Institute of Integrated Engineering, Itabira Campus, Federal University of Itajuba,
Itabira 35903-087, MG, Brazil; muriloasr95@gmail.com (M.A.S.R.); carlos.henrique@unifei.edu.br (C.H.d.O.)
[2] Institute of Pure and Applied Sciences, Itabira Campus, Federal University of Itajuba,
Itabira 35903-087, MG, Brazil; gabi@unifei.edu.br
[3] Institute of Science and Technology, Itabira Campus, Federal University of Itajuba,
Itabira 35903-087, MG, Brazil; rodrigobraga@unifei.edu.br
[4] Smart Campus Facens, Facens University, Sorocaba 18085-784, SP, Brazil; roberto.netto@facens.br
* Correspondence: anasantos@unifei.edu.br

Abstract: This work aims to implement and analyze the effect of the Single Minute Exchange of Die (SMED) implementation in the bean packaging operation in a company in east Minas Gerais, Brazil. Design/Methodology/Approach: The research methodology used was action research. Two cycles of action research were conducted; the first to carry out phase one of SMED, and the second to execute phases two and three. Originality/Research gap: There are few studies on the application of Lean Manufacturing tools in agroindustry. Some works present case studies, mainly in the food supply chain aiming to fill this gap. Regarding SMED applied in agribusiness, no work was found. Key statistical results: The implementation of this methodology allowed the reduction of setup time by around 58%, the distance travelled by operators in the process by approximately 50%, in addition to gains in a production capacity of 14%. Practical Implications: It is concluded that the application of the methodology caused an increase in the company's productivity, as it was possible to obtain gains in productive capacity without changing the amount of hours worked or the number of employees involved in the production process. Limitations of the investigation: This methodology was applied only once and the challenges encountered were not documented.

Keywords: agroindustry; lean manufacturing; single minute exchange of die; action research; reduce setup time; productivity gain

Citation: Ribeiro, M.A.S.; Santos, A.C.O.; de Amorim, G.d.F.; de Oliveira, C.H.; da Silva Braga, R.A.; Netto, R.S. Analysis of the Implementation of the Single Minute Exchange of Die Methodology in an Agroindustry through Action Research. *Machines* **2022**, *10*, 287. https://doi.org/10.3390/machines10050287

Academic Editors: Luís Pinto Ferreira, Paulo Ávila, João Bastos, Francisco J. G. Silva, José Carlos Sá and Marlene Brito

Received: 10 March 2022
Accepted: 13 April 2022
Published: 20 April 2022

Publisher's Note: MDPI stays neutral with regard to jurisdictional claims in published maps and institutional affiliations.

1. Introduction

The global market for dried beans (beans, peas, chickpeas, etc.) has been increasing, due to the awareness of their health benefits. In addition, as they are foods rich in micro nutrients, such as potassium, magnesium, folate, iron and zinc, they are considered important sources of protein in vegetarian diets [1].

Regarding the largest bean producers in the world, a survey carried out by The Food and Agriculture Organization (FAO) [2] between 2018 and 2020 indicates that Brazil, with 13% of world production, corresponds to the second-largest producer of this type of grain, behind only India [2], as shown in Table 1.

The high national potential for bean production, in addition to the recent rise in the domestic market, highlights the need to increase the productivity of industries related to this market.

Lean Manufacturing (LM) or Toyota Production System (TPS), according to [3], consists of a production system that aims at productivity gains and waste reduction. In the 90s, the LM principles were analyzed in more detail in the book "Lean Thinking" and, since then, its essence has been transferred from the efficiency of production to a certain type of organizational intervention and management focused on best practices and process improvement methodologies. In this way, efforts began to focus on increasing added value within the entire flow (from suppliers to end customers) and reducing waste in their processes [4].

Table 1. World bean production in tons from 2018 to 2020 [2].

Countries	2018	2019	2020	Average	Average %
India	6,220,000	5,310,000	5,460,000	5,663,333	25.46%
Brazil	2,916,365	2,908,075	3,053,012	2,953,243	13.28%
Myanmar	2,721,079	3,030,000	3,035,290	2,934,697	13.19%
China	1,324,407	1,322,508	1,281,586	1,309,500	5.89%
Tanzania	1,096,930	1,197,489	1,267,684	1,187,368	5.34%
U.S.A.	1,108,120	932,220	1,495,180	1,178,507	5.30%
Mexico	1,196,156	879,404	1,056,071	1,043,877	4.69%
Others	3,872,843	3,928,618	4,003,239	3,934,900	17.69%
Total	**22,706,612**	**21,461,027**	**22,669,231**	**22,278,957**	**100**

The number of LM implementations has grown and these have been found beyond the automotive sector (such as healthcare, construction, food processing) and various manufacturing processes (such as Product Development (PD), Supply Chain Management (SCM), accounting) [5]. Although agroindustry operates in a different context from small and medium-sized companies in which lean thinking is already applied, since 2010 there have been initiatives in countries such as Sweden, New Zealand and Japan to implement lean thinking in companies of this type of sector [6–8].

For LM to be fully introduced, the implementation of several tools is necessary, including the Single Minute Exchange of Die (SMED). SMED Setup is a set of activities that prepares a system for manufacturing a product, while setup time is the preparation period between the end of the last product made and the first product manufactured in the next process. Single-minute exchange of die (SMED) involves an installation performed in single-digit minutes (i.e., 9 min or less) [9]. References [9,10] states that the implementation of this tool allows for a series of improvements in the production system, such as productivity gains, reduction of stocks, gains in the flexibility of the production system and reduced lead time.

Shorter life cycles and highly customized products restrict the reach of the LM principles since they need changeability in production lines and laborious adjustments, buffer stocks and cycle times. However, the digital transformation and affordable hardware and software solutions promoted by the Industry 4.0 (I4.0) are allowing the boosting of LM tools using technologies such as RFID, cloud computing and sensors/actuators [11,12]. The systematic review in [13] contributes to the understanding of the junction between LM, I4.0 and manufacturing sustainability. The relationship between LP and I4.0 is explored by [14]. The proposed framework summarizes I4.0 technologies and LM practices in both directions. It helps understanding the requirements and effects of the application of I4.0/LM in a productive process. This allows the extension of concepts and practices to the agroindustry.

Therefore, considering LM and I4.0, there are two categories of work: (i) most of then are interested in a conceptual and theoretical discussion [13]; and (ii) those that focus on applications and case studies [14,15]. This work falls into the second category.

Additionally, the application of LM in the agroindustry has a literature gap. Some works such as [16–18] present case studies, aiming to fill this gap. Some research can be found in the food supply chain such as [19–22]. However, as discussed in [23], there is a lack of studies in this area, mainly in LM/SMED. The review presented in [9] points to action research methodology in various companies, improving sustainability and revisiting Shingo stages by monitoring and employee training [24].

This study addresses the implementation of the SMED tool in a company in eastern Minas Gerais, Brazil, that works with the processing, packaging, and distribution of beans. The packaging operation is the bottleneck in the process. It is what limits the company's production capacity. Therefore, this work aims to answer the following research problem: What is the effect of implementing SMED on the bean packaging operation in a company in east Minas Gerais, Brazil? Moreover, this work also try to answer: "can the LM/SMED tool be applied in a bean industry?" and, in a broader context, "can the use of LM empower agroindustry to Agro4.0?". A few studies have since been carried out in this context, our manuscript tries to fill these gaps by conducting an SMED in agroindustry through action research.

Thus, in terms of theoretical contribution, the relevance of this study is because although the international literature presents some work on the benefits of implementing SMED, there are few studies on its application in this type of company that consider the setup of a set of grain processing and packaging machines [9]. According to [25,26], sectors related to agriculture lack studies on the application of Lean Manufacturing and have the potential for developing the philosophy. In addition, this work also contributes with an empirical study on SMED in grain processing and packaging.

From the point of view of practical and managerial implications, all organizations that intend to implement SMED in a production line can benefit from this work, as the various aspects of SMED in operations management will be identified. Thus, knowledge of this information and encouragement to implement SMED in companies can help identify waste and consequently improve flexibility and add value to their production processes. The study of LM in this context also contributes to the dissemination of knowledge on the subject to current managers and professionals linked to the area of operation.

Concerning its relevance, the results of this research may contribute to the implementation of Lean in the Operation and thus create more opportunities for improvement so that companies can reach the status of excellence in operations. This work also contributes to the improvement of meeting the requirements of the market for bean packers and business strategies, as well as reducing costs and increasing the competitiveness in the market.

In addition to this Introduction, this manuscript is structured as follows. Section 2 provides the theoretical framework. Sections 3 and 4 describe the methodology and development, respectively. The results and discussions are carried out in Section 5. Finally, Section 6 presents the conclusion.

2. Theoretical Framework

2.1. Lean Manufacturing

Lean Manufacturing is a production system that emerged in Japan between the late 1940s and early 1950s, in the post-war period. Its birthplace was the Japanese company Toyota Motor Company and for this reason, it was first known as TPS—Toyota Production System [3,27,28]. TPS was developed exclusively by the Japanese, who wanted to create their own style of production, which would better suit their culture and surpass the conventional large-scale production system, producing a small amount of many different types of cars [3].

The production style was mainly due to two reasons: the first reason was the fact that the Japanese market, due to the context of the war in the 1930s, had a demand for small quantities of different types of cars and trucks, making large-scale production unfeasible [28].

The second reason was stated by the then president of Toyota, Kiichiro Toyoda, who in 1945 said that the Japanese auto industry had three years to reach the productivity of the US auto industry; otherwise, this branch of the Japanese industry would not survive [3]. Consequently, Toyota executives Eiji Toyoda and Taiichi Ohno, with the collaboration of Shingo Shingeo through consultancies, created a new production system in order to overcome the difficulties encountered in Japanese industry [28,29].

Ohno in [3] defines STP as "[...] a method to fully eliminate waste and increase productivity". In a production system, waste is related to every activity that generates costs without adding value to the product. In addition, the efforts to reduce waste also motivate the deployment of smart production systems by utilizing an autonomation policy [30,31].

The "TPS House" was developed in a way to explain TPS. The idea of representing TPS through a house occurred because a house can only remain stable if the foundation, pillars, and roof are strong and well connected [27]. According to [27], the roof represents the goals of this system, which are to improve quality, reduce costs and lead time. The foundation is composed of processes that guarantee the stability of the system, such as level production (heijunka), stable and standardized processes, visual management, and the Toyota Model philosophy.

The pillars that support this system are just-in-time (JIT) and production automation with a human touch or autonomation [30,31]. JIT means that the correct material will reach the assembly line of a flow process at the right time and in the right amount. On the other hand, automation with a human touch, unlike conventional automation, allows machines to automatically stop the production line whenever they identify a problem, using devices capable of identifying product failures. These pillars enabled the large-scale reduction of waste found in conventional car production [3].

As Toyota made greater profits than other companies did, even in times of crisis, the interest of researchers and other Japanese companies in this Toyota production model increased. TPS began to be studied and the ideas of this productive system began to be disseminated throughout the world. With the publication of the first edition of the book "The Machine that Changed the World" in 1990, TPS also became known as Lean Production or Lean Manufacturing [3,28].

According to [28], the researcher at the International Automotive Vehicles Program (IMVP), John Krafcik, explained why this production system is called "lean":

[...] by using smaller amounts of everything compared to mass production: half the factory worker effort, half the manufacturing effort, half the tooling investment, and half the planning hours to develop new products in half the time. It also requires far less than half of the current inventories at the manufacturing site, results in far fewer defects, and produces an ever-increasing range of products.

The total elimination of waste, that is, the elimination of activities that do not add value and generate costs, allows companies to reduce the efforts necessary to produce their products and consequently increase the efficiency of the production system and reduce costs [3,28]. The LM aims to produce products and services at the lowest cost and as fast as required by the customer [32].

LM concepts have been widely disseminated and studied throughout the world. Recently, this concept was expanded outside the production systems of factories, being called Lean Thinking. According to [28], "lean thinking is a way of specifying value, aligning the actions that create value in the best sequence, carrying out these activities without interruption whenever someone requests them and carrying them out in an increasingly effective way".

2.1.1. Waste and Value

Waste and value are two key concepts of Lean. As discussed in [33] seven or eight wastes are identified in the literature as well as the concept of waste as an activity that does not add value; defining waste, in general, as any input to the system that is not transformed into an output that represents value to the customer (customer satisfaction), at the right time and amount (just-in-time). As for the concept of value, Ref. [28] stated that value is something that can only be decided by the end customer, being expressed in a relevant way only when it comes to a particular product, whether it is a good, a service or both, which meets the needs of the end customer at a given time and at a specific price. It indicates that the value can be determined by what the customer is willing to pay.

2.1.2. The Eight Types of Waste

Taiichi Ohno was the pioneer in recognizing the existence of waste in production systems, as well as the one who dedicated himself to eliminating it. Ref. [3] states that the shop floor is the best source of manufacturing information, providing up-to-date and direct information about the management. In addition, it is where you can identify the waste present in manufacturing. Ref. [27] states that Toyota first identified seven wastes in production, but the author includes an eighth type of waste (Figure 1):

1. Defect: the production of defective items or the reprocessing of defective items, which generally result in wasted time and effort;
2. Overproduction: the production of items excessively, exceeding demand;
3. Wait: the time when the worker and machine are idle;
4. Non-utilized talent: the waste of employees' time, ideas and skills;
5. Transport or unnecessary motion: the movement of materials excessively or inefficiently;
6. Inventory: inventory that is larger than necessary, causing additional costs related to excessive transport and storage, obsolescence, damaged products, and longer lead times;
7. Motion: the movement of an employee, during work, to carry out activities that do not add value and are unnecessary;
8. Extra-processing or incorrect processing: processing too much of an item.

Figure 1. The eight types of wastes.

2.2. Quick Changeover

First, it is necessary to define a concept that will be broadly covered in this work, the concept of machine setup time, also known as setup. According to [34], setup is the change in production to manufacture another product in a machine, or in a sequence of interconnected machines, through the exchange of parts, dies, devices, molds, among others. Setup time is the time elapsed between the production of the last item of the completed cycle and the first item, of quality, of the cycle that has just been started.

SMED, also known as quick changeover, is a concept developed by Shigeo Shingo between 1950 and 1969 to reduce machine setup time. It consists of a methodology with the objective of reducing setup time to single digit minutes. In this way, it is possible to reduce setup times from several hours to less than 10 min [29,35–37].

SMED was developed to solve one of the biggest problems faced by companies: to produce in a diversified manner and in small batches. Believing that it was impossible to reduce setup times drastically, line managers were for a long time forced to adopt low-diversity and large-batch production to minimize the effect of machine setup time on production capacity. However, large-scale production contributes to the waste of overproduction, which leads to an increase in stock and consequently generates additional costs [29,37].

Reducing setup time was the strategy that enabled diversified production in small batches, without loss of production capacity, enabling a significant reduction in stocks [10,35,37].

In addition to the reduction in machine setup time, Ref. [10] highlights several positive effects of adopting the SMED methodology. The main ones are the improvement of production capacity and machine utilization rates, production without stocks, elimination of errors in the setup process, improved quality of products, improved safety in the setup process, improved organization of the tools needed in the setup process, motivation in relation to the occurrence of setup, reducing the necessary qualification of the operator who performs the setup and improvement of production flexibility.

2.3. SMED Implementation Steps

While conducting studies to reduce setup times, Ref. [29] realized that machine setup operations could be divided into two—internal setup and external setup. According to [10], the internal setup or internal setup time (IST) consists of the machine setup operations, which can only be performed while the machine is stopped. On the other hand, the external setup or external setup time (EST), according to the author, consists of the setup operations of the machines that can be performed while the machine is running. The stages of the SMED implementation process, according to [10], are divided into four:

- Initial phase: Setup time is not divided into internal setup and external setup;
- Stage 1: Setup division into internal setup and external setup;
- Stage 2: Transformation of internal setup into external setup;
- Stage 3: Simplification of internal setup and external setup operations.

The initial stage is the step in which the methodology implementation planning takes place. At this stage, a study of the current conditions of the setup process is carried out, identifying all activities and the time spent on each one of them and the displacement of operators [10,35].

Stage 1 is the step in which IST and EST are differentiated, which represents the most important point of the methodology. This stage ensures that activities that may take place while the machine is still running actually take place while the machine is still on. The focus of this stage is the activities related to the setup and functional verification of raw materials, tools, and fastening devices that must be performed while the machine is still on [10,35,36].

Stage 2, on the other hand, is a stage in which internal setup activities are converted, when possible, into external setup, minimizing the number of operations that are performed while the machine is not in operation [10,36].

Finally, Stage 3 is the stage in which the activities of internal setup and external setup are simplified as much as possible, with the objective of increasing the ease, security, and speed of setup. Setup activities are modified in order to make them faster and can reduce the workload. EST can be simplified, mainly through improvements in the storage and transport of the elements necessary for the machine setup process. IST, on the other hand, can be simplified through the implementation of parallel operations, the use of functional fixators, and the elimination of adjustments [10,36].

3. Methodology

The research method used in this study was action research. According to [38], action research consists of a scientific research methodology that can be used to solve social or organizational problems, with the participation of people who live with this problem, through actions in the object of study. The steps required to implement this research methodology are shown in Figure 2.

Figure 2. Steps for implementing action research (Adapted from [38]).

This research method was chosen because it is intended to make changes to the object of study. The authors plan to carry out the action research cycle twice. The first cycle to perform stage 1 of SMED, separating the internal setup from the external setup. The second cycle to execute stages 2 and 3 of SMED, transforming the internal setup into an external setup and simplifying the setup operations.

4. Development

4.1. Context and Proposal

The object of study of this work is a medium-sized company located in east Minas Gerais, Brazil. Its main activity is the processing, packaging, and distribution of beans, in addition to the distribution of rice and cassava flour. The main product sold by the company is carioca beans. The company also works with other varieties, black and red beans. The flow of operations of the processing and packaging of beans throughout the production process is shown in Figure 3.

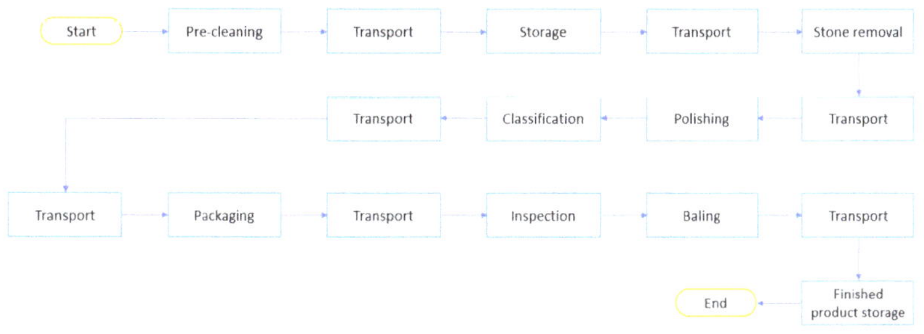

Figure 3. Flowchart of bean processing and packaging process.

After unloading the beans, the processing begins. In the operations of the processing, impurities from the harvest are removed, such as straw, clods, stones, and dust. In addition, bean husks and broken beans are also removed. All transport carried out in the processing is carried out by bucket elevators. The packing process, on the other hand, consists of bean packing and baling operations. In the packaging, the beans are packed in 1 kg packages. In the baling, these packages are again wrapped in 10 kg or 30 kg bales.

For the implementation of SMED, the setup chosen was the changeover of the 30 kg bale of red beans for the 10 kg bale of carioca beans. This setup was chosen because it is the most complete performed by the company since the improvements obtained in this process of setting up the machines can be used in all other setups of the packaging operation. In addition to the complete setup, the partial setup and the basic setup are also performed in this operation.

The complete setup consists of activities in the processing, packaging, and baling operations. The processing operator cleans all the machinery in this process to avoid mixing beans of different varieties. The packaging/processing operator, with the help of the repairperson, replaces the primary and secondary packaging, in addition to changing the baler die. The partial setup consists of the same actions in the processing and the replacement of primary packaging in the packaging operation. While the basic setup only consists of replacing the primary packaging in the packaging operation.

For the implementation of SMED, a team composed of four members was first assembled, with specific functions to monitor and map the execution of the setup. These members are: the observer, who is responsible for observing and recording the activities of the operators; the timekeeper, who is responsible for taking the time for each activity; the shadow, who has the function of recording the route and the number of steps of operators; and the Kaizen man, who has the function of identifying points of improvement in the execution of the setup.

In order to map the setup in the initial phase, the team filmed the setup of the machines to raise all the necessary steps for the implementation. Once the implementation steps were defined, the first data collection was carried out.

Through the obtained data, it could be verified that there were 84 activities necessary for carrying out the machine setup. Of these 84 activities, 15 were identified as external setup, as they could be performed while the machinery was still in operation. In addition, the total time for carrying out the activities corresponds to approximately 85 min.

In the data collection, the distance travelled by the operators was also measured through the number of steps, as well as the path travelled by the operators in the production system. The path was illustrated by the spaghetti diagram shown in Figure 4.

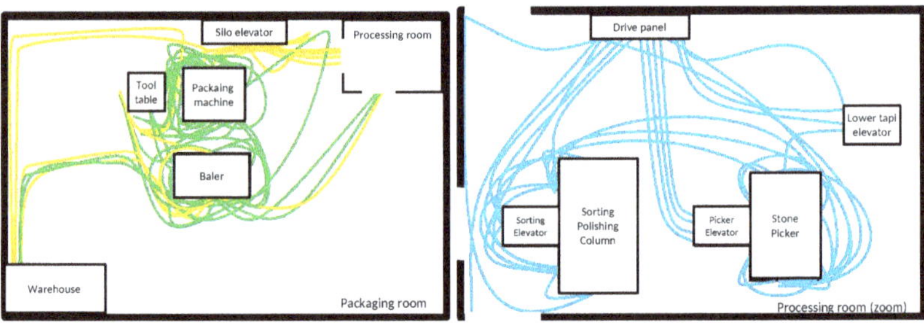

Figure 4. Spaghetti diagram of setup in the initial phase. Each color represents a different operator.

During the machine setup, the operator responsible for processing, shown in blue, covered 448 steps. In this work, each step of the operator was considered equivalent to 60 cm, so the processing operator covered about 269 m. The operator responsible for the packaging and baling operations, represented in green, travelled about 453 m. Finally, the repairperson, represented in yellow, moved approximately 143 m.

4.2. First Cycle of Action Research

From the first data collection, besides the analysis and classification of machine setup activities between external and internal setup, the time of use of the repairperson during machine adjustments was also observed in order to increase its participation in the process. The activities classified as EST started to be performed while the machinery was in operation. Furthermore, these activities, which were previously dependent on the operator, are now performed only by the repairperson.

Based on these setup changes, a new data collection was performed to assess the effect of EST elimination on setup time. With the elimination of external setup activities, the number of activities performed during the machine setup period dropped to 69 activities; in addition, the execution time was reduced to about 60 min, representing a reduction of approximately 30% compared to the previous time. The paths taken by the operators and the repairperson during the execution of setup underwent changes with the absence of activities considered EST. In general, there was a reduction in the path of operators, as shown in Figure 5.

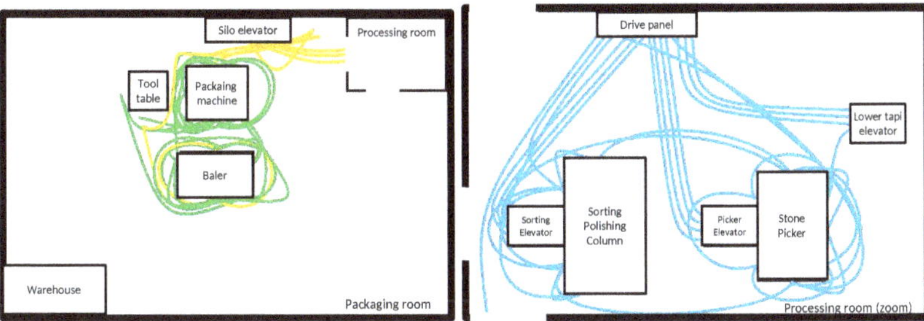

Figure 5. Spaghetti diagram of setup after cycle 1. Each color represents a different operator.

With the reduction in the number of activities during the machine setup process, the operator responsible for the improvement process, shown in blue, started to cover approximately 194 m. The operator of the packaging and baling operations, in green, and the repairperson, in yellow, moved approximately 250 and 51 m, respectively. Table 2 summarizes the results obtained with the implementation of the first cycle of action research.

Table 2. Summary of the first cycle of action research.

Data collection	List and timing of activities:
	- Internal Setup = 85 min
	- External Setup = 0 min
	Travelled distance:
	- Processing:
	    Operator = 269 m
	- Packaging/Balling:
	    Operator = 453 m
	    Repairperson = 143 m

Data Analysis	Classification of setup activities into internal setup and external setup; Repairperson usage time.
Implementation	Separation of the internal and external setup in the execution of the setup; Redefining the activities of operators; Redistribution of external setup activities to the repairperson.

Evaluation	Measurement of the new setup method:
	- Internal Setup = 60 min
	- External Setup = 25 min
	Travelled distance:
	- Processing:
	    Operator = 194 m
	- Packaging/baling:
	    Operator = 250 m
	    Repairperson= 51 m

4.3. Second Cycle of Action Research

Based on the information obtained in the data collection of the first action research cycle, an analysis was carried out to identify improvements in the IST activities, unnecessary activities, the feasibility of using more employees, improvements in the sequence of activities, opportunities for error reduction, layout improvement, and opportunities to reduce adjustment time and the number and size of screws removed during setup.

Based on data analysis, an action plan was developed using the 5W2H tool in order to obtain not only a reduction in setup time but also a reduction in the occurrence of errors, gains in product quality, and gains in the organization of tools needed. The action plan is detailed in Table 3.

Table 3. Action Plan.

What to Do? (What)	Why Do It? (Why)	How to Do It? (How)	When to Do It? (When)	Where to Do? (Where)	Who Is Going to Do? (Who)	How Much Does It Cost? (How Much)
Use the ratchet wrench for installation and removal of screws	To reduce the necessary adjustment time	Through the purchase of the ratchet wrench and operator training	September/October	In the baler	Packaging/Balling Operator	BRL 160.00
Reduce from 4 to 3 different screw sizes	To reduce the necessary adjustment time	By replacing the 13 mm drive belt screws with 14 mm screws	September/October	In the baler	Third-party company	BRL 100.00
Use a method to simplify installation of screws	To reduce the necessary adjustment time	By using a method that already existed in the baling die but which was unknown to employees	September/October	In the baler	Packaging/Balling Operator	BRL 0.00
Install compressed air in the packaging silo	To improve the quality of cleaning of the packaging silo, which was performed with a piassava broom since pieces of piassava can contaminate the product about to be packaged	By purchasing and installing the necessary materials	September/October	In the baler	Packing/baling coordinator and operator	BRL 50.00
Implement 5S on the tool table	To improve the organization of tools and reduce the time wasted in finding them	Through the application of the 5 senses of this methodology	September/October	Tool table	Researcher, coordinator and operator of the packaging/baling	BRL 0.00
Introduce a new repairperson and a new operator into the setup process	To share the necessary activities with more employees, reducing setup time and necessary displacement	Through training	September/October	In the processing and packaging sectors	Researcher, coordinator, operators, and assistants	BRL 0.00
Reorganization of setup activities	So that the setup in the packaging sector can start simultaneously with the setup in the processing sector	Through the redistribution of activities with the new employees involved in the setup	September/October	In the processing and packaging sectors	Researcher, director, coordinator, operators and assistants	BRL 0.00
Create a new standardized work sequence	To optimize the time needed to run the setup and avoid errors and rework	Through meetings and discussions	September/October	In the processing and packaging sectors	Researcher, director, coordinator, operators and assistants	BRL 0.00

The method to simplify the installation of the screws is to place slots in two holes of the baling die; this allows the installation of two screws before positioning this die in the baler. In addition, there were two screws of similar sizes (13 and 14 mm), which allowed the reduction from four to three different screw sizes by enlarging the smaller holes. Finally, a ratchet wrench was purchased for the installation and removal of the screws, which reduced the time for adjustments.

A new operator and a new repairperson were introduced to participate in the setup since there were already employees available to perform the activities (they used to perform activities with lower priority). The new operator was responsible for the activities of the packaging operation, allowing the former operator to carry out only the activities of the baling operation. The former repairperson started to help both the processing operator and the packaging operator. The new repairperson was responsible for part of the activities related to the processing.

Through the introduction of new employees in the setup, it was possible to reorganize the machine setup activities. This reorganization of activities allowed the actions related to the processing, the packaging operation, and the baling operation, which were previously carried out in sequence, to be started in parallel. With all these setup changes, a new standardized work sequence was developed. The standardization of this new work sequence contributes to the elimination of errors that could occur during the setup of the machines.

In order to assess the effect of the proposed improvements to the setup, set out in the action plan, a new data collection was carried out. With the changes proposed in the action plan, the setup time was reduced to approximately 36 min, which corresponds to the setup time of the baling operation activities, as it is the procedure that demands the most time of the entire setup operations. This reduction in setup time corresponds to around 40% in relation to the time obtained after the first cycle of the action research, and a reduction of around 58% in relation to the initial setup time.

As for the path taken by employees, there was a reduction in the path travelled and the distance travelled, mainly due to the introduction of the new operator and the new repairperson. The path taken by operators and repair people is illustrated in Figure 6.

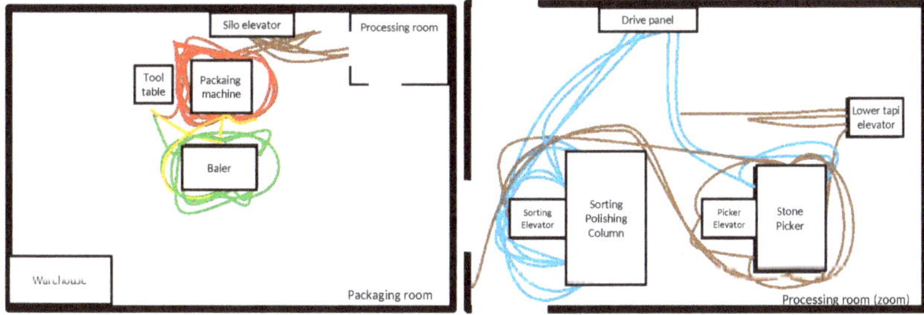

Figure 6. Spaghetti diagram of setup after cycle 2. Each color represents a different operator.

At the end of the second cycle of the action research, the processing operator, represented in blue, started to walk approximately 60 m, while the new repairperson, represented in brown, covered approximately 102 m. As for the new operator of the packaging operation, represented in red, and the operator of the baling operation, represented in green, they moved approximately 63 and 172 m, respectively. Finally, the repairperson for the packaging and baling operations, represented in yellow, started to travel approximately 35 m.

Table 4 summarizes the results obtained with the implementation of the first cycle of action research.

Table 4. Summary of the second cycle of action research.

Data collection	New setup method:
	- Internal Setup = 60 min
	- External Setup = 25 min
	Installation and removal of screws:
	- Number of screws: 14 screws
	Different sizes of screws: 4 sizes
	Distance travelled by operators:
	- Processing:
	    Operator = 194 m
	- Packaging:
	    Operator = 250 m
	    Repairperson = 51 m

Data Analysis	Improvements in internal setup activities; Identification of unnecessary activities; Feasibility of using more employees; Improvements in the sequence of activities; Opportunities to reduce errors during machine setup; Improved layout; Number and size of screws removed during setup; Opportunities to reduce adjustment time.
Implementation	Acquisition of a ratchet key; Reduction from 4 to 3 different screw sizes; Using a method to simplify the installation of the baling die screw; 5S Implementation on the tool table; Installation of compressed air for cleaning the packaging silo; Introduction and training of a new repairperson and a new operator in the setup process; Reorganization of setup activities; Creation and implementation of a standardized work sequence;

Evaluation	Measurement of the new setup method:
	- Internal Setup = 36 min
	- External Setup = 25 min
	Travelled distance:
	- Processing:
	    Operator = 60 m
	    Repairperson = 102 metros
	- Packaging/processing:
	    Packing operator = 63 m
	    Processing operator = 172 m
	    Repairperson = 35 metros
	Installation and removal of screws:
	- Number of screws: 14 screws
	- Different sizes of screws: 3 sizes

5. Results and Discussions

Both action research cycles carried out strongly contributed to the reduction of the necessary setup time, in addition to also reducing the displacement of the employees involved. Consequently, the reduction in the setup time of the machines contributed to the increase in the company's production capacity since part of the time required by the setup became available for production.

It is important to highlight that the improvement actions performed for the most complete setup also had a positive effect on the execution time of the partial setup and the basic setup, mentioned above.

Productive capacity gains, as well as the effect of complete setup improvements on partial and basic setup, are described in Table 5. At the end of each action-research cycle, there were reductions in setup time and employee displacement, in addition to an increase in the company's production capacity. The result obtained in each cycle is described in Figure 7.

As shown in Figure 7, at the end of the second cycle of the action research, setup time was reduced by approximately 58%, the displacement of employees, even with the increase from three to five employees, was reduced by around 50%. The productive capacity had an increase of approximately 14%. During the implementation of SMED, operators initially resisted, as they believed the changes would entail a greater workload. However, as the methodology was implemented, operators began to realize the benefits brought by SMED and that there would not be an increase in work. From that moment on, they started to support the changes brought about by the methodology.

Table 5. Effect of setup improvements on production capacity.

First Cycle of Action Research	

Effect on setup time	- Complete Setup: −30%
	    Average setup/month: 2
	- Partial Setup: −38%
	    Average setup/month: 32
	- Basic Setup: −45%
	    Average setup/month: 23

Effect on monthly productive capacity	- Operational availability $(2 \times 25 \text{ min}) + (32 \times 16.32 \text{ min}) + (23 \times 5.37 \text{ min}) = 695.52$ min/month - Capacity increase: approximately 32,661 kg of beans/month (9.33%) - Prior productive capacity: approx. 350,000 kg of beans/month - Production capacity afterward: approximately 382,661 kg of beans/month
Second Cycle of Action Research	

Effect on setup time	- Complete Setup: −40%
	    Average setup/month: 2
	- Partial Setup: −32%
	    Average setup/month: 32
	- Basic Setup: 0%
	    Average setup/month: 23

Effect on monthly productive capacity	- Operational availability $(2 \times 23.23 \text{ min}) + (32 \times 8.42 \text{ min}) + (23 \times 0 \text{ min}) = 315.26$ min/month - Approximately 14,823 kg of beans/month (3.87%) - Approximately 382,661 kg of beans/month - Production capacity afterward: approximately 397,484 kg of beans/month

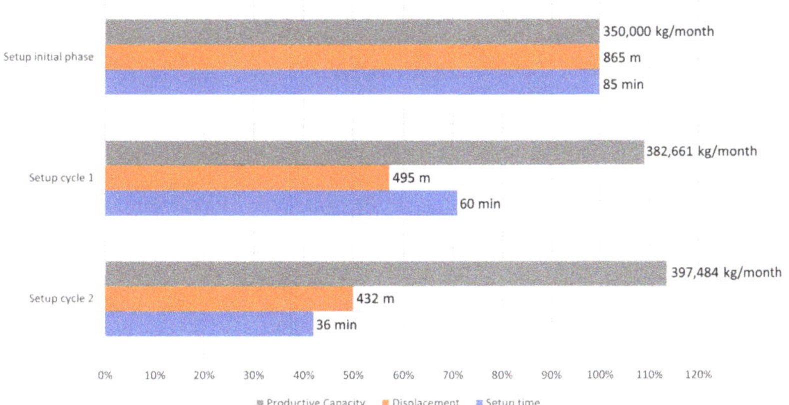

Figure 7. Result at the end of each action research cycle.

6. Conclusions

This work aimed to evaluate the effect of SMED implementation in the bean packaging operation in a company in eastern Minas Gerais, Brazil. With the implementation of this methodology, it was possible to obtain a reduction in setup time and, consequently, there was an increase in the company's productive capacity, in addition to a reduction in the displacement of employees. As for the increase in the productive capacity, there was no change in the workforce or working hours, so it can be concluded that the company achieved productivity gains.

In addition to the increase in productivity, quality gain, gains in the organization of tools, and the reduction in the occurrence of errors mentioned above, SMED also had the advantages of increased motivation regarding the occurrence of setup and increased production flexibility, as the effect of set-up time on production capacity has been reduced, allowing machine set-up to take place more times a day if necessary.

Finally, for the implementation of SMED, there was full support and encouragement from the company's management, which was already looking for alternatives that could expand the company's production capacity. Management was satisfied with the results obtained, mainly due to the low investment (approximately BRL 310.00).

In terms of a theoretical contribution, this work is one of the few studies that deal with the implementation of SMED in a company in the grain processing and packaging sector, showing the possible results obtained by the methodology and confirming its effectiveness in reducing setup time. This study shows that the LM/SMED tool can be applied in a bean industry and LM can empower the agroindustry to Agro4.0. Few studies have been carried out in this context, so our manuscript tries to fill these gaps by conducting an SMED in agroindustry through action research.

In terms of practical contribution and relevance to the sector, despite the little interaction of Lean tools with companies in this segment, this study showed that SMED proved to be an effective option for obtaining gains in productive capacity by reducing setup time, opening doors for new studies on the implementation of other Lean Manufacturing methodologies, such as line balancing and time planning.

As a limitation of the study, it is possible to highlight that the methodology was applied only once. The findings were also not fully documented. So, there is still room for improvement in the research method.

As future work, the relationship between LP and I4.0 and Agro4.0 can be explored. Practical applications including process-related systems and technologies of I4.0 [12], combined with scientific methodology, are an opportunity to fill the existing literature gap in the area.

Author Contributions: Conceptualization, M.A.S.R. and A.C.O.S.; methodology, A.C.O.S., G.d.F.d.A. and C.H.d.O.; software support, R.A.d.S.B. and R.S.N.; investigation, M.A.S.R. and A.C.O.S.; data curation, M.A.S.R., R.A.d.S.B. and R.S.N.; writing—original draft preparation, R.A.d.S.B. and A.C.O.S.; writing—review and editing, G.d.F.d.A. and C.H.d.O.; funding acquisition, R.A.d.S.B. and R.S.N. All authors have read and agreed to the published version of the manuscript.

Funding: The APC was funded by Federal University Itajuba and Institute of Science and Technology—ICT Unifei Itabira

Conflicts of Interest: The authors declare no conflict of interest.

References

1. Coêlho, J.D. Feijão: Produção e mercados. *Cad. Setorial ETENE* **2021**, 1–9. Available online: https://www.bnb.gov.br/s482-dspace/handle/123456789/1031 (accessed on 31 January 2021).
2. Food and Agriculture Organization. FAOSTAT. Available online: https://www.fao.org/ (accessed on 31 January 2021).
3. Ohno, T.; Bodek, N. *Toyota Production System: Beyond Large-Scale Production*; Productivity Press: New York, NY, USA, 2019.
4. Santos, A.C.O.; da Silva, C.E.S.; Braga, R.A.D.S.; Corrêa, J.É.; de Almeida, F.A. Customer value in lean product development: Conceptual model for incremental innovations. *Syst. Eng.* **2020**, *23*, 281–293. [CrossRef]
5. Danese, P.; Manfè, V.; Romano, P. A systematic literature review on recent lean research: State-of-the-art and future directions. *Int. J. Manag. Rev.* **2018**, *20*, 579–605. [CrossRef]
6. Melin, M.; Barth, H. Lean in Swedish agriculture: Strategic and operational perspectives. *Prod. Plan. Control* **2018**, *29*, 845–855. [CrossRef]
7. Doevendans, H.J.; Grigg, N.P.; Goodyer, J. Exploring Lean deployment in New Zealand apple pack-houses. *Meas. Bus. Excell.* **2015**, *19*, 46–60. [CrossRef]
8. Aoki, R.; Katayama, H. Heijunka Operation Management of Agri-Products Manufacturing by Yield Improvement and Cropping Policy. In *International Conference on Management Science and Engineering Management*; Springer: Berlin/Heidelberg, Germany, 2017; pp. 1407–1416.
9. da Silva, I.B.; Godinho Filho, M. Single-minute exchange of die (SMED): A state-of-the-art literature review. *Int. J. Adv. Manuf. Technol.* **2019**, *102*, 4289–4307. [CrossRef]
10. Shingo, S.; Dillon, A.P. *A Revolution in Manufacturing: The SMED System*; Routledge: London, UK, 2019.
11. Shahin, M.; Chen, F.F.; Bouzary, H.; Krishnaiyer, K. Integration of Lean practices and Industry 4.0 technologies: Smart manufacturing for next-generation enterprises. *Int. J. Adv. Manuf. Technol.* **2020**, *107*, 2927–2936. [CrossRef]
12. Dombrowski, U.; Richter, T.; Krenkel, P. Interdependencies of Industrie 4.0 & lean production systems: A use cases analysis. *Procedia Manuf.* **2017**, *11*, 1061–1068.
13. Jamwal, A.; Agrawal, R.; Sharma, M.; Giallanza, A. Industry 4.0 technologies for manufacturing sustainability: A systematic review and future research directions. *Appl. Sci.* **2021**, *11*, 5725. [CrossRef]

14. Ciano, M.P.; Dallasega, P.; Orzes, G.; Rossi, T. One-to-one relationships between Industry 4.0 technologies and Lean Production techniques: A multiple case study. *Int. J. Prod. Res.* **2021**, *59*, 1386–1410. [CrossRef]
15. Mrugalska, B.; Wyrwicka, M.K. Towards lean production in industry 4.0. *Procedia Eng.* **2017**, *182*, 466–473. [CrossRef]
16. Lermen, F.H.; Echeveste, M.E.; Peralta, C.B.; Sonego, M.; Marcon, A. A framework for selecting lean practices in sustainable product development: The case study of a Brazilian agroindustry. *J. Clean. Prod.* **2018**, *191*, 261–272. [CrossRef]
17. Cardozo, E.R.; Rodríguez, C.; Guaita, W. Small and medium size enterprises of the food and agriculture sector and sustainable development: Approach based on principles of lean manufacturing. *Inf. Tecnol.* **2011**, *22*, 39–48. [CrossRef]
18. Carrera, J.; Fernandez del Carmen, A.; Fernández-Muñoz, R.; Rambla, J.L.; Pons, C.; Jaramillo, A.; Elena, S.F.; Granell, A. Fine-tuning tomato agronomic properties by computational genome redesign. *PLoS Comput. Biol.* **2012**, *8*, e1002528. [CrossRef] [PubMed]
19. Taylor, D. Strategic considerations in the development of lean agri-food supply chains: A case study of the UK pork sector. *Supply Chain. Manag.* **2006**, *11*, 271–280. doi: [CrossRef]
20. De Steur, H.; Wesana, J.; Dora, M.; Pearce, D.; Gellynck, X. Applying Value Stream Mapping to reduce food losses and wastes in supply chains: A systematic review. *Waste Manag.* **2016**, *58*, 359–368. doi: [CrossRef]
21. Zarei, M.; Fakhrzad, M.; Jamali Paghaleh, M. Food supply chain leanness using a developed QFD model. *J. Food Eng.* **2011**, *102*, 25–33. doi: [CrossRef]
22. Cox, A.; Chicksand, D.; Palmer, M. Stairways to heaven or treadmills to oblivion?: Creating sustainable strategies in red meat supply chains. *Br. Food J.* **2007**, *109*, 689–720. doi: [CrossRef]
23. Satolo, E.G.; Hiraga, L.E.D.M.; Zoccal, L.F.; Goes, G.A.; Lourenzani, W.L.; Perozini, P.H. Techniques and tools of lean production: Multiple case studies in brazilian agribusiness units. *Gestão Produção* **2020**, *27*. . [CrossRef]
24. Culley, S.; Owen, G.; Mileham, A.; McIntosh, R. Sustaining changeover improvement. *Proc. Inst. Mech. Eng. Part B J. Eng. Manuf.* **2003**, *217*, 1455–1470. [CrossRef]
25. Dora, M.; Lambrecht, E.; Gellynck, X.; Van Goubergen, D. Lean Manufacturing to Lean Agriculture: It's about time. In *IIE Annual Conference, Proceedings of the 2015 Industrial and Systems Engineering Research Conference, Nashville, TN, USA, 30 May–2 June 2015*; Institute of Industrial and Systems Engineers (IISE): Honolulu, HI, USA, 2015; p. 633.
26. Reis, L.V.; Kipper, L.M.; Velásquez, F.D.G.; Hofmann, N.; Frozza, R.; Ocampo, S.A.; Hernandez, C.A.T. A model for Lean and Green integration and monitoring for the coffee sector. *Comput. Electron. Agric.* **2018**, *150*, 62–73. [CrossRef]
27. Liker, J. *The Toyota Way Fieldbook*; McGraw Hill: New York, NY, USA, 2005.
28. Womack, J.P.; Jones, D.T. Lean thinking—Banish waste and create wealth in your corporation. *J. Oper. Res. Soc.* **1997**, *48*, 1148. [CrossRef]
29. Shingo, S.; Dillon, A.P. *A Study of the Toyota Production System: From an Industrial Engineering Viewpoint*; CRC Press: Boca Raton, FL, USA, 1989.
30. Dey, B.K.; Sarkar, B.; Seok, H. Cost-effective smart autonomation policy for a hybrid manufacturing-remanufacturing. *Comput. Ind. Eng.* **2021**, *162*, 107758. [CrossRef]
31. Dey, B.K.; Pareek, S.; Tayyab, M.; Sarkar, B. Autonomation policy to control work-in-process inventory in a smart production system. *Int. J. Prod. Res.* **2021**, *59*, 1258–1280. [CrossRef]
32. Bhamu, J.; Sangwan, K.S. Lean manufacturing: Literature review and research issues. *Int. J. Oper. Prod. Manag.* **2014**, *34*, 876–940. [CrossRef]
33. Thürer, M.; Tomašević, I.; Stevenson, M. On the meaning of 'waste': Review and definition. *Prod. Plan. Control* **2017**, *28*, 244–255. [CrossRef]
34. Marchwinski, C.; Shook, J. *Lean Lexicon: A Graphical Glossary for Lean Thinkers*; Lean Enterprise Institute: Brookline, MA, USA, 2003.
35. Dave, Y.; Sohani, N. Single minute exchange of dies: Literature review. *Int. J. Lean Think.* **2012**, *3*, 27–37.
36. McIntosh, R.; Culley, S.; Mileham, A.R.; Owen, G. A critical evaluation of Shingo's' SMED' (Single Minute Exchange of Die) methodology. *Int. J. Prod. Res.* **2000**, *38*, 2377–2395. [CrossRef]
37. Carrizo Moreira, A.; Campos Silva Pais, G. Single minute exchange of die: A case study implementation. *J. Technol. Manag. Innov.* **2011**, *6*, 129–146. [CrossRef]
38. Coughlan, P.; Coghlan, D. Action research for operations management. *Int. J. Oper. Prod. Manag.* **2002**, *22*, 220–240. [CrossRef]

Article

Multi-Model In-Plant Logistics Using Milkruns for Flexible Assembly Systems under Disturbances: An Industry Study Case

Adrian Miqueo [1,*], Marcos Gracia-Cadarso [1], Marta Torralba [2], Francisco Gil-Vilda [3] and José Antonio Yagüe-Fabra [1]

1   Aragonese University Institute for Engineering Research, Universidad de Zaragoza, C/María de Luna 3, 50018 Zaragoza, Spain
2   Centro Universitario de la Defensa Zaragoza, Ctra. Huesca s/n, 50090 Zaragoza, Spain
3   LEANBOX SL, C/Juan de Austria 126, 08018 Barcelona, Spain
*   Correspondence: adrian.miqueo@unizar.es

Abstract: Mass customisation demand requires increasingly flexible assembly operations. For the in-plant logistics of such systems, milkrun trains could present advantages under high variability conditions. This article uses an industrial study case from a global white-goods manufacturing company. A discrete events simulation model was developed to explore the performance of multi-model assembly lines using a set of operational and logistics Key Performance Indicators. Four simulation scenarios analyse the separate effects of an increased number of product models and three different sources of variability. The results show that milkruns can protect the assembly lines from upstream process disturbances.

Keywords: milkrun; in-plant logistics; flexible assembly; simulation; high-mix low-volume; lean manufacturing

Citation: Miqueo, A.; Gracia-Cadarso, M.; Torralba, M.; Gil-Vilda, F.; Yagüe-Fabra, J.A. Multi-Model In-Plant Logistics Using Milkruns for Flexible Assembly Systems under Disturbances: An Industry Study Case. *Machines* **2023**, *11*, 66. https://doi.org/10.3390/machines11010066

Academic Editors: Luís Pinto Ferreira, Paulo Ávila, João Bastos, Francisco J. G. Silva, José Carlos Sá and Marlene Brito

Received: 15 December 2022
Revised: 29 December 2022
Accepted: 30 December 2022
Published: 5 January 2023

1. Introduction

Since the end of the 20th century, it is considered that demand trends are shifting from mass production towards mass customisation [1] and mass personalisation [2]. To address this situation, manufacturing companies need to produce an increasing number of different products, in smaller quantities each, without compromising on quality or price [3]. For consumer goods manufacturers, this means shifting from large batches of very similar products towards high-mix low-volume production. To gain an advantage or simply remain competitive, production flexibility, reconfigurability and resilience are key [4].

In a typical discrete production process—e.g., automobiles, white goods, home electronics, furniture, toys—the assembly stage taking place after manufacturing is also of capital importance [5]. Traditional assembly operations are performed in manual or semi-automated lines or cells, which are usually dedicated to one product or a small family of products closely related [6]. These products are assembled in batches to minimise the losses incurred due to product changeovers [7,8]. Looking at existing assembly operations approaches to build upon, Lean Manufacturing [9] proposes a methodology inherently oriented towards reduced batch sizes, frequent product changeovers, multi-product assembly cells and cross-functional operator teams [10,11]. In this context, it seems clear that traditional assembly lines face serious threats when confronted with the high-mix low-volume demand brought by the mass customisation paradigm. The main challenges include dealing with complexity, uncertainty and disturbances, successfully deploying disruptive digital technologies [12]—i.e., Industry4.0 [13] or smart manufacturing [14]—and further integrating the sub-systems related to assembly: supporting functions such as internal logistics [15], maintenance [16] or quality control [17].

Internal logistics is the supply chain function most closely related to the assembly operations since it is tasked with feeding components to the assembly line or cell without introducing production constraints [18,19]. Flexible assembly lines driven by mass customisation and featuring mixed- or multi-model production pose additional challenges to internal logistics [20], which impact directly on the classic Lean supply performance indicators [21]. In-plant milkruns [22] (*misuzumashi* [23], *tow-train* [18]) are one of the best available Lean tools for efficiently supplying parts to flexible multi-model assembly lines [24].

The brief literature review that will be presented in Section 2 shows that despite an increasing research depth on the topic of milkrun logistic systems for flexible assembly lines, there are still limited published works which include variability. Two papers are very closely related to our research topic: Korytkowski et al.'s [25] is great but features a single-model assembly line, while Faccio et al.'s [26] article considers mixed-model assembly lines, but the sources of variability considered there are limited to milkrun train capacity and refilling interval. This connects with the key avenues for future work identified by Gil-Vilda et al. [19], which point to including variability and disturbances to the study of milkrun systems.

In consequence, the goal of this article is to continue exploring the use of milkrun trains for the internal logistics of flexible assembly operations featuring multiple manual assembly lines. In particular, we aim to look at scenarios where demand presents mass customisation characteristics (i.e., high-mix low-volume). The work presented here aims to evaluate the performance of milkrun trains and assembly lines in this demand context by focusing on two main aspects, following the lines for further investigation detected by Gil-Vilda et al. [19], namely the product mix (multi-model in opposition to single-model assembly) and the impact of variability and stochastic disturbances.

To address the aforementioned objectives, the following research questions are formulated:

1. What is the effect on the operational and logistics Key Performance Indicators (KPIs) of producing multiple models in an assembly line compared to single-model production? Are there significant differences between mixed-model and multi-model production from the milkrun internal logistics point of view?
2. How is the milkrun-assembly lines system affected by variability? In particular, to what extent is it impacted by assembly process variability and supply chain disturbances?

To carry out this research, Discrete Events Simulation (DES) was the chosen tool. A real industrial study case from a global white-goods manufacturer site located in northern Spain is presented and used to provide the foundations of the different simulation scenarios analysed to address the research questions.

The structure of this article is the following: Section 2 presents a brief literature review on the topic, highlighting the key findings made by previous research and the open lines of research derived from them. Section 3 Methodology introduces the assumptions used to build the simulation model, details the study case data and the parameters as well as the performance indicators selected to define and assess the simulation scenarios. Section 4 Results presents the outcome of the simulation, which is then discussed in Section 5.

2. Literature Review

Feeding the components to assembly lines requires complex in-plant logistics to do so in an efficient, flexible and responsive manner. Although many feeding policies could be used [27], some have clear advantages when facing a demand situation of mass customisation or mass personalisation.

In the context of Lean logistics, milkruns (also named 'tow-trains' or shuttles) are defined as *'pickups and deliveries at fixed times along fixed routes'* [18]. Inbound and outbound milkrun delivery systems work analogously, sharing a key aspect: *'milkruns are round tours on which full and empty returnable containers are exchanged in a 1:1 ratio'* [22].

Several authors have proposed different approaches for classifying milkrun systems. For instance, Kilic et al. [28] proposed that the main problem for milkrun design is to determine the routes and time periods aiming to minimise total cost, which are composed of transportation and Work In Process (WIP) holding costs. Their framework classifies milkrun problems depending on the need to determine the time periods, the routes or both; for one- or multiple-routed milkruns; and considering either equally or differently timed routes. On the other hand, Mácsay et al. [29] described four milkrun-based material supply strategies, while Klenk et al. [30] modelled milkrun systems using Methods-Time Measurement (MTM) parameters and explored six major milkrun concepts.

Alnahhal et al. conducted a literature review in 2014 [31] that found a scarcity of studies looking at in-plant milkrun systems as a whole, and that there was a research tendency to drift away from Lean goals to look for optimality based on restrictive objectives in its stead. Later articles, however, addressed in-plant milkruns from multiple angles; in particular, for mixed-model assembly systems closely related to multi-model systems, which are the focus of this article. A plethora of study cases have also been published in recent years, helping to illustrate the benefits of milkruns and the production challenges they help to overcome. The following subsections look into some of them in further detail.

2.1. In-Plant Milkruns for Mixed-Model Assembly Lines

Alnahhal et al. [32] looked into using milkruns for mixed-model assembly lines from decentralised supermarkets. Variables such as train routing, scheduling and loading problems were considered, aiming to minimise the number of trains, loading variability route length variability and assembly line inventory costs. Different analysis tools were employed: analytical equations, dynamic programming and Mixed-Integer Programming (MIP). On the other hand, Golz et al. [33] used a heuristic solution in two stages to minimise the number of shuttle drivers, focusing on the automotive sector.

This sector was also the focal point of Faccio et al.'s work [26], in which they proposed a general framework using short-term (dynamic) and long-term (static) sets of decisions allowing to size up the feeding systems for mixed-model assembly lines composed of supermarkets, kanbans and tow-trains. In another article [34], Faccio et al. dived deeper into the subject by investigating kanban number optimisation. It was highlighted that traditional kanban calculation methods fell short under a multi-line mixed-model assembly systems.

Emde et al. also looked at optimising some aspects of mixed-model assembly lines, namely (1) the location of in-house logistics zones [35] and (2) the loading of tow-trains to minimise the inventory at the assembly and to avoid material shortages, using an exact polynomial procedure [36]. Discrete Events Simulation was used by Vieira et al. [37] in an automated way (using a tailored API on top of a DES commercial software) to model and analyse the costs of mixed-model supermarkets.

2.2. Other Aspects of In-Plant Milkruns

A few articles examined the performance evaluation of milkrun systems. Klenk et al. [38] evaluated milkruns in terms of cost, lead time and service level. Their article used real data from the automotive industry with a focus on dealing with demand peaks. Bozer et al. [39] presented a performance evaluation model used to estimate the probability of (1) exceeding the physical capacity of the milkrun train and (2) exceeding the prescribed cycle time. This model assumed a basic, single-train system and that assembly lines are never starved of components. It highlighted some of the milkrun advantages: low lead times, low variability and low line-side inventory levels. Other articles describe milkrun systems evaluation methods which employ cost efficiency [29] or the required number of tow-trains [40]. Many authors used discrete event simulation to evaluate the potential performance of milkrun systems as a tool for milkrun design [41], evaluating dynamic scheduling strategies [42] or digital twin verification and validation [43].

The Association of German Engineers (VDI—*Verein Deutscher Ingenierure*) proposed the standardisation guidelines VDI 5586 [44] for in-plant milkrun systems design and dimensioning. Schmid et al. [45] discussed the draft VDI norm and found several drawbacks. Their article states that algorithms can support the milkrun design process; however, this system's design cannot be formulated as a regular optimisation problem. In a later article, Urru et al. [46] highlighted that VDI 5586 was the only norm for milkrun logistics systems design and that it is only applicable under severe restrictions. A methodology was then proposed to complement the VDI guideline. Kluska et al. proposed a milkrun design methodology which includes the use of simulation as supporting tool [41].

Gyulai et al. [47] provided an overview of models and algorithms for treating milkrun systems as a Vehicle Routing Problem (VRP). This article introduced a new approach with initial solution generation heuristics and a local search method to solve the VRP.

Gil-Vilda et al. [19] focused on studying the surface productivity and milkrun work time of U-shaped assembly lines fed by a milkrun train using a mathematical model. This article established promising avenues for future research: (1) assessing the impact of the number of parts per container and (2) analysing the impact of variability.

On the topic of variability, two articles stand out. Korytkowski [25] posed the research question about *'how disturbances in the production environment and managerial decisions affect the milkrun efficiency'*. This work analyses a single-model assembly line by employing discrete events simulation including three variability parameters—assembly process coefficient of variability, probability of a delayed milkrun cycle start and the magnitude of such delay—in addition to other three parameters: WIP buffer capacity, TAKT time synchronisation, and the milkrun cycle time. The KPIs used were throughput, WIP stock, milkrun utilisation and workstation starvation. The key conclusions were that TAKT sync does not affect the KPIs, even in conjunction with limited WIP buffer capacity. It was also found that a higher milkrun cycle time decreases the milkrun utilisation and increases the assembly line stock. Finally, this article concluded that milkrun systems mitigate the impact of production variations, which implies that they do not require large safety times built into them. Faccio et al. [26] also introduced variability sources in their dynamic milkrun framework for mixed-model assembly lines. In particular, this article includes tow-train capacity variability (related to the number of parts per kanban container, which is linked to the stochastic demand considered) and refilling interval variability.

2.3. In-Plant Milkrun Study Cases

There is no scarcity of published articles featuring study cases of in-plant milkrun systems. However, there are not so many articles specifically focusing on milkruns feeding multi-model assembly lines, and only a few articles consider stochastic variables. It is also noteworthy that the majority of study cases on the topic belong to the automotive industry. Table 1 summarises the study case articles found in this brief review, which includes the articles mentioned previously as well as a few additional documents [48–52] which specifically present milkrun study cases.

Table 1 shows some noteworthy points. First of all, no article specifically shows study cases of multi-model assembly lines, although there are some articles on mixed-model systems. Secondly, very few articles present real industrial study cases outside of the automotive sector. Finally, variability has not been commonly considered by research articles on the topic so far. The work presented here aims to cover the three highlighted shortcomings.

Table 1. Key aspects of selected research articles on in-plant milkrun systems which include study cases.

Article	Analysis Tool	Objective	No. Lines	No. Vehicles	Product Mix	Variability	Real Industry Case	Sector
Aksoy [51]	MILP and heuristics	MR route optimisation	Multi	Multi	Single	No	Yes	Automotive
Alfonso [53]	Simulation	Ergonomy and material flow improvement	Multi	Single	Single	No	Yes	Automotive
Alnahhal [32]	MIP, DP and math modelling	Min WIP, variability, handling cost	**Multi**	Multi	**Mixed**	No	No	NS [1]
Coelho [43]	Simulation	Verify and validate digital twin framework for in-plant logistics	Multi	NS	NS	**Yes**	Yes	Automotive
Costa [52]	Simulation	Train loading	Multi	Single	Single	No	Yes	Electronics
Emde [48]	MIP and heuristics	Min WIP	**Single**	Single	**Mixed**	No	No	Automotive
Faccio [26]	Math model	Min no vehicles and WIP	**Multi**	Multi	**Mixed**	**Yes**	Yes	Automotive
Faccio [34]	Math model	Optimal no. kanbans	**Multi**	Multi	**Mixed**	**Yes**	Yes	Automotive
Gil-Vilda [19]	Math model	Max surface productivity	Single	Single	Single	No	Yes	Unknown
Golz [33]	MILP and heuristics	Min no. trains	**Multi**	Single	**Mixed**	**Yes**	No	Automotive
Gyulai [47]	Heuristics and local search method	Min no. vehicles	Multi	Multi	NS	No	NS	Automotive
Kilic [28]	Mixed Integer Programming (MIP)	Min cost (no vehicles × distance travelled)	Multi	Multi	NS	No	Yes	Automotive
Klenk [38]	Math model	Handling demand peaks	Multi	Single	NS	**Yes**	Yes	Automotive
Korytkowski [25]	Simulation	Effect of disturbances and management decisions	Single	Single	Single	**Yes**	No	NS
Pekarcikova [54]	Simulation	Improve logistic flows	Single	Single	Single	No	NS	Automotive
Rao [42]	Simulation	Improve material flow, reduce no. vehicles	Multi	Multi	Single	No	NS	NS
Satoglu [50]	Math model and heuristics	MR route to minimise handling and stock costs	Multi	Single	Single	No	Yes	Electronics
Simic [49]	Particle swarm optimisation	Min stock costs	Single	Single	Single	No	No	Automotive

[1] NS: Not Specified.

3. Materials and Methods

In this article, the operational performance of two assembly lines and the milkrun train that feeds them is evaluated under different conditions. The system consisting of assembly lines and internal logistics was studied by considering a set of inputs, a Discrete Events Simulation model and a set of output KPIs, as depicted in Figure 1. The model consists of two main parts: the assembly lines and the supply chain feeding the components to the Assembly Line (AL) in containers using a milkrun train. Simulation was chosen for building this model because it allows the introduction of stochastic elements [55], such as process or logistics variability, which is necessary to achieve this work's goals. The simulation tool employed was FlexSim® (2022.0, FlexSim Software Products, Inc., Orem, UT, USA). Several simulation scenarios are created by modifying different parameters and disturbances values to analyse desired aspects of the system behaviour. Section 3.1 details the modelling assumptions. Section 3.2 includes the notation and definitions employed, and Section 3.3 includes the input data used in the models, which are used for validation (Section 3.4) and the experiment design (Section 3.5).

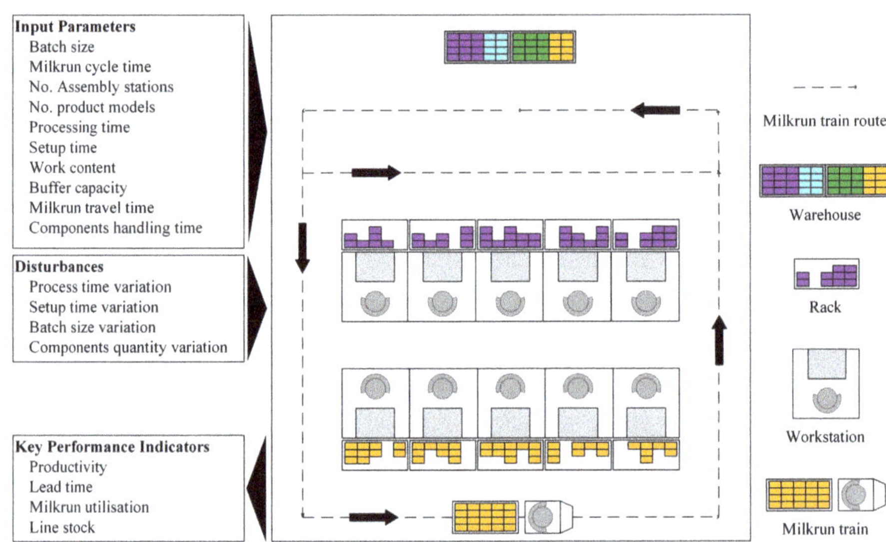

Figure 1. In-plant milkrun for multi-model assembly lines. Input parameters and disturbances are changed when analysing the performance of the system using simulation. Model output includes relevant operational and logistics Key Performance Indicators (KPIs) for evaluation.

3.1. Assumptions

The simulation model depicted in Figure 1 is made of two main subsystems: (1) two manual assembly lines, which feature operators, workstations, product buffers and components racks; and (2) internal logistics, which include a milkrun train, the components Points Of Use (POUs), a warehouse and the information flow necessary to ensure the assembly line receives the required components on time; see Figure 2.

Assembly lines: Figure 2a,b show the elements of the assembly lines used in this model, which feature the following assumptions following the classification of assembly systems by Boysen et al. [6]:

- The assembly systems are unpaced, buffered lines.
- These are fixed-worker assembly lines: operators are assigned to stations.
- There is manual assembly only (no semi- or fully automated work content).
- The number of workstations is constant. Each station can process only one product unit at a time.
- Operators need to gather all components specified by the Bill of Materials (BOM) to proceed to assemble at their stations; see Figure 2a.
- The demand mix is known and it continues for the whole simulation horizon.
- The assembly lines can be single-model, mixed-model or multi-model. Single-model lines only produce one product variant per AL. Mixed-model lines can produce more than one model, but there is no setup time between products. Multi-model lines are similar to mixed-model lines but they do incur setup time losses when changing over from one product model to another.
- Setup times, where present, are not dependent on the product sequence.
- The product sequence consists of an alternating pattern of batches of products. The batch size is stochastic, based on a discrete uniform distribution to represent the probability of a batch being released to the assembly line with fewer units than standard. This represents the disturbances caused by upstream manufacturing processes. The probability distribution is governed by the batch size coefficient of variability (CV_q).

- Processing and setup times are stochastic. They follow a lognormal distribution based on mean values and standard deviations, which are expressed by the coefficients of variability (CV_p, CV_s).
- Slightly different processing times on each station mean that these are unbalanced assembly lines, as shown in the 'Input' subsection.

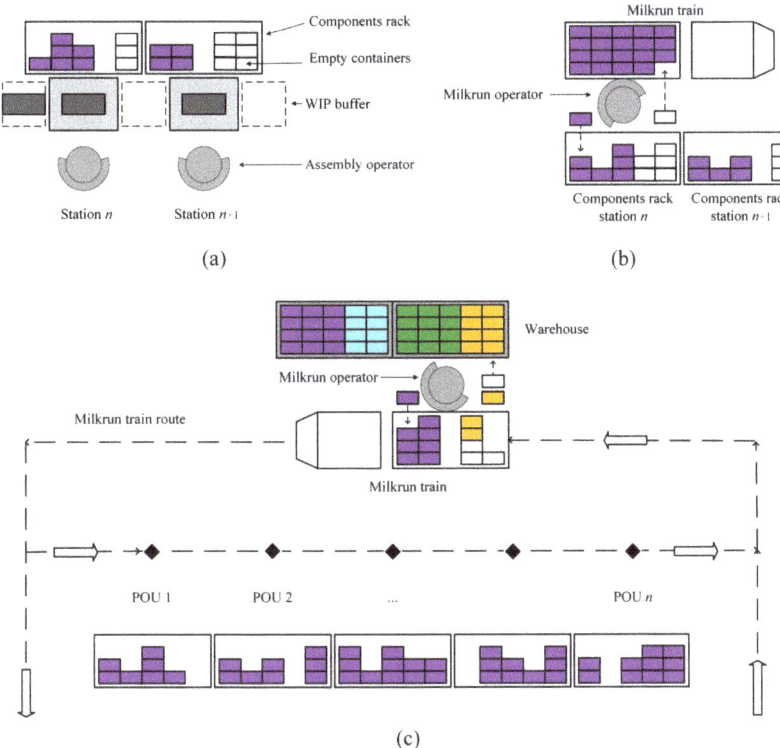

Figure 2. Simulation model subsystems interaction. (**a**) Assembly line stations; (**b**) Milkrun operator loading and unloading containers to assembly station; (**c**) Milkrun train picking at the warehouse, followed by the components replenishment cycle across all Points Of Use (POUs) of the route.

Internal Logistics: Figure 2 shows the main components of the internal logistics, which consists of four subsystems:

- Information flow between the assembly lines and the milkrun train, so that the milkrun picks up the right components for the product models that will be needed in the AL. This includes the calculations of the number of containers of each component N_i. This is worked out based on the expected consumption over the milkrun cycle time (d), the no. of pieces of component i per product unit (n_i) and the no. of pieces per container (q_i), with a minimum of 2, as shown in Equation (1). This minimum of 2 containers is required to prevent assembly line starvation, which could occur otherwise since the milkrun logic implies taking empty containers and replacing them with full ones on the next cycle.

$$N_i = \max\left(\left\lceil \frac{d_i \cdot n_i}{q_i} \right\rceil, 2\right) \tag{1}$$

- The number of pieces in each component container is stochastic, based on the standard number of pieces per container and a coefficient of variability (CV_c). A discrete uniform distribution is employed, which uses CV_c as the lower limit and the standard no. of pieces as the upper limit. This represents the probability of a certain number of pieces being non-conforming due to quality problems, inaccurate counting at the external suppliers' production site or incorrect re-packing at the in-plant warehouse, especially for components packed in bulk, such as nuts and bolts.

- Milkrun train picking at the warehouse (see Figure 2c) is modelled as a single POU. The milkrun train is emptied upon arrival, and it is thereafter filled again with the required containers for the next supply cycle.
- The milkrun transportation time from/to all POUs (Figure 2c) is based on historical time measurements from the industrial study case. Since the data show very little variability, the model assumes a deterministic transportation time given by the input parameter T_t.
- Supply chain operator loading and unloading of component containers to the assembly lines at each POU, as shown in Figure 2b. There are two possible situations: (1) Regular cycle (same product model): the operator replaces the empty boxes in the 'returns rack' with full boxes of the same component. The handling time is different for full and empty containers; see the input subsection. (2) Product changeover cycle (before the assembly line changeover): in which the milkrun operator firstly replaces any current product empty container to ensure that the current batch can be finished and then loads the next containers of the next product components so that they are available to the assembly operators when they finish the stations' changeover.

3.2. Notation

The following notations are introduced:
Input: Parameters

Q	Batch size
CT	Assembly cycle time
CT_{MR}	Milkrun cycle time
L	No. of assembly lines, index l.
K	No. of assembly workstations (no. POUs) per assembly line, index k.
M	No. of product models, index m.
T_p	Processing time
T_s	Setup time
WC	Work content (i.e., total process time)
WIP	No. of work in progress units between workstations
T_t	Milkrun transportation time to/from assembly line
T_h^e	Milkrun operator container handling time, empty container
T_h^f	Milkrun operator container handling time, full container

Input: Disturbances

CV_p	Process time coefficient of variation: $CV_p = \sigma_{Tp}/\mu_{Tp}$
CV_s	Setup time coefficient of variation: $CV_s = \sigma_{Ts}/\mu_{Ts}$
CV_c	Conforming units per container coefficient of variation
CV_q	Batch size coefficient of variation

Output: Key Performance Indicators

P	Productivity (units/operator-h): production rate of conforming units per assembly operator.
LT	Lead Time (min): average time for a batch of units to be finished from the moment the last unit of the previous batch is finished.
U	Milkrun Utilisation (%): fraction of total available time that the supply chain operator is busy (picking components at the warehouse, driving the milkrun train and handling containers to load/unload the components at the POUs).
S	Stock in the assembly line (units): average stock of components held in the assembly line measured in equivalent finished product units.

3.3. Input Data

The simulation model uses data provided by the industrial study case, which presents a common situation faced by plenty of manufacturing businesses globally. Table 2 shows the model parameters base, min and max values.

Table 2. Input parameters and disturbances base and range values.

Parameter	Units	Min	Max	Base Value
Q	units			48
CT	s			see Table 3
CT_{MR}	min			140
L	lines			2
K	stations			5
M	models	2	4	4
T_p	s			see Table 3
T_s	s			480
WC	s			see Table 3
WIP	units			1
T_t	min			4
T_h^e	s			1
T_h^f	s			2
CV_p		0	0.50	0.15
CV_s		0	0.50	0.15
CV_q		0	0.50	0.10
CV_c		0	0.20	0.00

The operations considered in this model include two manual assembly lines which assemble four product models, two on each line. The mean processing times for each model and station along with work content and cycle time is summarised in Table 3. These processing times were obtained from the industrial company standard operating procedures, which in turn are calculated using MTM.

Table 3. Product processing time input data.

Line	m	T_p (s)					CT (s)	WC (s)
		$k = 1$	$k = 2$	$k = 3$	$k = 4$	$k = 5$		
1	1	192.8	187.5	185.5	188.2	190.1	192.8	944.1
	2	214.3	210.2	215.4	212.0	210.7	215.4	1062.6
2	3	237.6	238.5	236.7	233.0	232.1	238.5	1177.9
	4	176.1	176.1	175.1	173.2	173.0	176.1	873.5

The products within a line share materials, technological features and general purposes, but they require different components, assembly fixtures and tooling. This calls for changeovers to adjust the workstations when a batch of a different product model is required. The parameter governing setup times is T_s, which takes each operator approximately 6 min (see Table 2), independently of the product sequencing.

Each product unit consists of many different components, as shown in Table 4. Most components are required only once per finished product unit, although some components, especially the smaller ones, may be required in larger numbers.

Table 4. Bill Of Materials summary data.

m	No. Components					Total No. Components	Total Pieces
	$k = 1$	$k = 2$	$k = 3$	$k = 4$	$k = 5$		
1	16	6	10	11	4	47	62
2	28	4	14	13	13	72	132
3	20	7	20	18	21	86	160
4	16	9	9	24	14	72	105

Components are transported to the POUs and then presented to the assembly operators in containers, i.e., boxes, trays or small trolleys. Each container carries a certain number of pieces of one component, typically a few dozens for middle- and large-size components, and about one hundred pieces for small components, such as bolts, screws and washers.

In this particular study case, an important number of components are packed in very large quantities per container compared to the number of pieces needed to feed the assembly line for the duration of the milkrun cycle. Note that the the milkrun cycle time is approximately similar to the time required to complete a production batch. To illustrate this fact, Table 5 shows the number of components of each product model that are packed in *large quantities*. Here, *large quantities* refers to the case in which one single container includes a number of pieces allowing to assemble more than two full batches of products—i.e., it is equivalent to the assembly line consumption of two milkrun cycles.

Table 5. Details of the high number of components served in large quantities [2] to the assembly lines.

Number of Components	Product Model				Avg
	$m = 1$	$m = 2$	$m = 3$	$m = 4$	
Total no. components	47	72	86	72	69
Packed in large quantities [2]					
No. components	13	25	29	37	26
Percentage components	28%	35%	34%	51%	28%

[2] Containers including a no. of pieces equivalent to the consumption of more than two milkrun cycles.

When the milkrun operator arrives at each POU, the containers are handled between the train and the back side of the POU racks. Based on measurements at the industrial partner facility, one second was estimated for handling empty containers and two seconds for containers full of components, as shown in Table 2. When walking from the milkrun train to the POU, the milkrun operator's speed was considered 1 m/s. The milkrun train speed in the assembly line area was found to be around 1 m/s, and the POU positions are separated approximately 2 m from each other, resulting in a 12 m long assembly line. Regarding the milkrun train travel from the warehouse to either assembly line, the industrial partner measurements showed little variability for an average travel time of approximately 4 min each way. The milkrun preparation time at the warehouse (picking time) was simulated considering the warehouse as a single picking point and treated as any POU of the assembly line.

The DES model takes into account the inherent variability of manual assembly operations by using lognormal distributions for processing and setup times, following the recommendations of [56]. The lognormal distribution is generated using the mean (μ) values of T_p and T_s—see Table 3—and the standard deviation (σ), which is given as a percentage of the mean by the coefficients of variation CV_p and CV_s. The base values for the coefficients were estimated from historical data provided by the industrial partner of this study case. The data allowed estimating CV_p and CV_s to be in the range of 0.15–0.20 for manual assembly lines. Note that since one of the goals of this work is to analyse the influence of processing and setup times variability on the internal logistics performance, CV_p and CV_s will take a range of values in certain simulation scenarios. Another two sources of variability, introduced in Section 3.1, are considered: the conforming units per container variability (CV_c) and the batch size quantity variability (CV_q). They are relevant along with the processing and setup variability because the logistic performance of the milkrun system is directly related to them.

3.4. Verification and Validation

The validation and verification of the simulation models were performed separately for assembly operations and internal logistics.

For the assembly operations section, historical production KPIs data were gathered and compared against the results of a simple parametric model and a discrete events simulation model. The results presented by the authors in [57] allowed the validation of both models by comparison against real industry study case data. It was also possible to verify the parametric model against the simulation model (considering no variability) because their results difference was smaller than 3.5% for any considered performance metric. In summary, the results indicated that both parametric and simulation models slightly underestimate total output and that they overestimate the production rate, labour productivity and line productivity. Both models were found to be reliable for the context considered here since the mean relative error was 1.63% and the max relative error was 4.9%.

Regarding the internal logistics part of the simulation model, the validation was carried out using measurements at the industrial partner assembly lines from June 2022. A total of 18 milkrun cycle measurements were registered, finding an average milkrun utilisation of 78.4%. This was compared with the equivalent simulation model results ($U = 71.6\%$) to calculate a relative error of 8.7%, slightly below 10%, which was considered satisfactory for the scope of this work.

3.5. Experiment Design

To address the research questions laid out in Section 1, several simulation scenarios were designed and then implemented on the simulation model by modifying the model's parameters. Table 6 summarises the parameters and range of values used to set up the simulation scenarios.

Table 6. Simulation scenarios.

Scenario	Parameter	Units	Range
i. Product mix	M	models	$\{2,4\}$
	T_s	s	$\{0,480\}$
ii. Process variability	CV_p, CV_s	per unit	$[0,0.50]$
iii. Batch size variability	CV_q	per unit	$[0,0.50]$
iv. Components quantity var.	CV_c	per unit	$[0,0.20]$

The first research question—*'(1) What is the effect on the operational and logistics KPIs of producing multiple models in an assembly line compared to single-model production? Are there significant differences between mixed-model and multi-model production from the milkrun internal logistics point of view?'*—is examined by changing the number of product models under demand (one model per assembly line for single-model, $M = 2$; two models per assembly line per mixed- and multi-model, $M = 4$) and the setup time duration parameter (T_s set to 0 s for mixed-model, 480 s for multi-model). For this *scenario i.*, process and batch quantity coefficients of variability take their base values (T_p and T_s 0.15, CV_q 0.10), and the conforming units per container coefficient of variability is set to 0, as stated in Table 2.

The second research question—*'(2) How is the milkrun-assembly lines system affected by variability? In particular, to what extent is it impacted by assembly process variability and supply chain disturbances?'*—will be decomposed into the three variability sources considered in the simulation model. Firstly, process variability is governed by parameters CV_p (assembly processing time variability) and CV_s (setup time variability). These parameters will take values ranging from 0 (no variability at all) up to 0.50 (high variability), making up *scenario ii.* Secondly, the batch size variability coefficient will be used to represent in-plant manufacturing issues leading to smaller-than-standard batches of products being released for assembly. Similarly to the previous scenario, in *scenario iii.* CV_q values will range from 0 to 0.50, covering from no disturbances up to half of the batches having fewer units than it was intended. Finally, *scenario iv.* looks into external supplier perturbations which are simulated using the components quantity coefficient of variability. CV_c will take values in the range of 0 to 0.20, meaning that each components container can have up to 20% fewer valid pieces in the less favourable case. The effect of the interactions between the variability parameters was not analysed because a preliminary two-level full factorial design of experiments showed that two-factor interactions were not significant for the KPIs under study in comparison to the effects of the variability parameters by themselves.

The following Section 4 Results shows the outcome of the simulation scenarios introduced here.

4. Results

This section includes the outcome of the simulations corresponding to *scenarios i.-iv.* Section 4.1 addresses the first research question, and Section 4.2 includes *scenarios ii.-iv.*, which jointly address the second research question.

The results shown here are obtained with a simulation horizon of 74 h with a warm-up time of 2 h (i.e., nine production shifts after the warm-up is finished). To account for the stochastic nature of the results, each simulation scenario is run 20 times. This number was chosen because it was found that using a larger number of runs did not affect the resulting output in a statistically significant manner. At the start of each simulation run, all assembly stations and buffers between them are empty as well as all the components racks and the milkrun train.

The results shown in this section are presented in boxplots where the upper and lower limit of the boxes corresponds to the first and third quartiles. The coloured line is the mean and the whiskers limits are set to 1.5 times the interquartile range. Outlier data points (beyond the whiskers) are marked by a circle. The charts scale has been kept constant across all simulation scenarios to facilitate comparison.

4.1. Single-Model vs. Mixed-Model, Multi-Model Assembly

The selected operational KPIs comparing the performance of the assembly lines under *scenario i.* demand conditions are shown in Figure 3 and summarised in Table 7.

The productivity of single- and mixed-model lines is significantly superior to multi-model lines, as is expected considering that the setup time becomes zero (from 480 s per batch of 48 units, which represents just below 5% of the time needed to complete the batch on average). The difference in productivity between single- and mixed-model lines is related to operator idle and blocked times following product model changeovers as a result of cycle time differences between the incoming and outgoing products. Said difference does not account for significant productivity results in this case. Batch lead time, as expected, is slightly larger for mixed- and multi-model lines compared to single-model lines.

On the internal logistics KPIs side, milkrun utilisation and assembly line stock show a clear differentiation between single-model assembly lines and the other two. Incorporating multiple product models increases greatly the utilisation (from 51% to 72%, a +44% increment). Note that this steep increase could be linked to the high percentage of components packed in large quantities. This will be examined in the next Section 5 Discussion.

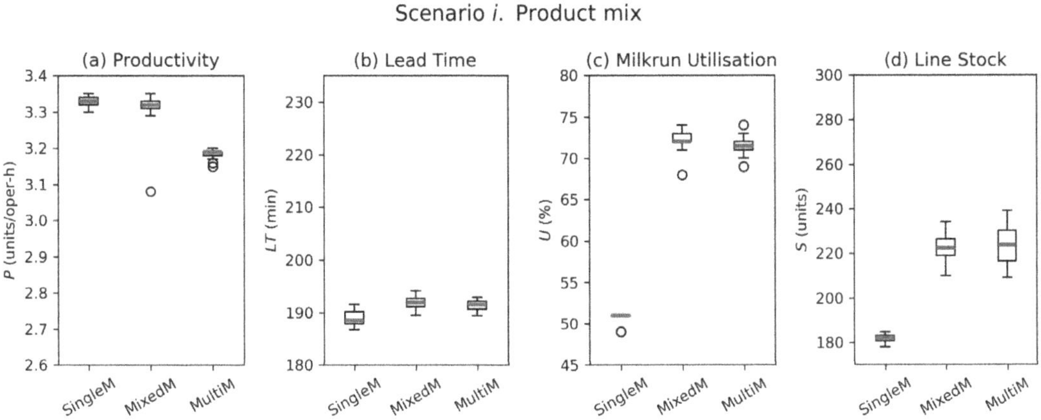

Scenario *i*. Product mix

Figure 3. *Scenario i.*: Mean and deviation values of KPIs for single-, mixed- and multi-model assembly lines. (**a**) Line productivity, (**b**) batch lead time, (**c**) milkrun utilisation and (**d**) assembly line stock levels.

The component stock in the assembly line also suffers an increase for mixed- and multi-model lines driven by the same reason: single-model assembly lines see their average component stock decrease as the containers with very large quantities of pieces are consumed over time. Contrarily, mixed- and multi-model lines are constantly fed with small component boxes full of pieces. In the case shown here, the difference is significant but not dramatic, at an approx. +22% increase (from 182 to 223 units).

In summary, increasing product mix negatively affects operational KPIs (reduces productivity, increases batch lead time), which was expected. It also increases greatly supply chain operator utilisation (+44% rise), although the magnitude of this sharp increase could be attributed to the high percentage of components packed in large quantities.

Table 7. *Scenario i*: Mean and standard deviation (SD) of main KPIs for single-, mixed- and multi-model assembly lines.

Product Mix	P (u/oper-h)		LT (min)		U (%)		S (u)	
	Mean	SD	Mean	SD	Mean	SD	Mean	SD
Single-model	3.33	0.015	188.9	1.5	50.60	0.82	181.7	1.9
Mixed-model	3.31	0.006	192.0	1.4	72.05	1.23	222.8	6.2
Multi-model	3.19	0.013	191.4	1.1	71.50	1.24	223.2	8.4

4.2. Variability and Disturbances

This subsection looks at how increasing levels of variability affect the operational (P, LT) and internal logistics KPIs (U, S). As described in Section 3.5, simulation experiments were set up to independently analyse the influence of assembly line process variability (CV_p and CV_s, scenario ii.), batch size variability (CV_q, scenario iii.) and conforming components variability (CV_c, scenario iv.).

4.2.1. Process Variability

To analyse the impact of the assembly line process and setup variability, the respective coefficients were modified increasingly from 0 up to 0.50 (the base value for the industrial case study is 0.15; see Table 2). Figure 4 shows the results of this simulation scenario, and Table 8 includes the results' numeric values for average and standard deviation.

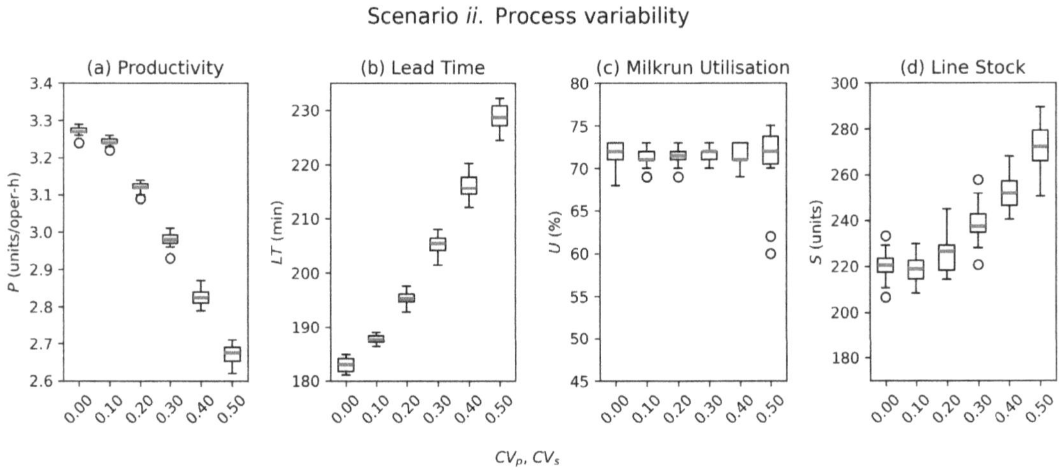

Figure 4. *Scenario ii.*: Mean and deviation values of KPIs for varying levels of process and setup coefficients of variation. (**a**) Productivity, (**b**) batch lead time, (**c**) milkrun utilisation, and (**d**) assembly line stock level.

In terms of operational KPIs, Figure 4a,b show that, as expected, an increase in process variability negatively the performance of the assembly line, especially considering that this lines' number of work-in-process units is limited to one. In particular, it can be seen that the productivity deteriorates greatly when CV_p and CV_s are greater than 0.20 both in terms of mean and standard deviation. Batch lead time follows the same trend.

Figure 4c shows that U does not suffer any changes, although its standard deviation increases slightly. On the other hand, the assembly line components' stock levels are severely impacted, rising from approx. 220 units for none or very small variability (CV_p and CV_s at 0–0.10) up to an average of approx. 270 units for CV_p, CV_s 0.50, which represents a noticeable +23% increase. Standard deviation also rises, but it remains small compared to the mean values of S, as shown in Figure 4d. In summary, only AL stock levels are affected by in-process variability, while the milkrun driver's workload remains unaffected.

Table 8. *Scenario ii.*: Mean and standard deviation of main KPIs for increasing values of process variability.

CV_p, CV_s	P (u/oper-h)		LT (min)		U (%)		S (u)	
	Mean	SD	Mean	SD	Mean	SD	Mean	SD
0.00	3.27	0.012	183.0	1.2	71.75	1.33	220.4	4.9
0.10	3.24	0.011	187.8	0.8	71.25	1.16	218.7	5.8
0.20	3.12	0.015	195.4	1.1	71.45	1.05	238.5	7.6
0.30	2.98	0.019	205.2	1.8	71.30	0.99	238.1	8.8
0.40	2.83	0.021	216.0	2.2	71.45	1.23	252.6	7.8
0.50	2.67	0.025	228.7	2.3	71.28	4.06	272.1	10.7

4.2.2. Batch Size Variability

To understand the impact that upstream manufacturing process issues would have on the assembly operational and internal logistics performance, *scenario iii.* was set up by changing the value of CV_q, which determines the probability of an assembly production batch smaller than standard. CV_q takes values between 0 (no disruption) and 0.50 (meaning that on average, half the batches released to the assembly lines have between 36 and 48 units.

The simulation results of *scenario iii.* are summarised in Figure 5, and average and standard deviation data are shown in Table 9.

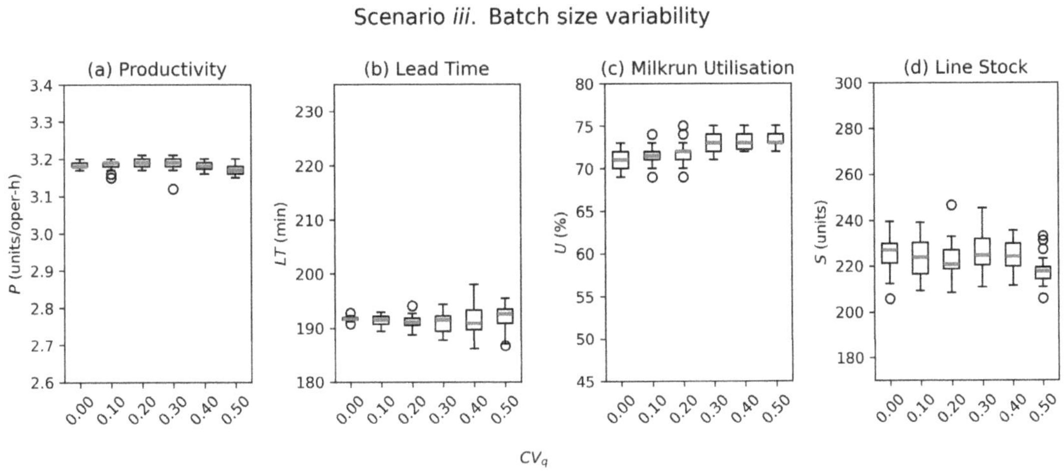

Figure 5. *Scenario iii.*: Mean and average values of KPIs for varying levels of batch size coefficients of variation. (**a**) Productivity, (**b**) batch lead time, (**c**) milkrun utilisation, and (**d**) assembly line stock.

Figure 5a,b shows that the average of both line productivity and lead time remains constant despite changes in CV_q. Although P standard deviation increases slightly, it remains very low at about 0.25–0.43% of the average value. The lead time StDev, on the other hand, does increase more than five-fold while remaining very low compared to average values (StDev of 0.24–1.39%). Therefore, the data show that batch size variability has no significant impact on the operational KPIs. Although variability rises as CV_q grows, it remains at very low levels in relative terms.

Figure 5c,d show very little impact on internal logistics KPIs as a result of an important rise in batch size variability. The milkrun utilisation average does increase slightly (from 71 to 73%, c.+4% rise), but the StDev reduction (from 1.25% to 0.82%) is not statistically significant. In a similar fashion, assembly line components stock decreases slightly in both average and standard deviation values, but none of these changes are statistically significant.

Table 9. *Scenario iii.*: Mean and standard deviation of main KPIs for increasing values of batch size variability.

CV_q	P (u/oper-h)		LT (min)		U (%)		S (u)	
	Mean	**SD**	**Mean**	**SD**	**Mean**	**SD**	**Mean**	**SD**
0.00	3.18	0.008	191.7	0.5	70.68	1.25	225.6	8.0
0.10	3.19	0.013	191.4	1.1	71.50	1.24	223.2	8.4
0.20	3.19	0.012	191.1	1.2	71.74	1.37	222.6	9.1
0.30	3.19	0.019	191.0	2.0	72.84	1.07	225.8	8.5
0.40	3.18	0.011	191.1	2.8	73.17	0.92	224.7	6.7
0.50	3.17	0.014	191.9	2.7	73.32	0.82	218.1	6.8

4.2.3. Components Quantity Variability

The goal of this subsection is to analyse the impact of the components quantity coefficient of variability CV_c. This coefficient is employed to represent disturbances within in-house or external suppliers' processes, resulting in a lower-than-standard number of conforming pieces in each component container. As explained in Section 3, the number of conforming pieces per container is simulated using a discrete uniform distribution which has the inferior limit set to CV_c percent of the nominal value. *Scenario iv.* considers CV_c values from 0 to 0.20, as shown in Table 10.

Figure 6a shows that productivity is affected negatively by an increase in CV_c, although the magnitude of the impact is very limited: only a -2.2% reduction from the base scenario when components containers have up to 20% less conforming pieces than expected. Similarly, lead time is impacted negatively by CV_c increase, as depicted in Figure 6b. The LT average rises slightly (c.+2%) and suffers a greater dispersion of results (StDev increases by +54%). All in all, even a substantial increase in components quantity variability does not affect the assembly lines' operational KPIs severely.

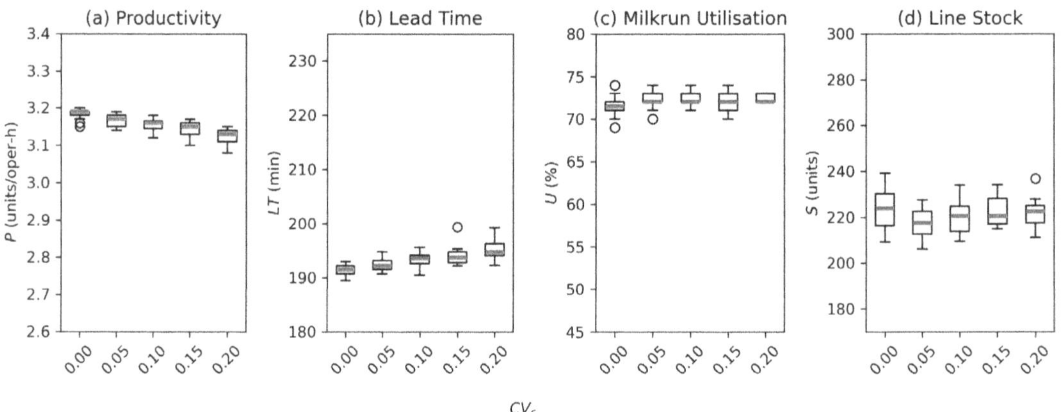

Figure 6. *Scenario iv.*: Mean and deviation values of KPIs for varying levels of components quantity coefficients of variation. (**a**) Productivity, (**b**) batch lead time, (**c**) milkrun utilisation, (**d**) assembly line stock.

Regarding internal logistics KPIs, Figure 6c,d show that an increase of CV_c has no significant impact on either milkrun utilisation or assembly line component stock levels.

Table 10. *Scenario iv.*: Mean and standard deviation of main KPIs for increasing values of component quantity variability.

	P (u/oper-h)		LT (min)		U (%)		S (u)	
CV_c	Mean	SD	Mean	SD	Mean	SD	Mean	SD
0.00	3.19	0.013	191.4	1.1	71.50	1.24	223.2	8.4
0.05	3.17	0.015	192.4	1.2	72.17	0.99	217.5	6.2
0.10	3.16	0.016	193.5	1.2	72.26	0.81	220.0	6.8
0.15	3.15	0.019	194.0	1.6	72.00	1.05	222.0	5.6
0.20	3.12	0.019	195.2	1.7	72.40	0.50	221.9	5.9

5. Discussion

The results shown in the previous section have been summarised in Table 11.

Table 11. Summary of KPI change trends resulting from each scenario considered.

Scenario	Productivity	Lead Time	Milkrun Utilisation	Line Stock
Goal	↑	↓	↓	↓
i. Product mix	↘	↗	↑↑	↑↑
ii. Process variability: CV_p, CV_s ↑	↓↓	↑↑	=	↑↑
iii. Batch size variability: CV_q ↑	=	=	≈	≈
iv. Components quantity variability: CV_c ↑	↘	↗	≈	≈

Increasing the product mix from single- to mixed- and multi-model assembly lines results in a moderate impact on operational performance (P, LT) but a very significant negative effect on internal logistics KPIs, which could have further implications. For instance, the rise of assembly line component stock would increase the required floor space and decrease the assembly line surface productivity.

It is important to note that according to the results shown in Section 3.1, the greatest factor affecting U is the product mix, with a remarkable +44% increase resulting from changing from single- to multi-model assembly.

This sharp increase in U is caused by the rising number of containers that need to be handled, which is due to two main reasons.

(1) First of all, the number of component containers to be handled is larger every time there is a product changeover, which is the case for almost every milkrun cycle under the assumption that the milkrun cycle time is approximately similar to the time required to complete a batch of products (cf. CT_{MR}, Q in Table 2 and CT in Table 3). The increased number of containers to be handled is due to the fact that the supply chain operator needs to take all the containers of the outgoing model from the POU racks regardless of how many component pieces are left and replace them with components for the incoming product model. During regular supply cycles, on the other hand, containers are only replaced if needed (empty boxes work as kanban signals).

(2) The second reason is related with the compound effects of the first reason and the fact that in this particular study case, we find a large number of components packed in large quantities (see Table 5). This fact means that for a significant percentage of the components, each milkrun train carries enough pieces to assemble more than four times the required amount of pieces. Furthermore, the milkrun train will need to take back to the warehouse a full container and a half-empty container every time a changeover is needed.

Thus, it seems reasonable to conclude that milkrun utilisation is higher on mixed- and multi-model lines compared to single-model assembly lines. However, the magnitude of the increase shown in the Results must be considered carefully, since it it would be strongly related to the container quantities of this particular industrial study case.

As a closing remark on this subject, two aspects could be looked at in order to reduce the milkrun utilisation for multi-model assembly lines. Firstly, if enough shop-floor space is available, small components packed in large quantities could be left by the workstations, forming an assembly line supermarket, independent of the regular milkrun cycles. For larger components, relaxing the rule of minimum two containers (see Equation (1)) could be considered. Secondly, packing components in smaller quantities (so that two containers cover approximately the consumption of a milkrun cycle) could also reduce the milkrun workload so that it is only slightly higher than for single-model assembly lines.

Production variability (CV_p, CV_s) is the most important disturbance factor affecting productivity, lead time and assembly line components stock. However, it does not affect supply chain operator utilisation because the productivity reduction implies a reduction of output rate (which slows down components consumption). The reason behind this is that the milkrun work logic establishes a fixed replenishment frequency (milkrun cycle time), resulting in a supply chain operator workload effectively unaffected by several minor variations over the course of a full replenishment cycle.

Despite the previous expectation that variability would always impact performance negatively, results from Sections 4.2.2 and 4.2.3 show that the internal logistics KPIs are not sensitive to disturbances originated by batch size and components quantity variability (CV_q and CV_c respectively). This implies that employing milkruns for the internal logistics of flexible multi-model assembly lines under high-mix low-volume demand is a way to shield this part of the supply chain from upstream disturbances, arriving from either external or internal processes.

It was also found that variability regarding batch size (CV_q) does not have any noticeable negative impact on operational performance, as shown in Figure 5c.

Note that as mentioned in Section 2, this article addresses a gap in the literature by specifically addressing in-plan logistics for multi-model assembly operations, including variability, and using a real study case—specially from an industry sector other than automotive.

The fact that the simulation model used in this work is based on a real industry study case provides valuable insight into the behaviour of similar assembly operations—internal logistics systems under increasingly hard conditions in terms of variability and product mix. However, it is important to note that this also limits the generalisation extent of the results obtained due to certain aspects listed below.

First of all, the case employed here considers only a relatively small product variation within each assembly line (ΔWC 13% and 34% for AL no.1 and AL no.2, respectively) and almost no difference in terms of average WC per model when comparing both lines (ΔWC c.2%). Understanding how much product variability affects the operational and internal logistics KPIs could be a potential avenue for further research to understand the extent of the potential benefits of employing milkruns for high-mix low-volume assembly.

Secondly, it could be argued that the number of conforming components coefficient of variability (CV_c) only modifies the number of pieces per container available to the assembly operator, but it does not realistically capture the possibility of components actually arriving at the assembly line and then causing quality control failures or unexpected assembly process time increases, which would imply additional productivity losses due to reasons such as product rework and idle/blocked assembly operators.

Thirdly, milkrun transportation time was considered deterministic because the industry case measurements indicated this time were consistent. However, for multi-train production sites, variability caused by occasional milkrun train traffic jams could be considered.

Finally, modelling the milkrun train as a single wagon could be slightly underestimating its utilisation despite the satisfactory validation results. Specifically, in potential scenarios featuring longer milkrun cycle times—note that the CT_{MR} parameter was unchanged through scenarios *i.* to *iv.*—this would entail a greater number of component containers and therefore potentially a greater number of required wagons leading to an increased walking time for the supply chain operator, which the current simulation model would not capture.

6. Conclusions

To address a mass customisation demand context that drives high-mix low-volume assembly operations, this article studied the implications of using milkrun trains for the internal logistics of multi-model assembly lines. Based on a real industrial study case from the white-goods sector, a discrete events simulation model was employed to set up four different scenarios which evaluate the effect of product mix and three different sources of variability. To measure such impact, a set of four Key Performance Indicators (KPIs) were used, two corresponding to assembly operations and two corresponding to supply chain efficiency.

It was found that multi-model lines increase significantly the milkrun utilisation and the assembly line components stock compared to single-model lines. However, the magnitude of this large increase could be partially attributed to particularities of the study case. Operational KPIs were also affected negatively but to a much lesser extent. Internal logistics performance is greatly affected by the variability of assembly line processing time, especially in terms of component stock. Other sources of variability, such as the ones affecting the number of units per production batch or the components quantity per container, have very limited impact on the selected KPIs. This would imply that employing milkruns for the internal logistics of flexible multi-model assembly lines under high-mix low-volume demand is a way to shield this part of the supply chain from upstream disturbances, arriving from either external or internal processes.

Two key limitations of this work are the relatively low product variability in terms of work content and the milkrun train physical features simplification.

Further research paths include exploring the implications of much greater product work content variability, incorporating more detailed physical models of the milkrun train and expanding the simulation model to include adjacent layers that could constrain the performance of the assembly system as a whole, such as quality (defects, reworks, quality controls) or breakdowns and maintenance.

Author Contributions: Conceptualization, A.M., M.T. and J.A.Y.-F.; methodology, A.M., M.T., F.G.-V. and J.A.Y.-F.; software, A.M. and M.G.-C.; validation, A.M. and M.G.-C.; formal analysis, A.M.; investigation, A.M.; resources, A.M.; data curation, A.M.; writing—original draft preparation, A.M.; writing—review and editing, A.M., M.G.-C., M.T., F.G.-V. and J.A.Y.-F., ; visualization, A.M.; supervision, M.T. and J.A.Y.-F.; funding acquisition, J.A.Y.-F. All authors have read and agreed to the published version of the manuscript.

Funding: This research work was undertaken in the context of DIGIMAN4.0 project ("DIGItal MANufacturing Technologies for Zero-defect Industry 4.0 Production", http://www.digiman4-0 .mek.dtu.dk/). DIGIMAN4.0 is a European Training Network supported by Horizon 2020, the EU Framework Programme for Research and Innovation (Project ID: 814225).

Data Availability Statement: The data presented in this study are available on request from the corresponding author. The data are not publicly available due to industrial company confidentiality agreement.

Acknowledgments: The authors would like to thank B/S/H/ Electrodomesticos España SA for its collaboration in this study.

Conflicts of Interest: The authors declare no conflict of interest.

Abbreviations

The following abbreviations are used in this manuscript:

AL	Assembly Line
BOM	Bill Of Materials
DES	Discrete Events Simulation
KPI	Key Performance Indicator
POU	Point Of Use
WC	Work Content
WIP	Work-In-Process

References

1. Koren, Y. *The Global Manufacturing Revolution: Product-Process-Business Integration and Reconfigurable Systems*; John Wiley & Sons: Hoboken, NJ, USA, 2010.
2. Hu, S.J. Evolving paradigms of manufacturing: From mass production to mass customization and personalization. *Procedia CIRP* **2013**, *7*, 3–8. [CrossRef]
3. Yin, Y.; Stecke, K.E.; Li, D. The evolution of production systems from Industry 2.0 through Industry 4.0. *Int. J. Prod. Res.* **2018**, *56*, 848–861. [CrossRef]
4. ElMaraghy, H.; Monostori, L.; Schuh, G.; ElMaraghy, W. Evolution and future of manufacturing systems. *CIRP Ann.* **2021**, *70*, 635–658.
5. Hu, S.; Ko, J.; Weyand, L.; Elmaraghy, H.; Lien, T.; Koren, Y.; Bley, H.; Chryssolouris, G.; Nasr, N.; Shpitalni, M. Assembly system design and operations for product variety. *CIRP Ann.-Manuf. Technol.* **2011**, *60*, 715–733. [CrossRef]

6. Boysen, N.; Fliedner, M.; Scholl, A. A classification of assembly line balancing problems. *Eur. J. Oper. Res.* **2007**, *183*, 674–693. [CrossRef]

7. Modrak, V.; Marton, D.; Bednar, S. The influence of mass customization strategy on configuration complexity of assembly systems. *Procedia CIRP* **2015**, *33*, 538–543. [CrossRef]

8. Tan, C.; Hu, S.J.; Chung, H.; Barton, K.; Piya, C.; Ramani, K.; Banu, M. Product personalization enabled by assembly architecture and cyber physical systems. *CIRP Ann.-Manuf. Technol.* **2017**, *66*, 33–36. [CrossRef]

9. Womack, J.P.; Jones, D.T.; Roos, D. *The Machine That Changed the World*; Macmillan Publishing Company: New York, NY, USA, 1990; p. 336. [CrossRef]

10. Horbal, R.; Kagan, R.; Koch, T. Implementing lean manufacturing in high-mix production environment. In *Lean Business Systems and Beyond*; Springer: Berlin/Heidelberg, Germany, 2008; pp. 257–267. [CrossRef]

11. Adnan, A.N.; Jaffar, A.; Yusoff, N.; Halim, N.H.A. Implementation of Continuous Flow System in Manufacturing Operation. *Appl. Mech. Mater.* **2013**, *393*, 9–14. [CrossRef]

12. Mourtzis, D.; Doukas, M. Design and planning of manufacturing networks for mass customisation and personalisation: Challenges and outlook. *Procedia Cirp* **2014**, *19*, 1–13. [CrossRef]

13. Rüßmann, M.; Lorenz, M.; Gerbert, P.; Waldner, M.; Justus, J.; Engel, P.; Harnisch, M. Industry 4.0: The future of productivity and growth in manufacturing industries. *Boston Consult. Group* **2015**, *9*, 54–89.

14. Nolting, L.; Priesmann, J.; Kockel, C.; Rödler, G.; Brauweiler, T.; Hauer, I.; Robinius, M.; Praktiknjo, A. Generating transparency in the worldwide use of the terminology industry 4.0. *Appl. Sci.* **2019**, *9*, 4659. [CrossRef]

15. Schmid, N.A.; Limère, V. A classification of tactical assembly line feeding problems. *Int. J. Prod. Res.* **2019**, *57*, 7586–7609. [CrossRef]

16. Guo, W.; Jin, J.; Hu, S.J. Allocation of maintenance resources in mixed model assembly systems. *J. Manuf. Syst.* **2013**, *32*, 473–479. [CrossRef]

17. Davrajh, S.; Bright, G. Advanced quality management system for product families in mass customization and reconfigurable manufacturing. *Assem. Autom.* **2013**, *33*, 127–138. [CrossRef]

18. Baudin, M. *Lean Logistics: The Nuts and Bolts of Delivering Materials and Goods*; CRC Press: Boca Raton, FL, USA, 2005.

19. Vilda, F.G.; Yagüe-Fabra, J.A.; Torrents, A.S. An in-plant milk-run design method for improving surface occupation and optimizing mizusumashi work time. *CIRP Ann.* **2020**, *69*, 405–408.

20. Banjo, O.; Stewart, D.; Fasli, M. Optimising Mixed Model Assembly Lines for Mass Customisation: A Multi Agent Systems Approach. In *Advances in Manufacturing Technology XXX*; Goh, K., Case, Y., Eds.; IOS Press: Amsterdam, The Netherlands, 2016; Volume 3, pp. 337–342. [CrossRef]

21. Tortorella, G.L.; Miorando, R.R.R.; Marodin, G. Lean supply chain management: Empirical research on practices, contexts and performance. *Int. J. Prod. Econ.* **2017**, *193*, 98–112. [CrossRef]

22. Meyer, A. *Milk Run Design: Definitions, Concepts and Solution Approaches*; KIT Scientific Publishing: Karlsruhe, Germany, 2017; p. 286. [CrossRef]

23. Chee, S.L.; Mong, M.Y.; Chin, J.F. Milk-run kanban system for raw printed circuit board withdrawal to surface-mounted equipment. *J. Ind. Eng. Manag.* **2012**, *5*, 382–405. [CrossRef]

24. Battini, D.; Gamberi, M.; Persona, A.; Sgarbossa, F. Part-feeding with supermarket in assembly systems: transportation mode selection model and multi-scenario analysis. *Assem. Autom.* **2015**, *35*, 149–159. [CrossRef]

25. Korytkowski, P.; Karkoszka, R. Simulation-based efficiency analysis of an in-plant milk-run operator under disturbances. *Int. J. Adv. Manuf. Technol.* **2016**, *82*, 827–837. [CrossRef]

26. Faccio, M.; Gamberi, M.; Persona, A.; Regattieri, A.; Sgarbossa, F. Design and simulation of assembly line feeding systems in the automotive sector using supermarket, kanbans and tow trains: A general framework. *J. Manag. Control* **2013**, *24*, 187–208. [CrossRef]

27. Battini, D.; Faccio, M.; Persona, A.; Sgarbossa, F. Design of the optimal feeding policy in an assembly system. *Int. J. Prod. Econ.* **2009**, *121*, 233–254. [CrossRef]

28. Kilic, H.S.; Durmusoglu, M.B.; Baskak, M. Classification and modeling for in-plant milk-run distribution systems. *Int. J. Adv. Manuf. Technol.* **2012**, *62*, 1135–1146. [CrossRef]

29. Mácsay, V.; Bányai, T. Toyota Production System in Milkrun Based in-Plant Supply. *J. Prod. Eng.* **2017**, *20*, 141–146. [CrossRef]

30. Klenk, E.; Galka, S.; Günthner, W.A. Analysis of Parameters Influencing In-Plant Milk Run Design for Production Supply. In Proceedings of the International Material Handling Research Colloquium, Gardanne, France, 25–28 June 2012; pp. 25–28.

31. Alnahhal, M.; Ridwan, A.; Noche, B. In-plant milk run decision problems. In Proceedings of the 2nd IEEE International Conference on Logistics Operations Management, GOL 2014, Rabat, Morocco, 5–7 June 2014; pp. 85–92. [CrossRef]

32. Alnahhal, M.; Noche, B. Efficient material flow in mixed model assembly lines. *SpringerPlus* **2013**, *2*, 1–12. [CrossRef] [PubMed]

33. Golz, J.; Gujjula, R.; Günther, H.O.; Rinderer, S.; Ziegler, M. Part feeding at high-variant mixed-model assembly lines. *Flex. Serv. Manuf. J.* **2012**, *24*, 119–141. [CrossRef]

34. Faccio, M.; Gamberi, M.; Persona, A. Kanban number optimisation in a supermarket warehouse feeding a mixed-model assembly system. *Int. J. Prod. Res.* **2013**, *51*, 2997–3017. [CrossRef]

35. Emde, S.; Boysen, N. Optimally locating in-house logistics areas to facilitate JIT-supply of mixed-model assembly lines. *Int. J. Prod. Econ.* **2012**, *135*, 393–402. [CrossRef]

36. Emde, S.; Fliedner, M.; Boysen, N. Optimally loading tow trains for just-in-time supply of mixed-model assembly lines. *IIE Trans. (Inst. Ind. Eng.)* **2012**, *44*, 121–135. [CrossRef]
37. Vieira, A.; Dias, L.S.; Pereira, G.B.; Oliveira, J.A.; Carvalho1, M.S.; Martins, P. Automatic simulation models generation of warehouses with milkruns and pickers. In Proceedings of the European Modeling and Simulation Symposium, Larnaca, Cyprus, 26–28 September 2016 .
38. Klenk, E.; Galka, S.; Günthner, W.A. Operating strategies for in-plant milk-run systems. *IFAC-PapersOnLine* **2015**, *28*, 1882–1887. [CrossRef]
39. Bozer, Y.A.; Ciemnoczolowski, D.D. Performance evaluation of small-batch container delivery systems used in lean manufacturing—Part 1: System stability and distribution of container starts. *Int. J. Prod. Res.* **2013**, *51*, 555–567. [CrossRef]
40. Fedorko, G.; Vasil, M.; Bartosova, M. Use of simulation model for measurement of MilkRun system performance. *Open Eng.* **2019**, *9*, 600–605. [CrossRef]
41. Kluska, K.; Pawlewski, P. The use of simulation in the design of Milk-Run intralogistics systems. *IFAC-PapersOnLine* **2018**, *51*, 1428–1433. [CrossRef]
42. Rao, Y.; Chen, Z.; Fang, W.; Guan, Z. Research on Simulation Based Dynamic Scheduling Strategy of the Logistics Distribution System for Assembly Lines in Engineering Machinery. *J. Phys. Conf. Ser.* **2021**, *2101*, 012046. [CrossRef]
43. Coelho, F.; Relvas, S.; Barbosa-Póvoa, A. Simulation-based decision support tool for in-house logistics: The basis for a digital twin. *Comput. Ind. Eng.* **2021**, *153*, 107094. [CrossRef]
44. VDI 5586 Blatt 2. *Routenzugsysteme—Planung und Dimensionierung (In-Plant Milk-Run Systems—Planning and Dimensioning)*; Standard, Vereine Deutscher Ingenieure: Düsseldorf, Germany, 2022.
45. Schmidt, T.; Meinhardt, I.; Schulze, F. New design guidelines for in-plant milk-run systems. In Proceedings of the 14th IMHRC Proceedings, Karlsruhe, Germany, 12–16 June 2016; p. 25.
46. Urru, A.; Bonini, M.; Echelmeyer, W. Planning and dimensioning of a milk-run transportation system considering the actual line consumption. *IFAC-PapersOnLine* **2018**, *51*, 404–409. [CrossRef]
47. Gyulai, D.; Pfeiffer, A.; Sobottka, T.; Váncza, J. Milkrun vehicle routing approach for shop-floor logistics. *Procedia CIRP* **2013**, *7*, 127–132. [CrossRef]
48. Emde, S.; Gendreau, M. Scheduling in-house transport vehicles to feed parts to automotive assembly lines. *Eur. J. Oper. Res.* **2017**, *260*, 255–267. [CrossRef]
49. Simic, D.; Svircevic, V.; Corchado, E.; Calvo-Rolle, J.L.; Simic, S.D.; Simic, S. Modelling material flow using the Milk run and Kanban systems in the automotive industry. *Expert Syst.* **2021**, *38*, e12546. [CrossRef]
50. Satoglu, S.I.; Sahin, I.E. Design of a just-in-time periodic material supply system for the assembly lines and an application in electronics industry. *Int. J. Adv. Manuf. Technol.* **2013**, *65*, 319–332. [CrossRef]
51. Aksoy, A.; Öztürk, N. A two-stage method to optimize the milk-run system. *Eur. J. Eng. Res. Sci.* **2016**, *1*, 7–11. [CrossRef]
52. Costa, B.; Dias, L.S.; Oliveira, J.A.; Pereira, G. Simulation as a tool for planning a material delivery system to manufacturing lines. In Proceedings of the 2008 IEEE International Engineering Management Conference, Estoril, Portugal, 28–30 June 2008; pp. 1–5.
53. Afonso, T.; Alves, A.C.; Carneiro, P.; Vieira, A. Simulation pulled by the need to reduce wastes and human effort in an intralogistics project. *Int. J. Ind. Eng. Manag.* **2021**, *12*, 274–285. [CrossRef]
54. Pekarcikova, M.; Trebuna, P.; Kliment, M.; Schmacher, B.A.K. Milk Run Testing through Tecnomatix Plant Simulation Software. *Int. J. Simul. Model.* **2022**, *21*, 101–112. [CrossRef]
55. Wang, Q.; Chatwin, C. Key issues and developments in modelling and simulation-based methodologies for manufacturing systems analysis, design and performance evaluation. *Int. J. Adv. Manuf. Technol.* **2005**, *25*, 1254–1265. [CrossRef]
56. Banks, J.; Chwif, L. Warnings about simulation. *J. Simul.* **2011**, *5*, 279–291. [CrossRef]
57. Miqueo, A.; Torralba, M.; Yagüe-Fabra, J.A. Models to evaluate the performance of high-mix low-volume manual or semi-automatic assembly lines. *Procedia CIRP* **2022**, *107*, 1461–1466. [CrossRef]

Article

Improving Urgency-Based Backlog Sequencing of Jobs: An Assessment by Simulation

Nuno O. Fernandes [1,2,*], Matthias Thürer [3], Mark Stevenson [4] and Silvio Carmo-Silva [2]

[1] Department of Industrial Engineering, Polytechnic Institute of Castelo Branco, 6000-767 Castelo Branco, Portugal
[2] Department of Production and Systems, University of Minho, 4710-057 Braga, Portugal
[3] Department of Intelligent System Science and Engineering, University of Jinan, Zhuhai 519070, China
[4] Department of Management Science, Lancaster University, Lancaster LA1 4YX, UK
* Correspondence: nogf@ipcb.pt

Abstract: When order release is applied, jobs are withheld in a backlog from where they are released to meet certain performance targets. The decision that selects jobs for release is typically preceded by a sequencing decision. It was traditionally assumed that backlog sequencing is only responsible for releasing jobs on time, whereas more recent literature has argued that it can also support load balancing. Although the new load-based rules outperform time-based rules, they can be criticized for requiring workload information from the shop floor and for delaying large jobs. While some jobs will inevitably be delayed during periods of high load, we argue that this delaying decision should be under control of management. A simulation study of a wafer fab environment shows that a time-based rule matches the performance of more complex load-based backlog sequencing rules that have recently emerged. The new rule realizes the lowest percentage of tardy jobs if the lower bound that distinguishes between early and urgent jobs is set appropriately. It provides a simpler means of improving release performance, allowing managers to delay jobs that have adjustable due dates.

Keywords: lean; order release; backlog sequencing; re-entrant flow shop; simulation

Citation: Fernandes, N.O.; Thürer, M.; Stevenson, M.; Carmo-Silva, S. Improving Urgency-Based Backlog Sequencing of Jobs: An Assessment by Simulation. *Machines* **2022**, *10*, 935. https://doi.org/10.3390/machines10100935

Academic Editor: Mosè Gallo

Received: 21 August 2022
Accepted: 8 October 2022
Published: 14 October 2022

Publisher's Note: MDPI stays neutral with regard to jurisdictional claims in published maps and institutional affiliations.

1. Introduction

Protecting throughput from variance is the key to achieving lean [1]. If order release is applied, jobs (or orders) are not immediately released onto the shop floor upon arrival at the production system, but release is controlled to meet certain performance targets, which creates a so-called backlog or pre-shop pool that effectively buffers the shop floor. Well-known approaches to order release include Constant Work-in-Process (ConWIP; e.g., [1–4]), Drum-Buffer-Rope (DBR, e.g., [5–7]), and Workload Control (e.g., [8–11]), among others. Meanwhile, a similar logic is also implemented in more recent manufacturing paradigms, such as cloud manufacturing (e.g., [12–17]).

The decision concerning which jobs to release is typically subdivided into a backlog sequencing decision, which determines the sequence in which jobs are considered for release, and a selection decision, which decides for each job, in sequence, whether it should be released given certain release criteria, e.g., applying a limit to the workload released to a station. Most studies on order release have focused on the selection decision, implicitly assuming that the sequencing decision should use some measure of urgency. Only recently it has been shown that incorporating load considerations in the sequencing decision can improve performance in the context of Workload Control (e.g., [18]), ConWIP (e.g., [19]) and DBR (e.g., [20]). However, including load considerations requires feedback from the shop floor and significantly increases the complexity of the backlog sequencing decision. The challenge is whether simple time-based rules can be developed that mimic the behavior of load-based rules but without requiring regular information being fed-back from the shop floor.

Taking a closer look at load-based backlog sequencing rules reveals that they gain an advantage by delaying large jobs during periods of high load. In other words, during periods when the incoming workload largely exceeds the available capacity, load-based rules focus on producing as many small jobs on time as possible to the detriment of some large jobs. It has long since been shown that this improves overall performance [21]. In this study, we mimic this behavior by proposing a new urgency-based rule that creates three classes of jobs: early (non-urgent), urgent, and very urgent. Using a simulation model of a re-entrant flow shop [22] we conjecture that prioritizing urgent over very urgent jobs leads to similar results to those obtained for load-based sequencing, but in a simpler way.

Re-entrant flow shops are of high practical relevance. They can be found, for example, in semiconductor wafer fab environments (see e.g., [23]) and in flexible manufacturing systems (see e.g., [24]). Moreover, the advent of lean management and industry 4.0 fostered a product-based view instead of a resource-based view [25]. This enables companies to streamline their production systems, so products can be manufactured using a fixed sequence of flexible resources [26]. However, re-entrant flow shops are also very challenging environments for order release methods [27] since jobs repeatedly pass through the same station (or stations) at different stages of processing. Our new rule not only provides a simpler means of improving order release performance in such contexts, but by using dedicated classes of jobs based on urgency it also facilitates the selection of those jobs that should be delayed in practice. The new rule mainly differs from load-based rules by not requiring feedback information from the shop floor.

The remainder of this paper is structured as follows. The literature is next reviewed in Section 2 to identify the order release method and the alternative backlog sequencing rules to be considered in our study. Section 3 then outlines the simulation model used to evaluate performance before the results are presented and discussed in Section 4. Finally, Section 5 puts forward a conclusion and outlines the managerial implications and limitations of the study.

2. Literature Review

This section identifies the order release method, and the backlog sequencing rules to be included in our study in Sections 2.1 and 2.2, respectively. A discussion of the literature is then presented in Section 2.3.

2.1. Order Release

Given its importance, several Workload Control release methods have been proposed in the literature, see e.g., the reviews by [28–31]. In this paper, we use a continuous version of Workload Control order release given its good performance in previous studies (see e.g., [32]). Workload Control restricts the load released to stations on the shop floor by load norms or load limits. Whenever a new job (or order) arrives to the production system, or whenever an operation is complete on the shop floor, the following steps are executed to decide on the release of jobs:

(1) Jobs waiting for release in the backlog are sorted according to a sequencing rule, and the job with the highest priority is considered for release first.
(2) If the load that results from releasing the job does not exceed the load norms for all stations in its routing, then the job is selected for release and released. Otherwise, the job remains in the backlog and the job with the next highest priority is considered for release.
(3) The release procedure stops when all jobs in the backlog have been considered for release or when there are no more jobs in the backlog to be released.

The load contribution to a station is calculated by dividing the processing time of the operation at a station by the station's position in a job's routing [33]. That is, $W_s := W_s + \frac{p_{ij}}{i}$ $\forall i \in R_j$, where W_s is the workload released to station s, i is the station position in the routing of the job, p_{ij} is the processing time of job j at the ith operation in its routing, and R_j the set of jobs awaiting release in the backlog.

2.2. Backlog Sequencing Rules

Many backlog sequencing rules have been applied in the literature. Most of this literature has assumed that backlog sequencing is only responsible for ensuring that orders are released on time. Consequently, most studies have applied a time-oriented backlog sequencing rule. For example, refs. [18,29] used the First-Come-First-Served (FCFS) rule, a time-oriented rule that sequences jobs according to their time of arrival to the backlog. Another time-oriented rule that is widely applied, since it reflects the existence of different due date allowances across jobs, is the Earliest Due Date (EDD) rule. This rule was used by, e.g., [18,34], among others. Finally, the Planned Release Date (PRD) rule is a time-oriented rule that also reflects differences in the number of operations and the operations throughput times in the routing of jobs. It sequences jobs according to planned release dates given by Equation (1) below. Two variants of this rule are used in the literature, where either waiting times or operation throughput times are treated as a constant. This rule was used, e.g., by [11,18,29].

$$\tau_j = \delta_j - \sum_{i \in R_j} \left(a_i + p_{ij} \right) \text{ or } \tau_j = \delta_j - \sum_{i \in R_j} b_i \tag{1}$$

τ_j = planned release date of job j
δ_j = due date of job j
a_i = estimated waiting time at the ith operation in the routing of a job
b_i = estimated throughput time at the ith operation in the routing of a job

More recently, Thürer et al. [18] questioned the assumption that backlog sequencing is mainly responsible for ensuring orders are released on time. The authors introduced a MODified Capacity Slack (MODCS) rule that combines load-balancing and urgency considerations. MODCS considers two classes of jobs, urgent and non-urgent, depending on whether jobs have a planned release date (refer to Equation (1)) that has already been passed or not. Urgent jobs have priority over non-urgent jobs. Urgent jobs are sequenced according to the Capacity Slack CORrected (CSCOR) rule. Non-urgent jobs are sequenced according to the PRD rule. CSCOR is a load-oriented rule that sequences jobs according to a capacity slack ratio, as given by Equation (2).

$$S_j = \frac{\sum_{i \in R_j} \left(\frac{\frac{p_{ij}}{i}}{N_s - W_s} \right)}{n_j} \tag{2}$$

The capacity slack ratio integrates three elements into one priority measure: the workload contribution of a job to a station, $\frac{p_{js}}{i}$; the load gap, N_s-W_s at the station that processes the ith operation of a job, and the number of operations in the routing of the job, i.e., the routing length n_j. The latter is used to average the ratio between workload contribution to stations and load gap at stations across the routing of the job. The lower the capacity slack ratio, S_j, the higher the priority of job j. This capacity slack rule was originally proposed by [35] and later used, e.g., by [29].

2.3. Discussion

Studies comparing the performance of different sequencing rules (e.g., [18,29]) showed that load-based rules (i.e., CS and MODCS) outperform time-based rules. However, there are two major issues. First, the CSCOR rule is rather complex and requires detailed information from the shop floor. Second, the CSCOR rule gains advantage by delaying larger jobs. This behavior is maintained also within MODCS during periods of high load when many jobs tend to become urgent. In this study we seek to address both issues by asking:

> *Can a new backlog sequencing rule be designed that matches the high performing MODCS rule whilst only considering job information and enabling a more controlled decision to be taken on which jobs to delay?*

In response, we propose a new time-oriented rule. This rule creates three classes of jobs based on urgency given by the PRD of the job, as illustrated in Figure 1: early (non-urgent), urgent, and very urgent. We then conjecture that prioritizing urgent over very urgent jobs provides similar results to those obtained for MODCS, despite the relative simplicity of the new rule. The new rule differs from load-based rules as it does not require feedback information from the shop floor. It only requires a PRD to be determined for each job to create three classes of urgency that then are used to prioritize jobs. By using dedicated classes, the new rule also facilitates the selection of jobs that should be delayed in practice. In general, it should be noted that, as per the definition, some jobs will almost inevitably be delivered slowly during periods of high load [36]. The simulation model used to test our conjecture and the experimental design is outlined next in Section 3.

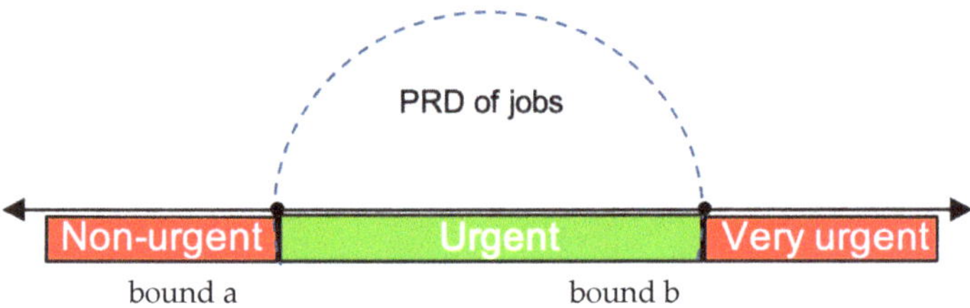

Figure 1. Classes of jobs based on urgency.

3. Methods

3.1. Simulation

A simulation model of a re-entrant flow shop was implemented using ARENA® software. In the model job inter-arrival times, processing times and due dates are stochastic variables. The re-entrant flow shop considered here is balanced and consists of three stations, each with two machines preceded by a single input buffer. Job routings are based on the six-step Mini-Fab model of Kempf [37], as depicted in Figure 2. That is, the production of each job is completed following a sequence of six processing steps.

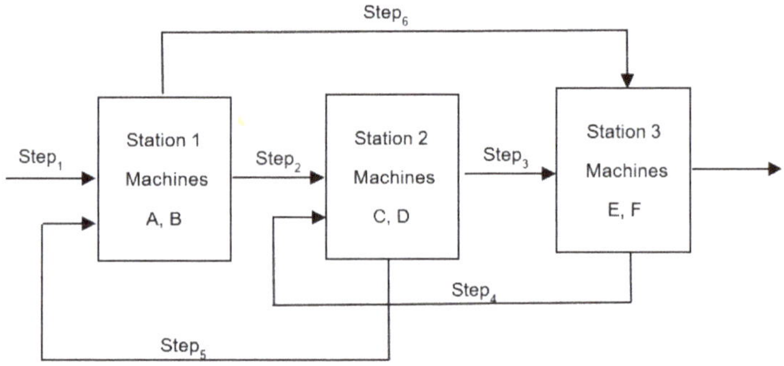

Figure 2. Illustration of the re-entrant flow shop under study.

In this study it is assumed that all materials for job processing are available and all the required information regarding job routings, job processing times, etc., is known at order arrival. This is in line with previous simulation studies on order release control (e.g., [10,29,38,39]). Processing times at machines follow a lognormal distribution with a mean of one time unit. Processing time variability of jobs is considered an experimental factor. We consider three levels for the coefficient of variation (CV) of the processing times, namely: 0.1, 0.2, and 0.4. We further assume that set-up times are sequence independent and considered to be part of the operation processing times. To avoid moving away from the focus of the research question and to avoid confounding factors, batch processing and unreliable machines were not considered. The inter-arrival time of jobs to the shop follows an exponential distribution. The mean of this distribution was defined to result in a system steady-state utilization rate of 90%.

Due dates are assumed to be set exogenously, i.e., by the customer. A random allowance, set between 15 and 30 time units, was added to the job entry time. Values were chosen so that releasing jobs onto the shop floor immediately upon arrival yields a percentage of tardy jobs close to 10% for an intermediate level of CV of the processing time (i.e., 0.2). Table 1 summarizes the simulated shop and job characteristics.

Table 1. Summary of simulated shop and job characteristics.

Shop and Job Characteristics	
Routing Variability	Re-entrant flows; same sequence for all jobs
No. of Stations	3
No. of Machines	6 (two per station)
Station Capacities	All equal
Station Utilization Rate	90%
No. of Operations per Job	6
Operation Processing Times	Log-normally distributed; mean = 1
Due Date Determination	Due Date = Entry Time + d; $d \sim U\,[15, 30]$
Inter-Arrival Times	Exp. Distribution; mean = 1.111

3.2. Order Release and Backlog Sequencing

Six settings of the load norms are considered, namely, 4, 5, 6, 7, 8, and 9 time units. As a baseline setting, immediate release (IMR) of jobs to the shop floor is also considered, i.e., without controlled order release. Since EDD and PRD backlog sequencing are equivalent in shops with a single routing for all jobs, only PRD is considered. We consequently consider three backlog sequencing rules from the literature: FCFS, PRD, and MODCS. Four versions of our new rule are considered, where we change the *bound a* (see Figure 1) that distinguishes early (non-urgent) and urgent jobs at four levels: PRD minus an allowance of 1, 2, 5, and 10 time units, respectively. These rules are referred to as NEW (-1), NEW (-2), NEW (-5), and NEW (-10), respectively. The *bound b* (see Figure 1) that distinguishes between urgent and very urgent jobs is set to the PRD plus an allowance of 10 time units for all four rules based on preliminary simulation experiments, i.e., very urgent jobs are those for which the PRD is more than 10 time units in the past. Finally, the allowance for the operation throughput time at each station is set based on the cumulative moving average realized during simulation experiments.

3.3. Dispatching

Jobs waiting in the input buffer of a station are prioritized according to operation due dates. The operation due date for the last operation in the routing of a job is equal to the due date of the job, as given by Equation (1), while the operation due date of each preceding operation is determined by successively subtracting the estimated station waiting time and processing time from the operation due date of the next operation. In this study, estimated station waiting times are given by the cumulative moving average, i.e., the average of all station waiting times realized until the current simulation time.

3.4. Experimental Design and Performance Measures

Our study considers three experimental factors tested at different levels: (i) three levels of processing time variability (CV = 0.1, 0.2, and 0.4); (ii) six levels of the workload norm (4, 5, 6, 7, 8, and 9 time units); and (iii) seven backlog sequencing rules. A full factorial design was used with 126 ($3 \times 6 \times 7$) experimental scenarios, where each scenario was run for 13,000 time units following a warm-up period of 3000 time units. Each experimental scenario was replicated 100 times. These simulation conditions allow for obtaining stable results with small (i.e., precise) confidence intervals for the performance measures.

The key performance measures considered are the total throughput time, the percentage of tardy jobs, and the mean tardiness. The total throughput time refers to the time that elapses between job entry to the system and job completion. The percentage of tardy jobs refers to the percentage of jobs completed after their due date. Tardiness is defined to be zero if the job is on time or early and it is equal to the completion date minus the due date if the job is late. In addition, we also measure the average shop floor throughput time, which is used as an instrumental performance variable. While the total throughput time includes the time that jobs wait before being released, i.e., the backlog waiting time, the shop floor throughput time refers to the time that elapses between job release to the shop floor and job completion.

4. Results and Discussion

An initial statistical analysis of simulation results was conducted using an ANOVA (Analysis of Variance). ANOVA results are presented in Table 2. All main effects and most of the two-way interactions were found to be statistically significant. There were no significant three-way interactions.

Table 2. ANOVA results.

	Source of Variance	Sum of Squares	Degrees of Freedom	Mean Squares	F-Ratio	*p*-Value
Percentage Tardy	Coefficient of Variation (CV)	18,370.15	2	9185.07	1142.61	0.00
	Norm	17,638.49	5	3527.70	438.84	0.00
	Pool Sequencing (S)	34,028.43	6	5671.40	705.51	0.00
	CV × Norm	2114.88	10	211.49	26.31	0.00
	CV × S	1706.67	12	142.22	17.69	0.00
	Norm × S	7315.81	30	243.86	30.34	0.00
	CV × Norm × S	415.26	60	6.92	0.86	0.77
	Error	100,266.43	12,473	8.04		
Mean Tardiness	Coefficient of Variation (CV)	418.28	2	209.14	1270.57	0.00
	Norm	408.90	5	81.78	496.84	0.00
	Pool Sequencing (S)	229.63	6	38.27	232.51	0.00
	CV × Norm	161.29	10	16.13	97.99	0.00
	CV × S	0.48	12	0.04	0.24	1.00
	Norm × S	130.72	30	4.36	26.47	0.00
	CV × Norm × S	3.90	60	0.07	0.40	1.00
	Error	2053.07	12,473	0.16		
Total Throughput Time	Coefficient of Variation (CV)	5585.58	2	2792.79	3435.25	0.00
	Norm	400.46	5	80.09	98.52	0.00
	Pool Sequencing (S)	18.12	6	3.02	3.71	0.00
	CV × Norm	124.71	10	12.47	15.34	0.00
	CV × S	9.12	12	0.76	0.93	0.51
	Norm × S	3.00	30	0.10	0.12	1.00
	CV × Norm × S	1.00	60	0.02	0.02	1.00
	Error	10,140.30	12,473	0.81		
Shop Floor Throughput Time	Coefficient of Variation (CV)	1865.34	2	932.67	6782.66	0.00
	Norm	18,268.67	5	3653.73	26,571.01	0.00
	Pool Sequencing (S)	3.40	6	0.57	4.12	0.00
	CV × Norm	256.49	10	25.65	186.53	0.00
	CV × S	0.80	12	0.07	0.49	0.92
	Norm × S	0.24	30	0.01	0.06	1.00
	CV × Norm × S	0.08	60	0.00	0.01	1.00
	Error	1715.14	12,473	0.14		

As somewhat expected from the choice and the design of our backlog sequencing rules, the backlog sequencing rules have the strongest impact on the percentage tardy and mean tardiness. This can also be observed from the results for the Scheffé multiple comparison procedure, which was applied to obtain a first indication of the direction and size of the performance differences. Table 3 gives the 95% confidence interval. When this interval includes zero, performance differences are not considered to be statistically significant. We can observe significant performance differences for all pairs for at least one performance measure except for between FCFS and PRD, for which performance is statistically equivalent. This is explored further in Sections 4.1 and 4.2 where detailed performance results are presented and their robustness evaluated.

Table 3. Results for Scheffé multiple comparison procedure.

Rule (x)	Rule (y)	Percentage Tardy		Mean Tardiness		Total Throughput Time		Shop Floor Throughput Time	
		Lower [1]	Upper	Lower	Upper	Lower	Upper	Lower	Upper
NEW (−2)	NEW (−1) [2]	−0.92	−0.25	−0.03 *	0.07	−0.12 *	0.10	−0.05 *	0.04
NEW (−5)	NEW (−1)	−2.12	−1.45	0.08	0.18	−0.14 *	0.07	−0.06 *	0.03
NEW (−10)	NEW (−1)	−2.49	−1.82	0.34	0.43	−0.15 *	0.06	−0.08 *	0.01
MODCS	NEW (−1)	0.17	0.84	−0.07 *	0.03	−0.12 *	0.09	−0.04 *	0.05
PRD	NEW (−1)	1.85	2.52	−0.06 *	0.04	−0.05 *	0.16	−0.02 *	0.06
FCFS	NEW (−1)	2.06	2.73	−0.03 *	0.06	−0.05 *	0.17	−0.05 *	0.04
NEW (−5)	NEW (−2)	−1.54	−0.87	0.06	0.16	−0.13 *	0.08	−0.06 *	0.03
NEW (−10)	NEW (−2)	−1.90	−1.23	0.32	0.41	−0.14 *	0.07	−0.08 *	0.01
MODCS	NEW (−2)	0.76	1.43	−0.09 *	0.01	−0.11 *	0.10	−0.03 *	0.05
PRD	NEW (−2)	2.43	3.11	−0.08 *	0.01	−0.04 *	0.17	−0.02 *	0.07
FCFS	NEW (−2)	2.65	3.32	−0.05 *	0.04	−0.04 *	0.18	−0.04 *	0.04
NEW (−10)	NEW (−5)	−0.70	−0.03	0.20	0.30	−0.12 *	0.10	−0.06 *	0.03
MODCS	NEW (−5)	1.96	2.63	−0.20	−0.10	−0.09 *	0.13	−0.02 *	0.07
PRD	NEW (−5)	3.63	4.31	−0.19	−0.10	−0.02 *	0.20	−0.01 *	0.08
FCFS	NEW (−5)	3.85	4.52	−0.16	−0.07	−0.01 *	0.20	−0.03 *	0.06
MODCS	NEW (−10)	2.33	3.00	−0.45	−0.36	−0.08 *	0.14	0.00	0.08
PRD	NEW (−10)	4.00	4.67	−0.44	−0.35	−0.01 *	0.21	0.01	0.10
FCFS	NEW (−10)	4.22	4.89	−0.42	−0.32	0.00	0.21	−0.01 *	0.08
PRD	MODCS	1.34	2.01	−0.04 *	0.06	−0.04 *	0.18	−0.03 *	0.06
FCFS	MODCS	1.55	2.23	−0.01 *	0.08	−0.03 *	0.18	−0.05 *	0.04
FCFS	PRD	−0.12 *	0.55	−0.02 *	0.08	−0.10 *	0.11	−0.07 *	0.02

[1] 95% confidence interval; [2] allowance for bound that distinguishes early jobs; * not significant at 0.05.

4.1. Performance Assessment

Results for a coefficient of variation (CV) of the processing times of 0.1 together with the 95% confidence intervals are given in Table 4. Results for different levels of the CVs are presented as part of our robustness analysis in Section 4.2.

Table 4. Simulation results for jobs CV = 0.1.

	Load Norm	STT [1]	TTT [2]	T [3]	P [4] (%)
IMR	None	13.35 ± 0.18	13.35 ± 0.18	0.39 ± 0.07	7.85 ± 0.73
FCFS	4	09.30 ± 0.02	12.62 ± 016	0.42 ± 0.07	7.20 ± 0.61
	5	10.34 ± 0.04	12.75 ± 0.16	0.34 ± 0.06	6.60 ± 0.59
	6	11.14 ± 0.06	12.99 ± 0.16	0.33 ± 0.06	6.80 ± 0.62
	7	11.67 ± 0.07	13.12 ± 0.16	0.34 ± 0.06	7.00 ± 0.63
	8	12.02 ± 0.09	13.18 ± 0.17	0.35 ± 0.06	7.17 ± 0.65
	9	12.26 ± 0.10	13.22 ± 0.17	0.36 ± 0.06	7.31 ± 0.66

Table 4. *Cont.*

	Load Norm	STT [1]	TTT [2]	T [3]	P [4] (%)
PRD	4	09.30 ± 0.02	12.58 ± 016	0.34 ± 0.06	6.25 ± 0.60
	5	10.36 ± 0.04	12.77 ± 0.16	0.31 ± 0.06	6.17 ± 0.61
	6	11.17 ± 0.06	13.00 ± 0.16	0.32 ± 0.06	6.65 ± 0.63
	7	11.70 ± 0.08	13.14 ± 0.16	0.34 ± 0.06	7.02 ± 0.65
	8	12.06 ± 0.09	13.21 ± 0.17	0.35 ± 0.06	7.26 ± 0.66
	9	12.31 ± 0.10	13.25 ± 0.17	0.36 ± 0.06	7.42 ± 0.67
MODCS	4	09.30 ± 0.02	12.56 ± 0.15	0.36 ± 0.06	4.25 ± 0.39
	5	10.35 ± 0.04	12.74 ± 0.15	0.30 ± 0.06	5.22 ± 0.51
	6	11.16 ± 0.06	12.98 ± 0.16	0.31 ± 0.06	6.27 ± 0.60
	7	11.69 ± 0.07	13.11 ± 0.16	0.32 ± 0.06	6.82 ± 0.63
	8	12.05 ± 0.09	13.18 ± 0.16	0.34 ± 0.06	7.13 ± 0.65
	9	12.31 ± 0.10	13.23 ± 0.17	0.35 ± 0.06	7.35 ± 0.66
NEW (−10) [1]	4	09.29 ± 0.02	12.64 ± 0.15	1.17 ± 0.12	3.74 ± 0.16
	5	10.32 ± 0.04	12.74 ± 0.15	0.88 ± 0.09	3.06 ± 0.15
	6	11.14 ± 0.06	12.98 ± 0.16	0.72 ± 0.08	2.75 ± 0.15
	7	11.68 ± 0.07	13.11 ± 0.16	0.60 ± 0.07	2.93 ± 0.21
	8	12.03 ± 0.09	13.17 ± 0.16	0.51 ± 0.07	3.45 ± 0.29
	9	12.28 ± 0.10	13.21 ± 0.17	0.45 ± 0.07	4.32 ± 0.38
NEW (−5)	4	09.29 ± 0.02	12.58 ± 0.17	0.57 ± 0.09	2.24 ± 0.17
	5	10.34 ± 0.04	12.74 ± 0.16	0.44 ± 0.07	2.18 ± 0.18
	6	11.14 ± 0.06	12.98 ± 0.16	0.39 ± 0.07	3.07 ± 0.29
	7	11.68 ± 0.07	13.11 ± 0.16	0.36 ± 0.06	4.50 ± 0.43
	8	12.04 ± 0.09	13.18 ± 0.16	0.36 ± 0.06	5.64 ± 0.51
	9	12.30 ± 0.10	13.22 ± 0.17	0.36 ± 0.06	6.46 ± 0.59
NEW (−2)	4	09.30 ± 0.02	12.58 ± 0.17	0.43 ± 0.09	2.59 ± 0.21
	5	10.35 ± 0.04	12.74 ± 0.16	0.35 ± 0.07	3.21 ± 0.31
	6	11.15 ± 0.06	12.98 ± 0.16	0.33 ± 0.07	4.84 ± 0.47
	7	11.68 ± 0.07	13.11 ± 0.16	0.32 ± 0.06	6.03 ± 0.56
	8	12.04 ± 0.09	13.18 ± 0.16	0.34 ± 0.06	6.67 ± 0.61
	9	12.30 ± 0.10	13.22 ± 0.17	0.35 ± 0.06	7.10 ± 0.64
NEW (−1)	4	09.30 ± 0.02	12.58 ± 0.17	0.39 ± 0.07	3.24 ± 0.31
	5	10.35 ± 0.04	12.74 ± 0.16	0.32 ± 0.06	4.25 ± 0.46
	6	11.15 ± 0.06	12.98 ± 0.16	0.32 ± 0.06	5.72 ± 0.55
	7	11.69 ± 0.07	13.12 ± 0.16	0.33 ± 0.06	6.54 ± 0.60
	8	12.04 ± 0.09	13.19 ± 0.16	0.34 ± 0.06	6.96 ± 0.63
	9	12.30 ± 0.10	13.23 ± 0.17	0.35 ± 0.06	7.26 ± 0.65

[1] Shop Throughput Time; [2] Total Throughput Time. [3] Tardiness [4] Percentage of Tardy Jobs.

For IMR, the percentage of tardy jobs is about 8%. When jobs are not retained in a pool before release, the shop and the total throughput times are identical (13.35 time units). However, when controlled job release is applied and workload norms are tightened, the workload on the shop floor is restricted, and the shop throughput time is reduced. For example, shop throughput time is reduced to 9.30 time units when the load norm is restricted to 4 time units under FCFS, i.e., a reduction of about 30%. This has a positive impact not only on total throughput times, but also on the percentage of tardy jobs if the workload norm is set appropriately. However, when norms are set too tightly, there may be an increase in the percentage of tardy jobs because the time in the pool offsets the reduction in shop floor throughput times, and there is an increase in sequencing deviations. That is, when norms are set too tightly some jobs may be delayed for long periods in the pre-shop pool (or backlog) before being released to the shop floor, which increases the mean tardiness. Results also confirm previous literature in the sense that PRD outperforms FCFS at tighter norms, and MODCS outperforms PRD. Since the superior performance of PRD over FCFS only occurs at tighter norms, it was found not to be statistically significant in our ANOVA.

Most importantly, our new backlog sequencing rule has the potential to outperform existing backlog sequencing rules if the lower bound that distinguishes between early and urgent jobs is set appropriately. The results further highlight that there is no best rule for all performance measures. NEW (-5) allows the lowest percentage of tardy jobs to be obtained, while NEW (-1) approaches the tardiness values of MODCS and PRD. Meanwhile, if the lower bound is set to be too large/loose (i.e., PRD minus 10 time units), then we increase the set of urgent jobs to include some jobs that are still early. As a result, the number of very urgent jobs is likely to increase, resulting in worse performance. This can be observed from Table 5, which provides more detailed information on the tardiness of jobs for a workload norm level of 4 time units.

Table 5. Distribution of the number of tardy jobs for a workload norm of 4 time units.

Lateness	FCFS	PRD	MODCS	New (-10)	New (-5)	New (-2)	New (-1)
]0,10]	497	399	298	155	120	180	212
]10,20]	50	48	23	63	29	23	33
]20,30]	1	3	10	46	24	16	16
]30,40]	0	0	6	24	23	12	14
]40,50]	0	1	3	20	8	10	5
]50,60]	0	0	3	16	2	2	2
]60,∞]	0	0	3	23	15	4	2
Total	548	451	346	347	221	247	284

The results in Table 5 highlight that: (i) both MODCS and our new rule improve overall performance by delaying some jobs and (ii) it is important for our new rule to capture only the jobs that are at risk of becoming tardy and still have a chance of being delivered on time as part of the urgent class of jobs that are released first.

4.2. Robustness Analysis

The relative performance of the different backlog sequencing rules is also not affected by the CV of the processing times. This can be observed from Tables 6 and 7, which give the results for a CV of 0.2 and 0.4, respectively. The main effect of the CV is on the best-performing norm level. The best-performing norm level in terms of the mean tardiness remains at five time units, but for the percentage of tardy jobs, the best-performing norm level increases with the coefficient of variation since we also observe less of an impact on the total throughput time and percentage of tardy jobs with an increase in the CV. A higher variability in processing times increases the load balancing opportunities for the release method, which in turn provides 'less room' for the sequencing rules.

Table 6. Simulation results for CV = 0.2.

	Load Norm	STT [1]	TTT [2]	T [3]	P [4] (%)
IMR	None	14.09 ± 0.20	14.09 ± 0.20	0.52 ± 0.08	9.96 ± 0.85
FCFS	4	09.49 ± 0.03	13.58 ± 0.21	0.74 ± 0.09	9.96 ± 0.70
	5	10.68 ± 0.04	13.52 ± 0.18	0.51 ± 0.08	8.55 ± 0.70
	6	11.56 ± 0.06	13.68 ± 0.18	0.46 ± 0.07	8.55 ± 0.71
	7	12.17 ± 0.08	13.81 ± 0.18	0.46 ± 0.07	8.80 ± 0.74
	8	12.60 ± 0.10	13.90 ± 0.18	0.47 ± 0.07	9.08 ± 0.76
	9	12.92 ± 0.11	13.95 ± 0.18	0.48 ± 0.07	9.34 ± 0.77
PRD	4	09.48 ± 0.03	13.49 ± 0.20	0.64 ± 0.09	8.85 ± 0.70
	5	10.70 ± 0.04	13.51 ± 0.18	0.46 ± 0.07	8.08 ± 0.71
	6	11.59 ± 0.06	13.69 ± 0.18	0.44 ± 0.07	8.43 ± 0.74
	7	12.20 ± 0.08	13.82 ± 0.18	0.45 ± 0.07	8.80 ± 0.75
	8	12.63 ± 0.10	13.91 ± 0.18	0.47 ± 0.07	9.16 ± 0.76
	9	12.95 ± 0.11	13.97 ± 0.19	0.48 ± 0.07	9.41 ± 0.78
MODCS	4	09.47 ± 0.03	13.40 ± 0.19	0.71 ± 0.09	5.23 ± 0.33
	5	10.69 ± 0.04	13.44 ± 0.17	0.46 ± 0.07	5.40 ± 0.44
	6	11.57 ± 0.06	13.62 ± 0.17	0.42 ± 0.06	6.70 ± 0.57
	7	12.18 ± 0.08	13.77 ± 0.17	0.42 ± 0.06	7.86 ± 0.65
	8	12.60 ± 0.09	13.86 ± 0.17	0.44 ± 0.07	8.59 ± 0.70
	9	12.94 ± 0.11	13.93 ± 0.18	0.45 ± 0.07	9.11 ± 0.75
NEW (−10) [1]	4	09.44 ± 0.02	13.42 ± 0.20	1.60 ± 0.14	4.85 ± 0.18
	5	10.64 ± 0.04	13.40 ± 0.17	1.09 ± 0.10	3.78 ± 0.17
	6	11.53 ± 0.06	13.58 ± 0.17	0.85 ± 0.09	3.29 ± 0.17
	7	12.14 ± 0.07	13.72 ± 0.17	0.66 ± 0.08	3.63 ± 0.25
	8	12.58 ± 0.09	13.83 ± 0.17	0.59 ± 0.07	4.94 ± 0.40
	9	12.90 ± 0.11	13.87 ± 0.18	0.53 ± 0.07	6.27 ± 0.53
NEW (−5)	4	09.47 ± 0.03	13.42 ± 0.19	1.07 ± 0.11	3.91 ± 0.19
	5	10.66 ± 0.04	13.41 ± 0.17	0.72 ± 0.09	2.97 ± 0.18
	6	11.55 ± 0.06	13.60 ± 0.17	0.58 ± 0.07	3.29 ± 0.25
	7	12.16 ± 0.08	13.74 ± 0.17	0.51 ± 0.07	4.81 ± 0.40
	8	12.60 ± 0.09	13.84 ± 0.17	0.48 ± 0.07	6.45 ± 0.55
	9	12.91 ± 0.11	13.90 ± 0.18	0.47 ± 0.07	7.67 ± 0.64
NEW (−2)	4	09.48 ± 0.03	13.43 ± 0.20	0.81 ± 0.10	4.00 ± 0.24
	5	10.68 ± 0.04	13.44 ± 0.17	0.53 ± 0.07	3.74 ± 0.30
	6	11.56 ± 0.06	13.63 ± 0.17	0.47 ± 0.07	5.05 ± 0.44
	7	12.17 ± 0.08	13.77 ± 0.17	0.45 ± 0.07	6.75 ± 0.58
	8	12.61 ± 0.09	13.86 ± 0.17	0.45 ± 0.07	7.94 ± 0.66
	9	12.92 ± 0.11	13.92 ± 0.18	0.45 ± 0.07	8.70 ± 0.71

Table 6. *Cont.*

	Load Norm	STT [1]	TTT [2]	T [3]	P [4] (%)
NEW (−1)	4	09.47 ± 0.03	13.44 ± 0.19	0.74 ± 0.09	4.45 ± 0.28
	5	10.68 ± 0.04	13.45 ± 0.17	0.49 ± 0.07	4.56 ± 0.38
	6	11.56 ± 0.06	13.63 ± 0.17	0.44 ± 0.07	6.05 ± 0.51
	7	12.18 ± 0.08	13.78 ± 0.17	0.44 ± 0.07	7.46 ± 0.63
	8	12.62 ± 0.10	13.87 ± 0.18	0.44 ± 0.07	8.36 ± 0.69
	9	12.93 ± 0.11	13.93 ± 0.18	0.46 ± 0.07	8.95 ± 0.74

[1] Shop Throughput Time; [2] Total Throughput Time. [3] Tardiness [4] Percentage of Tardy Jobs.

Table 7. Simulation results for CV = 0.4.

	Load Norm	STT [1]	TTT [2]	T [3]	P [4] (%)
IMR	None	16.36 ± 0.26	16.36 ± 0.26	1.10 ± 0.13	17.97 ± 1.00
FCFS	4	09.20 ± 0.03	17.50 ± 0.44	2.95 ± 0.29	18.09 ± 1.05
	5	11.38 ± 0.05	15.52 ± 0.24	1.27 ± 0.12	11.73 ± 0.75
	6	12.53 ± 0.07	15.49 ± 0.20	0.95 ± 0.10	11.79 ± 0.80
	7	13.41 ± 0.09	15.66 ± 0.21	0.87 ± 0.10	13.03 ± 0.87
	8	14.08 ± 0.11	15.83 ± 0.22	0.86 ± 0.10	14.49 ± 0.95
	9	14.58 ± 0.13	15.96 ± 0.22	0.90 ± 0.10	15.50 ± 0.98
PRD	4	09.90 ± 0.03	17.38 ± 0.41	2.79 ± 0.27	17.13 ± 1.03
	5	11.39 ± 0.05	15.50 ± 0.24	1.21 ± 0.12	11.43 ± 0.77
	6	12.56 ± 0.07	15.49 ± 0.22	0.92 ± 0.10	11.78 ± 0.82
	7	13.45 ± 0.09	15.68 ± 0.22	0.87 ± 0.10	13.33 ± 0.91
	8	14.11 ± 0.11	15.85 ± 0.22	0.88 ± 0.10	14.66 ± 0.97
	9	14.61 ± 0.13	15.98 ± 0.22	0.91 ± 0.10	15.67 ± 0.99
MODCS	4	09.87 ± 0.03	16.92 ± 0.36	2.87 ± 0.29	09.82 ± 0.38
	5	11.36 ± 0.05	15.29 ± 0.22	1.26 ± 0.12	06.54 ± 0.33
	6	12.52 ± 0.07	15.32 ± 0.20	0.91 ± 0.09	07.53 ± 0.46
	7	13.40 ± 0.09	15.51 ± 0.20	0.80 ± 0.08	10.23 ± 0.65
	8	14.06 ± 0.11	15.70 ± 0.20	0.79 ± 0.09	12.64 ± 0.81
	9	14.55 ± 0.13	15.85 ± 0.21	0.82 ± 0.09	14.42 ± 0.89
NEW (−10) [1]	4	09.81 ± 0.03	16.99 ± 0.39	4.12 ± 0.33	08.22 ± 0.24
	5	11.27 ± 0.04	15.19 ± 0.22	1.89 ± 0.15	05.81 ± 0.20
	6	12.42 ± 0.06	15.19 ± 0.19	1.30 ± 0.11	04.98 ± 0.22
	7	13.31 ± 0.08	15.41 ± 0.19	1.02 ± 0.09	06.21 ± 0.35
	8	13.98 ± 0.11	15.61 ± 0.20	0.89 ± 0.09	09.36 + 062
	9	14.48 ± 0.12	15.78 ± 0.20	0.86 ± 0.09	12.19 ± 0.79

Table 7. *Cont.*

	Load Norm	STT [1]	TTT [2]	T [3]	P [4] (%)
NEW (−5)	4	09.85 ± 0.03	16.91 ± 0.37	3.41 ± 0.29	07.68 ± 0.26
	5	11.32 ± 0.04	15.21 ± 0.21	1.55 ± 0.13	05.16 ± 0.21
	6	12.47 ± 0.06	15.24 ± 0.19	1.09 ± 0.10	05.04 ± 0.26
	7	13.35 ± 0.09	15.46 ± 0.19	0.90 ± 0.09	07.45 ± 0.47
	8	14.01 ± 0.11	15.66 ± 0.20	0.84 ± 0.09	10.75 ± 070
	9	14.51 ± 0.12	15.81 ± 0.21	0.84 ± 0.09	13.11 ± 0.84
NEW (−2)	4	09.87 ± 0.03	16.98 ± 0.37	3.03 ± 0.27	08.35 ± 0.31
	5	11.34 ± 0.05	15.28 ± 0.22	1.35 ± 0.12	05.48 ± 0.25
	6	12.50 ± 0.07	15.29 ± 0.20	0.97± 0.07	05.19 ± 0.39
	7	13.38 ± 0.09	15.50 ± 0.20	0.84 ± 0.09	09.03 ± 0.59
	8	14.04 ± 0.11	15.70 ± 0.20	0.81 ± 0.09	12.00 ± 079
	9	14.54 ± 0.13	15.85 ± 0.21	0.83 ± 0.09	14.05 ± 0.88
NEW (−1)	4	09.78 ± 0.03	17.00 ± 0.37	2.93 ± 0.26	08.99 ± 0.35
	5	11.36 ± 0.05	15.31 ± 0.22	1.30 ± 0.12	05.97 ± 0.30
	6	12.51 ± 0.07	15.32 ± 0.20	0.94± 0.09	06.86 ± 0.43
	7	13.38 ± 0.09	15.51 ± 0.20	0.82 ± 0.09	09.63 ± 0.63
	8	14.05 ± 0.11	15.71 ± 0.20	0.81 ± 0.09	12.43 ± 0.80
	9	14.55 ± 0.13	15.86 ± 0.21	0.83 ± 0.09	14.30 ± 0.89

[1] Shop Throughput Time; [2] Total Throughput Time. [3] Tardiness [4] Percentage of Tardy Jobs.

5. Conclusions

Load-based backlog sequencing rules were recently highlighted as an important means to improve order release performance. However, they rely on feedback information from the shop floor and they delay jobs with long processing times during periods of high loads. In answer to our research question—*Can a new backlog sequencing rule be designed that matches high performing MODCS rule whilst only considering job information and enabling a more controlled decision to be taken on which job to delay?*—we have shown that similar performance can be achieved in our simulations by simply subdividing orders in the backlog into early, urgent, and very urgent orders, and then releasing urgent before very urgent orders.

Our new, purely time-oriented rule not only provides a simpler means of improving order release performance, using dedicated classes, but it also facilitates the selection of jobs that are desirable to delay. This allows managers in practice to delay specific jobs for which customer due dates can be adjusted. It puts the control of which jobs to delay in the hand of managers. In fact, it may be an alternative explanation for the Workload Control paradox, which recognizes that order release methods often perform better in practice than expected given their simulation results [40]. Managers in practice are likely to show exactly this behavior–agreeing on new due date allowances for jobs that are otherwise 'hopelessly' delayed.

A main limitation of our study is that we have only focused on one order release method. Although this is justified by Workload Control arguably being the best order release method for high-variety contexts, future research could assess the impact of our new rule on the performance of alternative release methods, such as ConWIP and DBR. Another limitation is that the study only considered the specific case of a re-entrant flow shop. Future research could consider other shop configurations, e.g., flow shops and job shops. Meanwhile, a main advantage of our new method is that it uses dedicated job classes to determine which jobs should be delayed. This allows for delaying jobs with flexible customer due dates as part of the due date negotiation process.

Author Contributions: Conceptualization, N.O.F. and M.T; methodology, N.O.F. and M.T.; software, N.O.F.; validation, N.O.F. and M.T.; formal analysis, N.O.F. and M.T.; investigation, N.O.F., M.T., M.S. and S.C.-S.; resources, N.O.F., M.T., M.S. and S.C.-S.; data curation, N.O.F., M.T., M.S. and S.C.-S.; writing—original draft preparation, N.O.F. and M.T.; writing—review and editing, N.O.F., M.T., M.S. and S.C.-S.; visualization, N.O.F., M.T., M.S. and S.C.-S.; supervision, N.O.F., M.T., M.S. and S.C.-S.; project administration, N.O.F.; funding acquisition, N.O.F. All authors have read and agreed to the published version of the manuscript.

Funding: This research was funded by FCT—Fundação para a Ciência e Tecnologia within the R&D Unit Project Scope UIDB/00319/2020.

Institutional Review Board Statement: Not applicable.

Conflicts of Interest: The authors declare no conflict of interest.

References

1. Hopp, W.J.; Spearman, M.L. To pull or not to pull: What is the question? *Manuf. Serv. Oper. Manag.* **2004**, *6*, 133–148. [CrossRef]
2. Spearman, M.L.; Woodruff, D.L.; Hopp, W.J. CONWIP: A pull alternative to Kanban. *Int. J. Prod. Res.* **1990**, *28*, 879–894. [CrossRef]
3. Prakash, J.; Chin, J.F. Modified CONWIP systems: A review and classification. *Prod. Plan. Control* **2015**, *26*, 296–307. [CrossRef]
4. Jaegler, Y.; Jaegler, A.; Burlat, P.; Lamouri, S.; Trentesaux, D. The CONWIP production control system: A systematic review and classification. *Int. J. Prod. Res.* **2018**, *56*, 5736–5757. [CrossRef]
5. Goldratt, E.M.; Cox, J. *The Goal: Excellence in Manufacturing*; North River Press: Great Barrington, MA, USA, 1984.
6. Simons, J.V.; Simpson, W.P., III. An exposition of multiple constraint scheduling as implemented in the goal system (formerly disaster). *Prod. Oper. Manag.* **1997**, *6*, 3–22. [CrossRef]
7. Watson, K.J.; Blackstone, J.H.; Gardiner, S.C. The evolution of a management philosophy: The theory of constraints. *J. Oper. Manag.* **2007**, *25*, 387–402. [CrossRef]
8. Cigolini, R.; Portioli-Staudacher, A. An experimental investigation on workload limiting methods with ORR policies in a job shop environment. *Prod. Plan. Control* **2002**, *13*, 602–613. [CrossRef]
9. Portioli-Staudacher, A.; Tantardini, M. A lean-based ORR system for non-repetitive manufacturing. *Int. J. Prod. Res.* **2011**, *50*, 3257–3273. [CrossRef]
10. Thürer, M.; Stevenson, M.; Silva, C.; Land, M.J.; Fredendall, L.D. Workload control (WLC) and order release: A lean solution for make-to-order companies. *Prod. Oper. Manag.* **2012**, *21*, 939–953. [CrossRef]
11. Haeussler, S.; Netzer, P. Comparison between rule-and optimization-based workload control concepts: A simulation optimization approach. *Int. J. Prod. Res.* **2020**, *58*, 3724–3743. [CrossRef]
12. Delaram, J.; Houshamand, M.; Ashtiani, F.; Fatahi Valilai, O. Development of public cloud manufacturing markets: A mechanism design approach. *Int. J. Syst. Sci. Oper. Logist.* **2022**, 1–27. [CrossRef]
13. Delaram, J.; Houshamand, M.; Ashtiani, F.; Valilai, O.F. A utility-based matching mechanism for stable and optimal resource allocation in cloud manufacturing platforms using deferred acceptance algorithm. *J. Manuf. Syst.* **2021**, *60*, 569–584. [CrossRef]
14. Wu, S.; Cao, H.; Zhao, H.; Hu, Y.; Yang, L.; Yin, H.; Zhu, H. A softwarized resource allocation framework for security and location guaranteed services in B5G networks. *Comput. Commun.* **2021**, *178*, 26–36. [CrossRef]
15. Liu, Y.; Xu, X.; Zhang, L.; Wang, L.; Zhong, R.Y. Workload-based multi-task scheduling in cloud manufacturing. *Robot. Comput. Integr. Manuf.* **2017**, *45*, 3–20. [CrossRef]
16. Soualhia, M.; Khomh, F.; Tahar, S. Task scheduling in big data platforms: A systematic literature review. *J. Syst. Softw.* **2017**, *134*, 170–189. [CrossRef]
17. Xu, X.; Cao, L.; Wang, X. Adaptive task scheduling strategy based on dynamic workload adjustment for heterogeneous Hadoop clusters. *IEEE Syst. J.* **2014**, *10*, 471–482. [CrossRef]
18. Thürer, M.; Land, M.J.; Stevenson, M.; Fredendall, L.D.; Godinho Filho, M. Concerning Workload Control and Order Release: The Pre-Shop Pool Sequencing Decision. *Prod. Oper. Manag.* **2015**, *24*, 1179–1192. [CrossRef]
19. Thürer, M.; Fernandes, N.O.; Stevenson, M.; Qu, T. On the Backlog-sequencing Decision for Extending the Applicability of ConWIP to High-Variety Contexts: An Assessment by Simulation. *Int. J. Prod. Res.* **2017**, *55*, 4695–4711. [CrossRef]
20. Thürer, M.; Stevenson, M. On the Beat of the Drum: Improving the Flow Shop Performance of the Drum-Buffer-Rope Scheduling Mechanism. *Int. J. Prod. Res.* **2018**, *56*, 3294–3305. [CrossRef]
21. Land, M.J.; Su, N.P.B.; Gaalman, G.J.C. In search of the key to delivery improvement. In Proceedings of the 16th International Working Seminar on Production Economics, Innsbruck, Austria, 1–5 March 2010; Volume 2, pp. 297–308.
22. Graves, S.C.; Meal, H.C.; Stefek, D.; Zeghmi, A.H. Scheduling of Re-Entrant Flow Shops. *J. Oper. Manag.* **1983**, *3*, 197–207. [CrossRef]
23. Muhammad, N.A.; Chin, J.F.; Kamarrudin, S.; Chik, M.A.; Prakash, J. Fundamental simulation studies of CONWIP in front-end wafer fabrication. *J. Ind. Prod. Eng.* **2015**, *32*, 232–246. [CrossRef]
24. Rifai, A.P.; Dawal, S.Z.M.; Zuhdi, A.; Aoyama, H.; Case, K. Reentrant FMS scheduling in loop layout with consideration of multi loading-unloading stations and shortcuts. *Int. J. Adv. Manuf. Technol.* **2016**, *82*, 1527–1545. [CrossRef]

25. Fullerton, R.R.; Kennedy, F.A.; Widener, S.K. Lean manufacturing and firm performance: The incremental contribution of lean management accounting practices. *J. Oper. Manag.* **2014**, *32*, 414–428. [CrossRef]
26. Kundu, K.; Land, M.J.; Portioli-Staudacher, A.; Bokhorst, J.A.C. Order review and release in make-to-order flow shops: Analysis and design of new methods. *Flex. Serv. Manuf. J.* **2021**, *33*, 750–782. [CrossRef]
27. Neuner, P.; Haeussler, S. Rule based workload control in semiconductor manufacturing revisited. *Int. J. Prod. Res.* **2021**, *59*, 5972–5991. [CrossRef]
28. Wisner, J.D. A review of the order release policy research. *Int. J. Oper. Prod. Manag.* **1995**, *15*, 25–40. [CrossRef]
29. Fredendall, L.D.; Ojha, D.; Patterson, J.W. Concerning the theory of workload control. *Eur. J. Oper. Res.* **2010**, *201*, 99–111. [CrossRef]
30. Bagni, G.; Godinho Filho, M.; Thürer, M.; Stevenson, M. Systematic Review and Discussion of Production Control Systems that emerged between 1999 and 2018. *Prod. Plan. Control*, 2021; *in print*. [CrossRef]
31. Gómez Paredes, F.; Godinho Filho, M.; Thürer, M.; Fernandes, N.O.; Jabbour, C.J.C. Factors for choosing Production Control Systems in Make-To-Order Shops: A Systematic Literature Review. *J. Intell. Manuf.* **2022**, *33*, 639–674. [CrossRef]
32. Fernandes, N.O.; Carmo-Silva, S. Workload Control under continuous order release. *Int. J. Prod. Econ.* **2011**, *131*, 257–262. [CrossRef]
33. Oosterman, B.; Land, M.J.; Gaalman, G. The influence of shop characteristics on workload control. *Int. J. Prod. Econ.* **2000**, *68*, 107–119. [CrossRef]
34. Philipoom, P.R.; Steele, D.C. Shop floor control when tacit worker knowledge is important. *Decis. Sci.* **2011**, *42*, 655–688. [CrossRef]
35. Philipoom, P.R.; Malhotra, M.K.; Jensen, J.B. An evaluation of capacity sensitive order review and release procedures in job shops. *Decis. Sci.* **1993**, *24*, 1109–1133. [CrossRef]
36. Land, M.J.; Stevenson, M.; Thürer, M.; Gaalman, G.J.C. Job Shop Control: In Search of the Key to Delivery Improvements. *Int. J. Prod. Econ.* **2015**, *168*, 257–266. [CrossRef]
37. Kempf, K. Intel Five-Machine Six Step Mini-Fab Description. 1994. Available online: https://aar.faculty.asu.edu/research/intel/papers/fabspec.html (accessed on 30 April 2020).
38. Thürer, M.; Fernandes, N.O.; Ziengs, N.; Stevenson, M. On the meaning of ConWIP cards: An assessment by simulation. *J. Ind. Prod. Eng.* **2019**, *36*, 49–58. [CrossRef]
39. Fernandes, N.O.; Martins, T.; Carmo-Silva, S. Improving materials flow through autonomous production control. *J. Ind. Prod. Eng.* **2018**, *35*, 319–327. [CrossRef]
40. Bertrand, J.W.M.; van Ooijen, H.P.G. Workload Control order release and productivity: A missing link. *Prod. Plan. Control* **2002**, *13*, 665–678. [CrossRef]

Article

A Novel Robotic Manipulator Concept for Managing the Winding and Extraction of Yarn Coils

Rúben Costa [1], Vitor F. C. Sousa [2], Francisco J. G. Silva [1,2,*], Raul Campilho [1,2], Arnaldo G. Pinto [1], Luís P. Ferreira [1,2] and Rui Soares [1]

[1]   ISEP—School of Engineering, Polytechnic of Porto, 4200-072 Porto, Portugal
[2]   INEGI—Institute of Science and Innovation in Mechanical and Industrial Engineering, 4200-465 Porto, Portugal
*   Correspondence: fgs@isep.ipp.pt

Abstract: Wire rope manufacturing is an old industry that maintains its place in the market due to the need for products with specific characteristics in different sectors. The necessity for modernization and performance improvement in this industry, where there is still a high amount of labor dedicated to internal logistics operations, led to the development of a new technology method, to overcome uncertainties related to human behaviour and fatigue. The removal of successive yarn coils from a twisting and winding machine, as well as cutting the yarn and connecting the other end to the shaft in order to proceed with the process, constitutes the main problem. As such, a mobile automatic system was created for this process, due to its automation potential, with a project considering the design of a 3D model. This novel robotic manipulator increased the useful production time and decreased the winding coil removal cycle time, resulting in a more competitive, fully automated product with the same quality. This system has led to better productivity and reliability of the manufacturing process, eliminating manual labor and its cost, as in previously developed works in other industries.

Keywords: yarn winding; reels; yarn coil extraction; robotic manipulator; robotic arm

Citation: Costa, R.; Sousa, V.F.C.; Silva, F.J.G.; Campilho, R.; Pinto, A.G.; Ferreira, L.P.; Soares, R. A Novel Robotic Manipulator Concept for Managing the Winding and Extraction of Yarn Coils. *Machines* **2022**, *10*, 857. https://doi.org/10.3390/machines10100857

Academic Editor: Dan Zhang

Received: 28 July 2022
Accepted: 20 September 2022
Published: 26 September 2022

Publisher's Note: MDPI stays neutral with regard to jurisdictional claims in published maps and institutional affiliations.

1. Introduction

The rope industry is a very old sector that encompasses three subsectors, namely, synthetic ropes, net ropes and sisal ropes, each of which has a different production process [1]. The production of yarns (spinning) and the textile industry itself may be located in the same company, but they comprise completely different manufacturing processes and equipment, requiring that the yarns are wound into reels after their manufacture, which are then incorporated into the production equipment of the textiles. Thus, the yarn and rope industries need winding and unwinding operations, from the production to their application in the textile product, to be manufactured. The process of manufacturing yarn reels is governed by the final product, depending on whether it is a cable, rope or yarn. However, for any of these products, there is a need to include a system for unwinding, another for tensioning and, usually, accumulating the product to be treated, and yet another for winding, that is, a feeder, a dancer and a winder [2]. Even if automation is perfectly implemented in the main textile yarn production equipment and in the weaving, the initiation and finalization phases of the reels still require human intervention, which prevents this industry from being able to integrate the intended Industry 4.0 (I4.0) principles [3,4]. This situation can still be identified more intensely in the manufacture of synthetic ropes. This was the main motivation for the development of this work, with a view to also automate this type of operations. Nonetheless, it will be necessary to analyze whether the solution should be based on automation only, robotics only, or a hybrid solution. Based on this doubt, a study in the recent literature was carried out in order to consolidate the theoretical foundations that can serve as a support for the development of the solution.

Robotics has assumed a particular relevance in many recent studies, some of which seek to summarize different models for the development of future solutions in this area [5]. Chemweno and Tor [6] developed a model consisting of five steps that allows one to identify opportunities for robotization of manufacturing processes, which incorporates steps such as the decomposition and analysis of tasks aggregated to a given process, distributing, and allocating tasks between the operator and the robot, as well as design and simulate the operation of the production cell. At the same time, an evaluation criterion should also be developed at each of the stages; however, Säfsten et al. [7] argue that it is necessary to have a perfectly defined strategy to make the decision to robotize a given manufacturing process. This strategy involves previously carrying out of some refinements to the processes, properly analyzing the human factors, and only then moving on to the robotization of the processes. Process refinement aims to adjust operations to the robots' way of operating, while human factors intend to analyze operator safety aspects, necessary training and human–robot interaction capacity. Zheng et al. [8,9] carried out interviews having as target SME's (Small and Medium Enterprises) managers trying to identify the main difficulties felt in adopting robotic manufacturing systems. This study had as output an approach able to offer the possibility of switching the design of a labor-intensive manufacturing system to a robotic one. A similar study was also performed by Simões et al. [10] regarding the factors influencing managers in adopting collaborative robots in the industry, having identified 39 different factors, of which 12 have not previously been identified in the literature. The developed framework helps managers to deal with manufacturing evolution needs, helping in decision-making processes regarding the use of collaborative robots. Gultekin et al. [11] and Gadaleta et al. [12] argue that a rigorous study of the movement of robots is mandatory in order to avoid unnecessary energy consumption, which affects the sustainability of production operations. In addition to the numerous models developed by different researchers, some very interesting applications have also been studied. Silva et al. [13] studied the application of a robotic system to a cell for manufacturing suspension mats and cushions usually used in metallic car seat structures. With a hybrid use of robotics and automation, it was possible to develop an integrated inspection and packaging system for that type of product, based on a robot equipped with a gripper specially designed for this purpose. In addition, an artificial vision system was also developed, which allowed one to control different critical zones of the product, validating the previous manufacturing processes, and allowing the later organized packaging of the product in six stacks. This application made the process significantly more productive and efficient, requiring no human intervention, thanks to the automation of feeding and collecting the boxes where the product was deposited by the robot. On the other hand, Castro et al. [14] recovered a robot that was already out of service and adapted it to a welding cell for metallic structures for buses, giving new life to a robot that, through reprogramming and adaptation to a track for movement, still had all the necessary features necessary to meet the requirements imposed on this type of application. Daniyan et al. [15] studied the use of robotic systems in the railcar industry. This study involved a CAD/CAE software to determine the precision of movements when the robotic arm was subjected to the loads imposed by the process, verifying that the robotic system was capable of performing the functions for which it had been selected with the rigour required by the process, significantly improving the overall performance of the production system. Eriksen et al. [16] developed a robotic polishing system for forming tools, which works together with traditional tools, allowing for the texturing of surfaces with a high degree of repeatability. Sujan et al. [17] have developed a robotic system capable of grasping the extremely complex shapes presented by polycrystalline silicon nuggets, which need to be packaged in a fragile fused silica crucible. The system used a processing method consisting of a triple gripper with suction, which is handled by a seven-axis SCARA robot and an optical 3D vision application based on laser triangulation, capable of reconciling the complex shape of the nuggets with the shape of the crucible. It is thus possible to verify that several studies have been carried out in the context of raising awareness and increasing the adoption of robotic systems in manufacturing processes, and additional studies of particular applications of robotic systems to solve specific problems can also be found. Nevertheless, it is clear that robotics also need automation, both in peripheral systems and in the handling of parts and tools. Therefore, coexistence in production systems is recorded quite often, which is why it is also justified to carry out a literature search focused on automation-based solutions.

Magalhães et al. [18] developed a system for collecting, sorting, electronic inspection and repositioning for the next process at the end of automatic yarn bending machines. The system presents a return on investment perfectly affordable regarding the automotive industry standards and eliminates unnecessary manual operations, increasing the profitability of the process. The next process of these bent yarns is the over-injection. Faced with the need to feed and extract already bent yarns to/from plastic over-injection moulds, Silva et al. [19] developed an automatic vertical yarn feeding system, which automatically distributes these yarns on a tray, feeding them to a gripper specifically designed for this purpose. It then rotates with the yarns into the mould, placing them in the correct position in its interior. After over-injection, the same gripper has the proper geometry to extract the yarns already over-injected onto a conveyor belt, which dumps the parts into a finished product container. This system allowed for remarkable productivity gains. Martins et al. [20] and Moreira et al. [21] developed an automatic piece of equipment devoted to the manufacture of Bowden cables, which presents a series of small sequential operations. The aggregation of different stages of the manufacturing process has drastically improved the level of productivity and the reproducibility of the processes, increasing quality and radically reducing the flow of semi-products within the factory layout. Araújo et al. [22] developed an automatic system with application in guillotines for the cutting of thin metal sheets. Through a pneumatic compensation system, it allowed one to eliminate problems of bending thin metal sheets in the final zone of the cut, when they are still attached to the part from which they are being cut, where their weight promotes the deformation of that area. Nunes and Silva [23] developed a new concept of an automatic assembling system for components used in car windshield wipers, which was based on an indexable table provided with stirrer automatic feeding and positioning systems, increasing the flexibility of the process a lot and eliminating some tedious manual operations. Nevertheless, the developed solution needed an operator to feed the main/initial component. Based on a very similar problem, Costa et al. [24] used an in-line approach, completely different from that used in [23], more complex, but eliminating the need for manual feeding of the main component, i.e., the solution developed is fully automatic. These two approaches show how different strategies can be applied to fit an identical problem [25]. Strategic options can have an expressive meaning in several aspects, such as an increase in production, an improvement in quality or a decrease in breakdowns. Santos et al. [26], by substituting the type of cylinders in an APEX tire component manufacturing equipment, significantly improved breakdowns (−62%), also improving equipment efficiency by 9%. Araújo et al. [27] modified the movement strategy of cushions and suspension mats, which are being manufactured along a manufacturing line that still incorporated two manual workstations, eliminating these workstations and achieving an efficiency gain of 40%, requiring a relatively modest investment. Vieira et al. [28] and Figueiredo et al. [29], working on different problems usually detected in the manufacture of Bowden cables, successfully solved these problems through solutions based on conventional automation, reducing quality problems, increasing repeatability and productivity, as well as concentrating disperse operations in a single piece of equipment. The elimination of serious quality problems was also the main motivation of the work developed by Costa et al. [30], who developed a new concept for the assembly of gear subsets used in the movement of windshield wipers, based only on ingenious mechanical solutions and conventional automation, still achieving a productivity gain of 19%. Veiga et al. [31] have also successfully developed a new automatic system for assembling electrical connectors used in automobiles based solely on conventional automation. The literature is practically consensual in mentioning that the automation of industrial processes requires a very rigorous diagnostic phase, which must be followed by a phase of selection of the best solution that takes into account flexibility, rigour of results and a cost/benefit ratio [32]. It is thus clear that conventional automation can be perfectly integrated into more complex robotic solutions, without jeopardizing the total flexibility of the system, and even increasing that flexibility.

This work aimed to develop a new concept of a robotic system for the initiation and completion of the textile yarn winding operation, eliminating the need for operators connected to the process and allowing a manufacturing solution that permits an evolution to an integrated process, following the basic concepts of Industry 4.0.

2. Materials and Methods

The focus of this work stands on a new concept for the removal of successive coils of yarn from a twisting and winding machine. Afterwards, it cuts the yarn and connects the other end to the shaft, so as to continue the process. Originally, this was performed manually, but the need to eliminate human labor and improve the process led to its automatization.

Initially, an evaluation of the previous concept was performed, where the problems related to it and opportunities of improvement were identified. Then, the defined objectives/requirements were described, after which the new concept was proposed. Finally, the automated equipment's project was introduced, revealing the main stages of development and the final results achieved with it.

2.1. Analysis of the Previous Concept

The process described in this work consists in the treatment of a coil yarn and the respective logistics operation of the product. This process arose from the need for an improvement in the rope-making industry, which still needs a high volume of labor, and, therefore, to reduce costs by introducing a more automated system able to integrate the Smart Manufacturing concept in the near future.

The initial process to obtain this product requires a lot of labor, which is described in the diagram present in Figure 1.

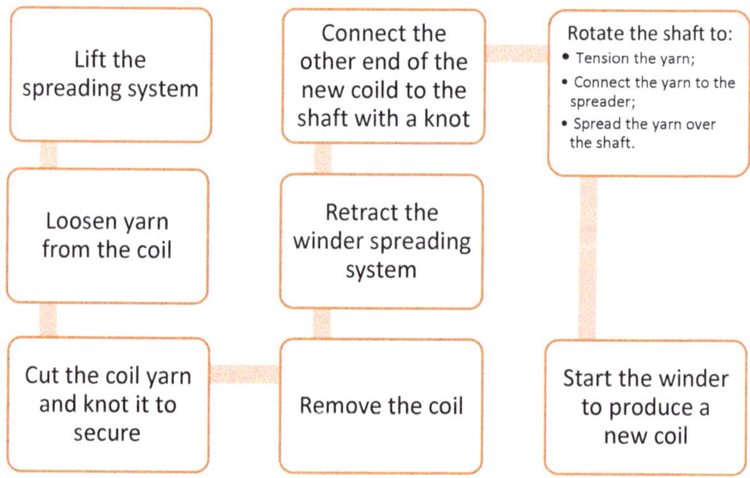

Figure 1. Manual process for obtaining a yarn coil.

In the rope-making industry, intensive labor is still predominant. In the particular case considered for this work, the removal of yarn coils (final product) from a winder and their storage is performed in a completely manual way. This project aims to provide a new concept for the automatization of this process, reducing the labor needed (ideally to make the process fully automatic), and also reducing the rate of defective products, increasing production speed and, in the long term, reducing costs for the company, resulting in a much more profitable process. In this way, the company can also start dedicating these saved man-hours to other, more important activities, in order to improve the final product and meet customer needs, adapting to market demand and always staying one step ahead regarding competitors.

The problem is that, due to the complexity of this process, as shown above, it was necessary to deconstruct the manufacturing process step by step, in order to understand its weak points and where there was room for improvement by automating it, so that no human labor was needed. After splitting into assemblies and subassemblies, a careful analysis of the parts and inserts was carried out, being the machine optimised in design, and being made a careful selection of materials, as well as a deep stress analysis using the CAE method. This is a procedure which involves the continuous work of a specialised engineering team to achieve the best possible result.

2.2. Defined Objectives/Requirements

Some conditions were established previously as requested by the market, which must be respected. Most of these conditions are an addition to the initial requirements defined for the functionalities initially defined.

Hence, the set of requirements determined are as follows:

- Coil dimensions can vary from Ø250 to Ø450 mm;
- Coil longitudinal dimensions can vary between 300 and 500 mm;
- Cycle time must be as low as possible, without exceeding one minute per coil.

The process to obtain the product should also follow the sequence of requirements defined as described in Figure 2.

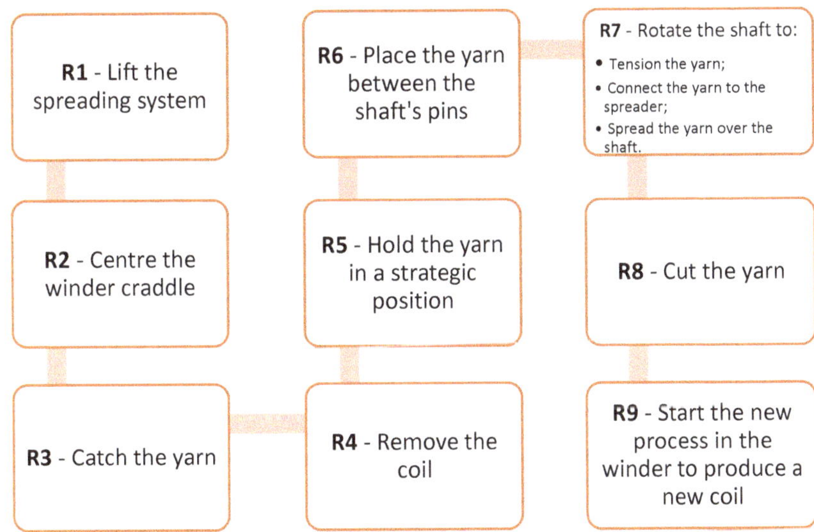

Figure 2. Diagram with the sequence of actions needed for the coil removal.

2.3. Proposed Concept

Taking into account the need to know which was the best approach for the development of this procedure, the strategy followed consisted in:

- A brainstorming process, which was carried out for a better preparation and coordination of the path to follow;

- An evaluation of the possible concepts, which could be adopted for the novel robotic simulator;
- A weighing of the proposed options, in order to choose the best one;
- The development of the selected system, through a mechanical design and an automation programming of the equipment, for it to perform the intended tasks.

During the brainstorming session, considering the defined objectives, the need to address the programming was verified, so a parallel reasoning process was also carried out, where it was determined which types of automation would be feasible. Figure 3 shows that process.

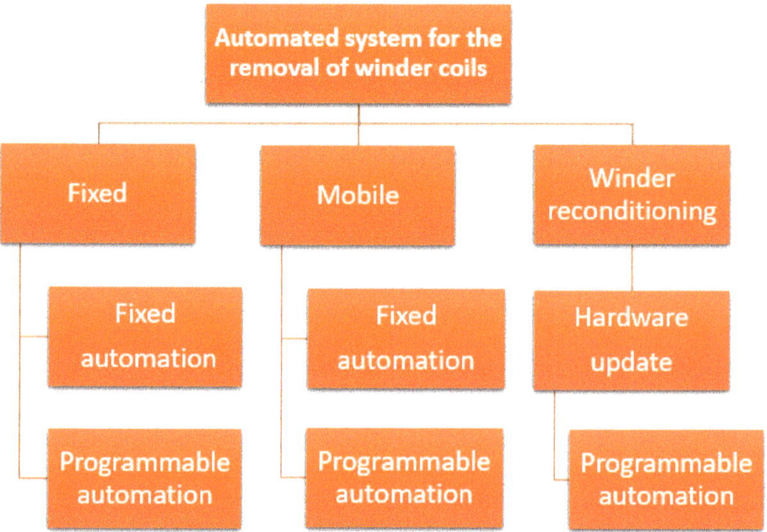

Figure 3. Brainstorming results.

Given the necessary requirements and the results of the brainstorming, a SWOT analysis was performed for each presented idea:

- The fixed solution brings greater freedom in the dimensions imposed on the project and allows for a reduction in cycle time, decreasing the cost of the process and the product; however, it is a solution that already exists on the market, it occupies a large area on the shop floor and each winder requires its own machine.
- The solution of reconditioning the winder is solid and, because it has programmable automation, it allows for the treatment of coils with different dimensions. This type of automation is achievable due to a hardware update, a stage of utter importance to the winder reconditioning. On the other hand, it is a complex project and the machine is limited in terms of space.
- The mobile solution is more dynamic, can treat several winders simultaneously and is an innovative solution, being an introductory method for the I4.0 concept. In spite of this, it is a complex project, with a high cost and design time.

Accordingly, weighing up the advantages and disadvantages, the conclusion was reached that the idea with the greatest potential for innovation is the Mobile System, due to the fact that it allows for continuous twenty-four-hours-a-day work on the coil production line, with little manpower involved.

3. Results

This section will include the design of the final equipment, detailing its most important components, as well as how they function. After the design, the development and implementation of the new production process will be described, as well as the results obtained after implementing it.

3.1. Equipment Presentation

Through the development of this project, it was decided to subdivide the EOAT (End of Arm Tool) assemblies by their functions in the tool and/or by where this assembly will be connected. The subdivision of the assemblies was carried out as can be seen in Figure 4.

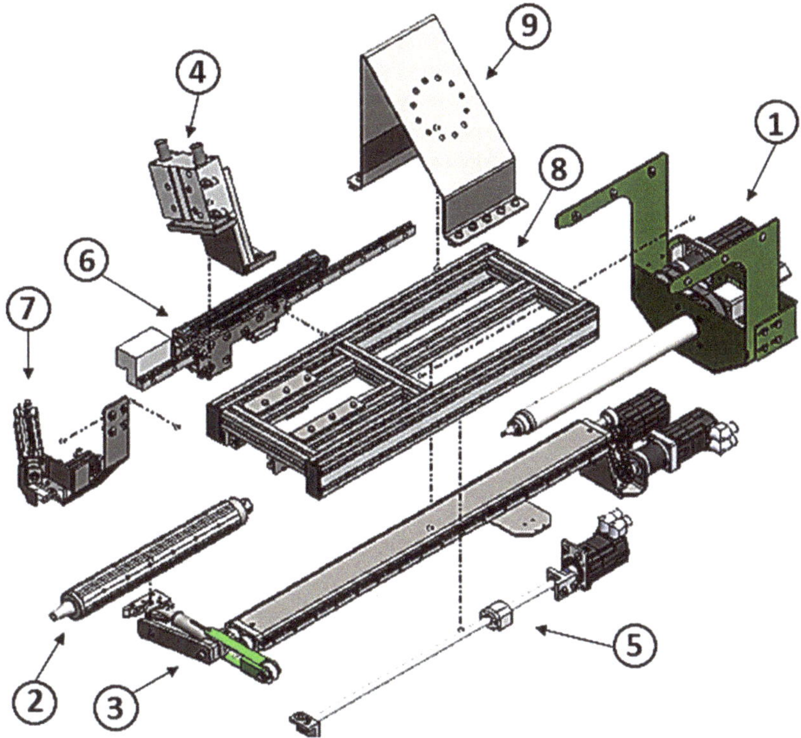

Figure 4. EOAT exploded view.

In Table 1, it is possible to observe the caption for Figure 4.

Table 1. EOAT exploded view's caption.

1—Coil Support Assembly	2—Expandable Pneumatic Shaft
3—Extensible Arm Assembly	4—Nozzle Activation Assembly
5—Ball Screw Assembly	6—Cradle Alignment Assembly
7—Handle Cut Assembly	8—Base Structure Assembly
9—Robotic Arm Linking Assembly	

3.1.1. Mechanical Project

For a better understanding of the various EOAT assemblies functioning, these will be deconstructed. The methodology for the sub-assemblies' presentation is by association with the product in chronological order and the other sub-assemblies dependent on it. This way, all the sub-assemblies that constitute the tool will be exposed. The diagram of

Figure 5 gives a better perspective of the assemblies to be presented and which are their dependences.

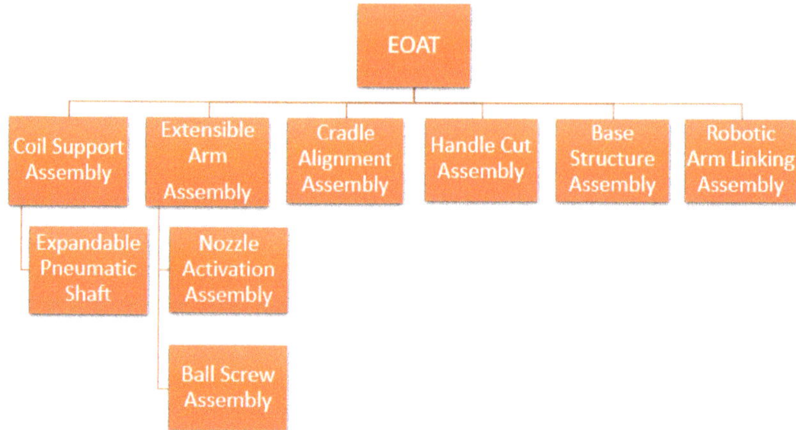

Figure 5. Diagram of EOAT's sub-assemblies.

Coil Support Assembly

The coil support assembly (Figure 6) is responsible for supporting the coil and every element attached to it.

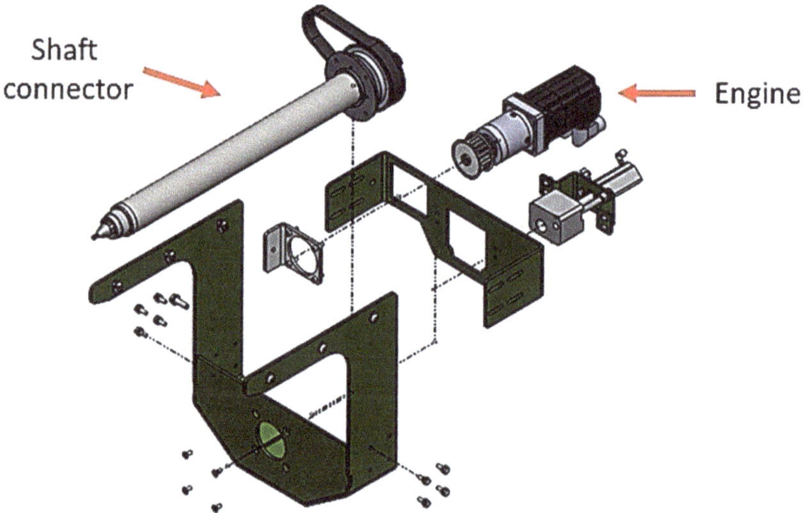

Figure 6. "Coil Holder" assembly's exploded view.

Within this assembly, there is a sub-assembly that needs to be explored more rigorously, depicted in Figure 7. This sub-assembly has four functions, namely:

- Connect and maintain the link between the EOAT and the Expandable Pneumatic Shaft;
- Support the coil in the EOAT;
- Transmit rotation to the Expandable Pneumatic Shaft;
- Activate the valve inside the Expandable Pneumatic Shaft.

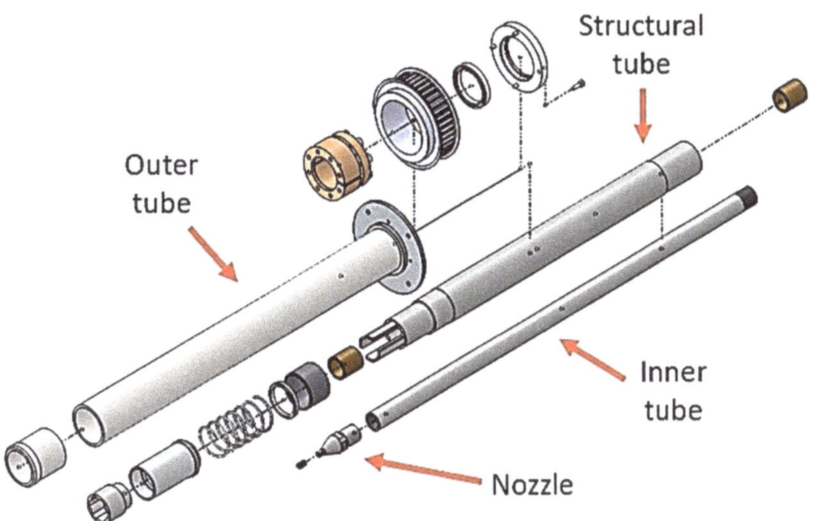

Figure 7. "Shaft connector" sub-assembly's exploded view.

In Figure 8, the mechanism that was developed in order to fulfil the functions described above is represented, where the nozzle has the purpose of activating the valve present in the expandable pneumatic shaft. The outer tube's task is to support the coil and protect the mechanism, and it is made of AISI 304 stainless steel in order to prevent damage to the product when they contact each other. The connection function is carried out by the box wrench, which has a mechanical linear movement performed by a spring. The nozzle, together with the inner tube, also serves to inflate the expandable pneumatic shaft. The linear movement is provided by a cylinder, and the mechanism responsible for this movement is also shown in Figure 8.

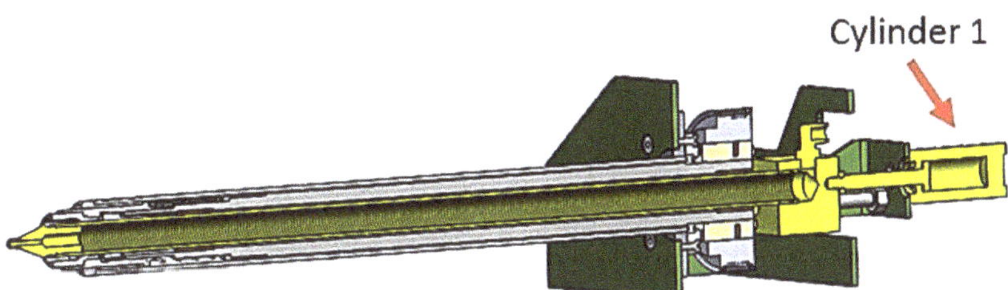

Figure 8. Sectional view of the mechanism responsible for picking up and expanding the expandable pneumatic shaft.

For the spring present in the "Shaft connector", the biggest restriction is its dimension, as it is limited both by the rotation transmission tube to the expandable pneumatic shaft and by the structural tube supporting the coil. Thus, a compression spring is required to ensure that the affected elements are always forward. The elements in question are two machined parts, one acting as a mating part in the hexagonal section of the expandable pneumatic shaft (which is threaded), while the other acts as a stop for the spring and also as an advance limiter.

A more pragmatic method was followed, where a spring was made for testing, obtaining optimum results, which is the reason why this concept has been adopted. This spring has its working displacement limited by stops, so it will not lose its quality. In addition, this shaft also has a mechanism for transmitting rotation to the expandable pneumatic shaft, shown in Figure 9. This mechanism should be able to prevent the inertia of the expandable pneumatic shaft and all the elements of which it is comprised.

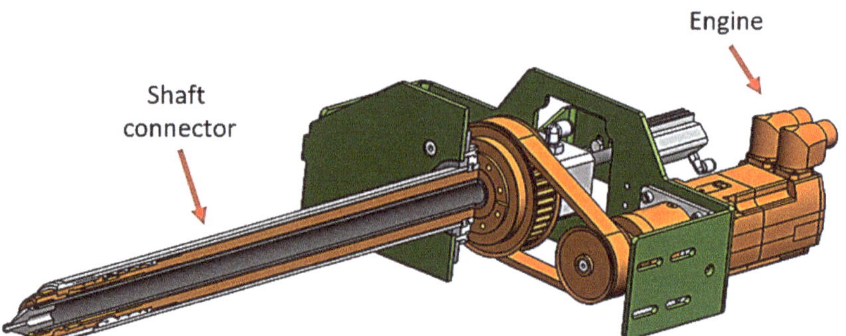

Figure 9. Sectional view of the mechanism responsible for rotating the expandable pneumatic shaft.

The torque, angular velocity, and the time for one revolution of the expandable pneumatic shaft have been calculated so that, with any future changes/improvements to the tool, the designer can be able to perform the calculations for the tool's modification without difficulty.

The expandable pneumatic shaft shown in Figure 10 has been sized to have an expansion of 10 mm in diameter, so that it can expand from Ø50 to Ø60 mm, with a valve at the inlet to ensure that this effect is possible.

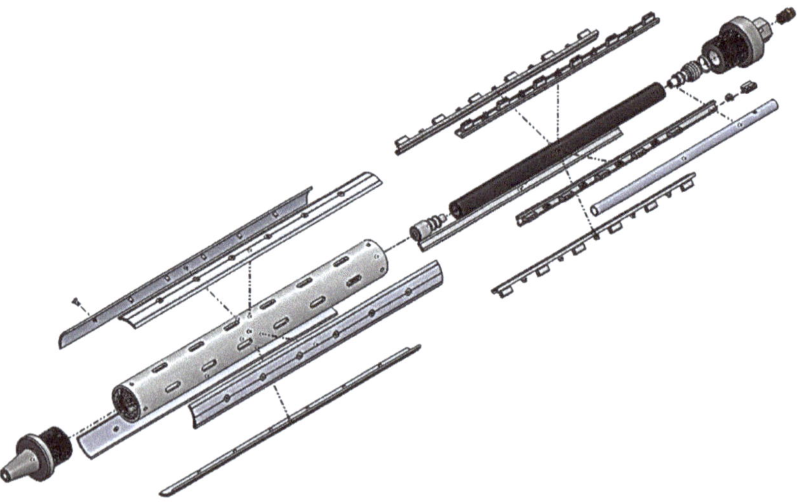

Figure 10. Expandable pneumatic shaft's exploded view.

The nipple valve present in the mechanism of Figure 11, which is located on its right side (brown), is activated by the nozzle present in the shaft connector, emptying the pneumatic shaft so that it is possible to remove the yarn coil. It should also be noted that it is the same valve that will allow the opposite process to be carried out before a new winding process begins. Figure 11 also shows, through the centre of the shaft, the air flow (in blue) from the valve to the expandable rubber sleeve, which is responsible for expanding the diameter of the shaft.

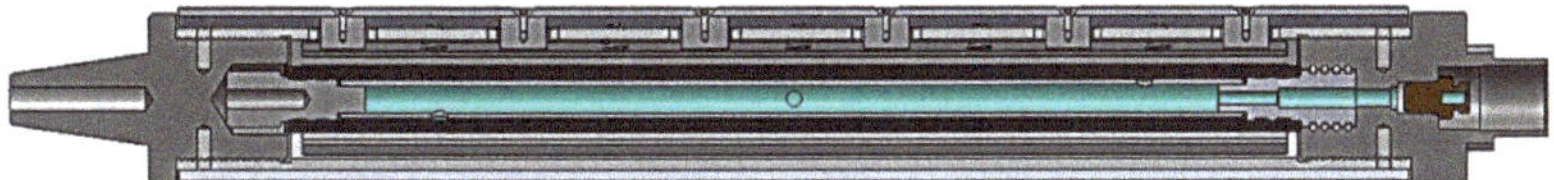

Figure 11. Sectional view of the YZ plane of the expandable pneumatic shaft's mechanism.

The airless shaft remains retracted by the action of conical springs, as depicted in Figure 12.

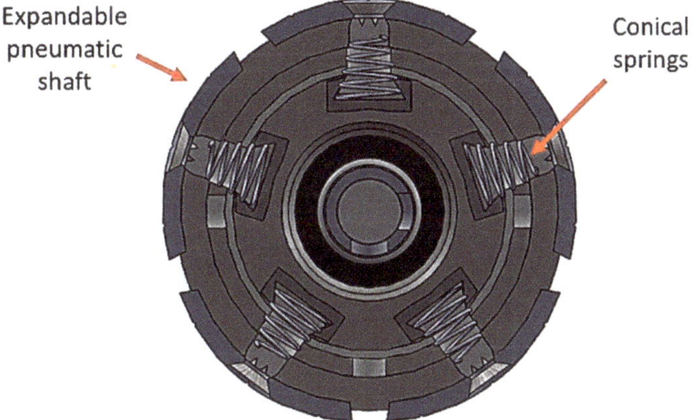

Figure 12. Sectional view of the XY plane of the expandable pneumatic shaft's mechanism.

The winder's pneumatic shaft spring is held inside a U-shaped aluminum profile, with the flaps more closed, and it is through these flaps that the spring remains fixed.

Extensible Arm Assembly

In order to meet the requirements, a method is needed to take the coil, pick up the yarn and transport it to some strategic points without damaging it.

Firstly, it is necessary to remove the coil without damaging the product, where all the yarn of the coil is tensioned and, when the pneumatic shaft is contracted, the coil contracts a little, a resistance that the bar must be able to withstand. A welded reinforcement has been added to this sub-assembly, and the plate has also been oversized so that it does not wrap. This sub-assembly shown in Figure 13 should also activate a lever that lowers a spreader present inside the winding machine.

The coil bar sub-assembly must have a rotation movement in order to reach certain necessary positions inside the winding machine. This movement is made by the mechanism represented in Figure 13, the extensible bar assembly.

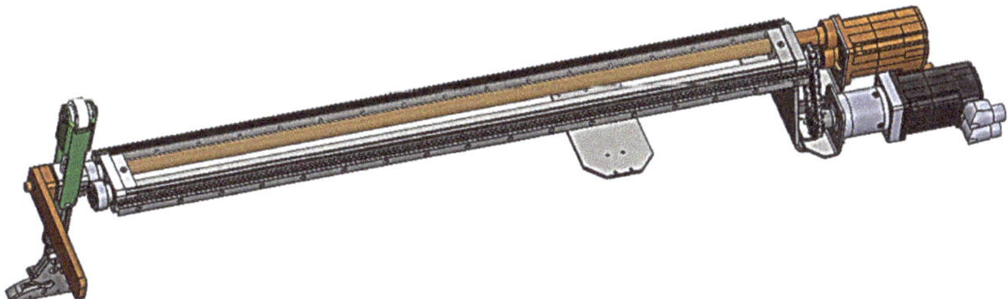

Figure 13. Mechanism responsible for rotating the reel bar.

The first sub-assembly in Figure 14 is named "coil bar", the second "yarn arm", and finally the third "extensible bar". The latter has the function of housing the entire mechanism for transmitting movements to the other two sub-assemblies.

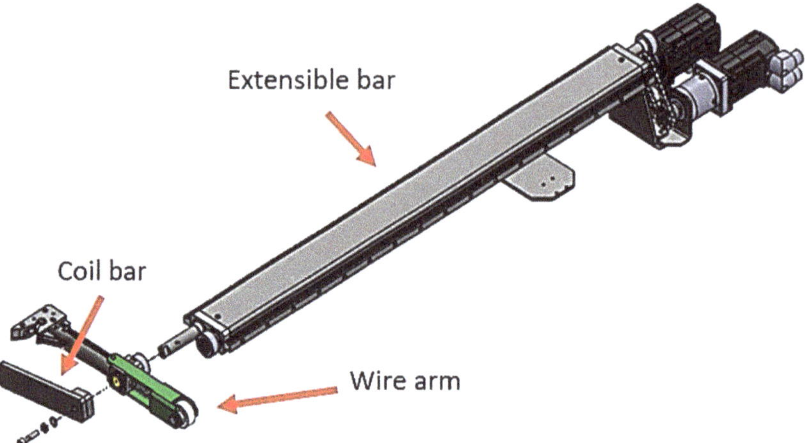

Figure 14. Exploded view of the three sub-assemblies present in the extensible arm.

Besides the rotational motion, the yarn arm sub-assembly has a crucial movement in allowing for a better accomplishment of the requirements, which is a linear motion, depicted in Figure 15.

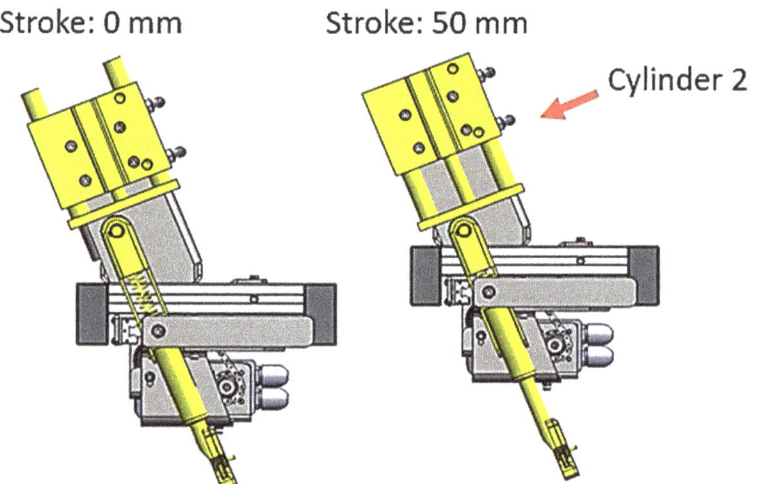

Figure 15. Linear movement of the yarn arm sub-assembly and its dependence.

The linear movement of the yarn arm sub-assembly is guided, with a 50 mm stroke. The cylinder exposed in Figure 15 is embedded in the nozzle activation system and has the visible inclination so that there is not much loss of stroke due to the angle at which it must act, thus allowing it to act in a straight line with the yarn arm sub-assembly.

The yarn arm stroke has the purpose to fulfil the requirement of attaching the yarn to the pneumatic shaft. This way, when the need arises to attach the yarn to the expandable pneumatic shaft, it is necessary to operate this mechanism together with a 30° rotation movement in all the EOAT, by moving the robotic arm. This 30° angle is essential, so that the yarn can reach the pins of the pneumatic shaft, since the stroke of the yarn arm sub-assembly is limited in size. As such, the mechanism is able to catch the yarn.

As the cylinder will exert a force on the yarn arm sub-assembly, and as the tension amount exerted on the arm is not known, a stock cylinder was used to carry out tests and get a better idea of the force required. Nonetheless, in the first instance, it is necessary to estimate the force it must overcome.

Within the yarn arm sub-assembly, another mechanism also exists, responsible for catching the yarn and keeping it attached, so that it is manoeuvrable; this being the "hand" of this tool. The mechanism is represented in Figure 16.

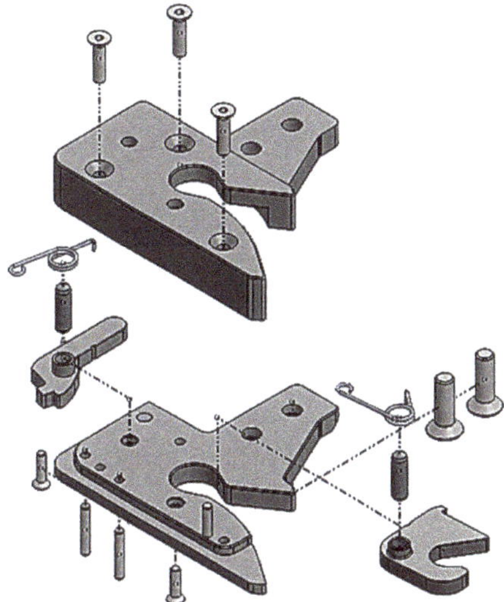

Figure 16. Gripper mechanism's exploded view.

This mechanism has three important positions: the rest, engagement and disengagement positions. Figure 17 shows the working positions of the gripper mechanism, and in the coupling situation, the pawl/knob locks the coupling part in the pictured position.

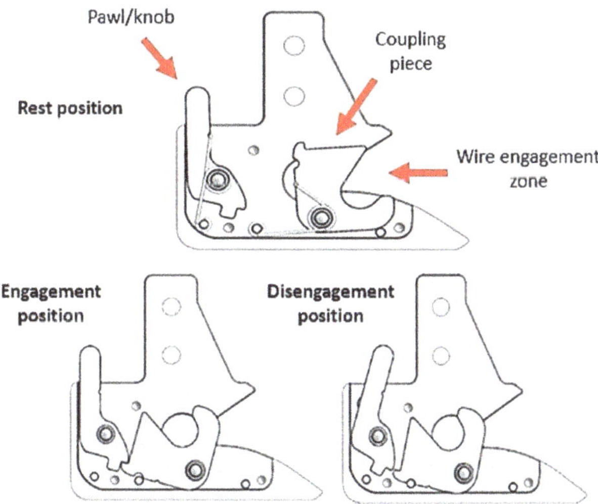

Figure 17. Working positions of the gripper mechanism.

In the situation where it is necessary to disengage for a new process, the pawl is activated, and the springs must act to return the mechanism to its rest position. Finally, the linear movement of the arm is carried out by the assembly shown in Figure 18.

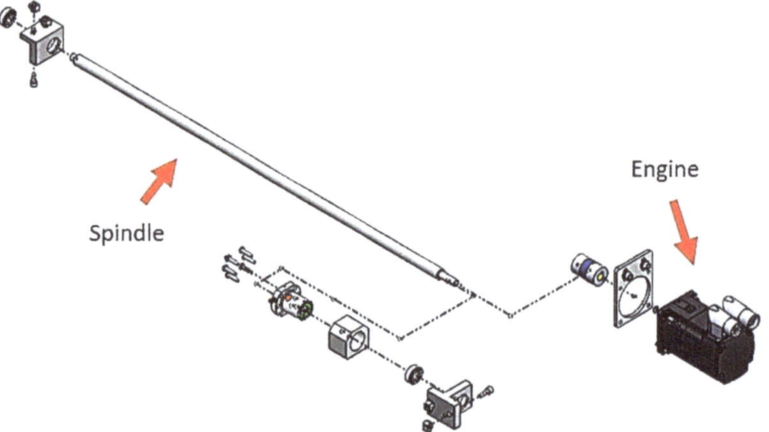

Figure 18. Spindle assembly's exploded view.

The assembly represented in Figure 19 is responsible for providing an axial movement to the extensible arm, allowing it to move inside the winding machine. The axial linear movement executed by the spindle allows the mechanism to perform what is demonstrated in Figure 19.

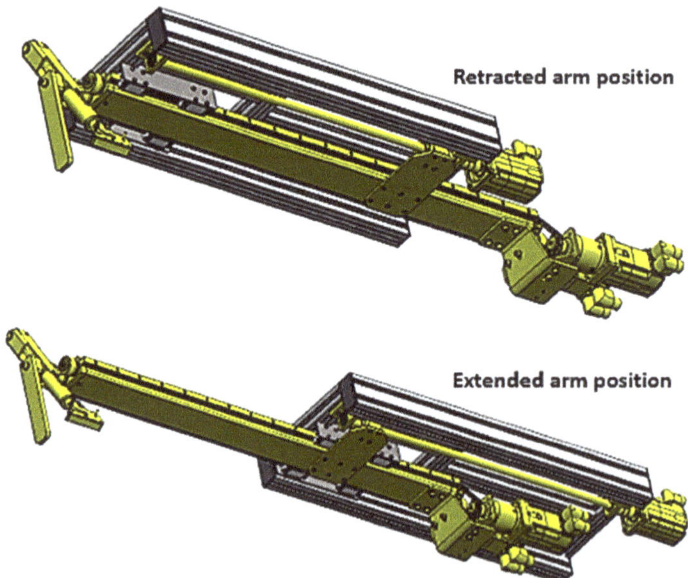

Figure 19. Linear movement of the extensible arm assembly and its dependence.

Cradle Alignment Assembly

The following assembly has a very simple function: to keep the cradle parallel to the floor. This procedure is aided with the movement of the robotic arm, being performed by a cylinder. This assembly was positioned at a strategic distance from the connection shaft (see Figure 20), so that, with the cradle aligned, the coupling between the connection shaft and the expandable pneumatic shaft, present in the winding machine, can be performed. The cradle alignment assembly is shown in Figure 20 and is fitted with a guide so that the pneumatic cylinder does not have to bear the strain exerted by the cradle.

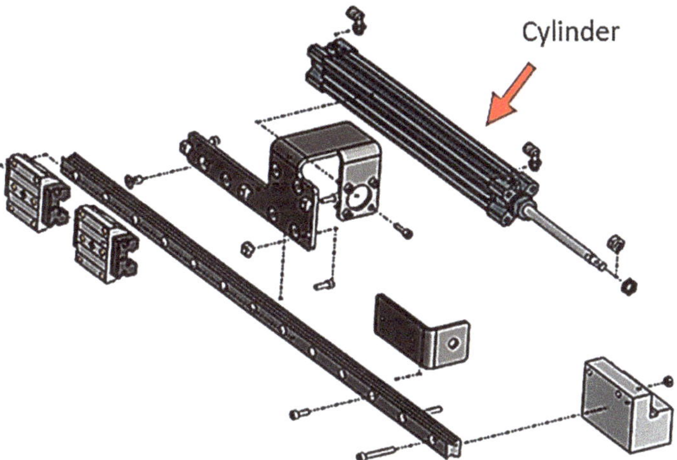

Figure 20. Cradle alignment kit's exploded view.

The pneumatic cylinder of this assembly must be able to overcome the frictional force generated by the linear guide and the acceleration force.

Handle Cut Assembly

The assembly represented in Figure 21, besides guaranteeing the cut, must assure that the yarn remains located in a strategic position, because the heated blade alone does not guarantee a clean cut of the yarn.

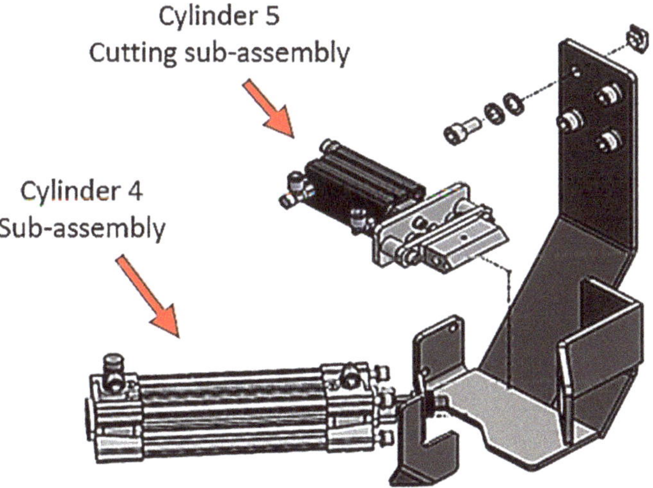

Figure 21. Handle cut assembly's exploded view.

By combining the two sub-assemblies in this mechanism, the plate creates a fixed and flat surface where the yarn will be enclosed for the proper functioning of the cutting blade. This system is represented in Figure 22.

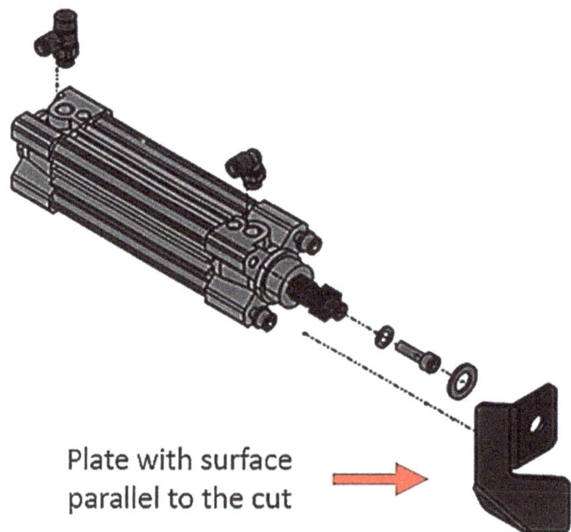

Figure 22. Handle sub-assembly's exploded view.

In this system, it is necessary to place the yarn in a certain position in order to achieve the cut, and the plate attached to the cylinder of the yarn pulling system has its own design for this purpose. Additionally, providing a flat area for the cutting system, it also ensures that once the yarn is pulled, it does not escape. On the other hand, in order to guarantee that the position of the sheet is not altered, a cylinder with an anti-rotation system (hexagonal rod) was selected.

The cutting system, as can be seen in Figure 23, comprises a blade that is heated through a resistance, and a probe to control the temperature. However, as most of the materials are metallic, this mechanism was made to avoid the temperature from damaging the cylinder, using Celeron to decrease the thermal transmission between the fixing materials to the cylinder.

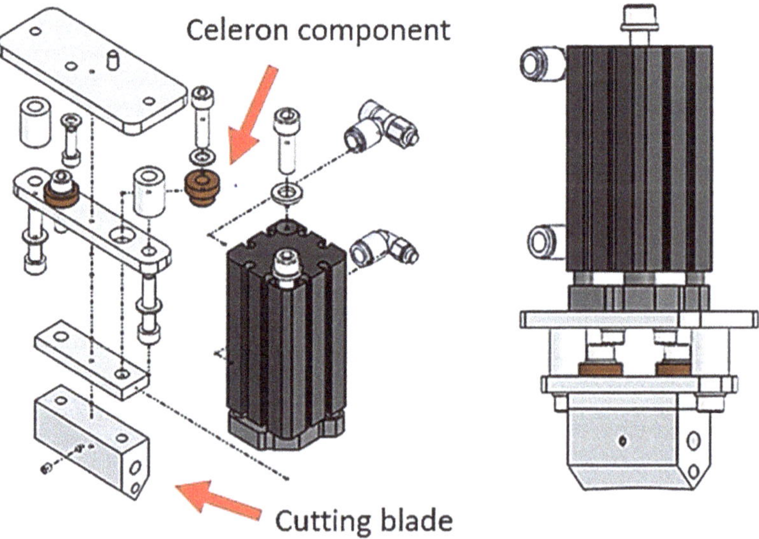

Figure 23. Cutting sub-assembly's exploded and assembled view.

Base Structure Assembly

All the developed assemblies need a common base that will serve as support and connection to the robotic arm. This assembly was developed in an aluminum profile, as shown in Figure 24, as it allows for greater freedom for adjustments if necessary.

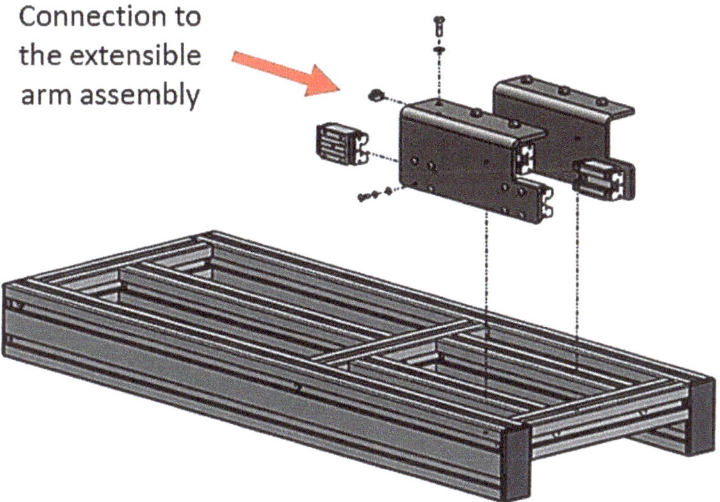

Figure 24. Aluminum profile frame with connection and support function.

The assembly that connects the base structure assembly and the robotic arm is the robotic arm linking assembly, and is depicted in Figure 25.

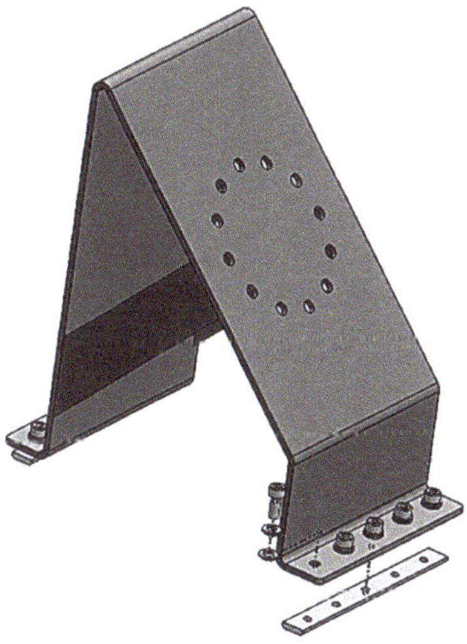

Figure 25. Robotic arm linking assembly.

This system possesses the function of supporting the entire EOAT, including the coil. Accordingly, as it is a critical point that must ensure the integrity of the entire system without collapsing, it must be studied to verify if its structural functionality has to be redesigned.

After a finite element analysis was performed, the simulation showed that it was necessary to introduce a reinforcement in the part's structure, as the EOAT was not safe in any position. Figure 26 displays the optimised design.

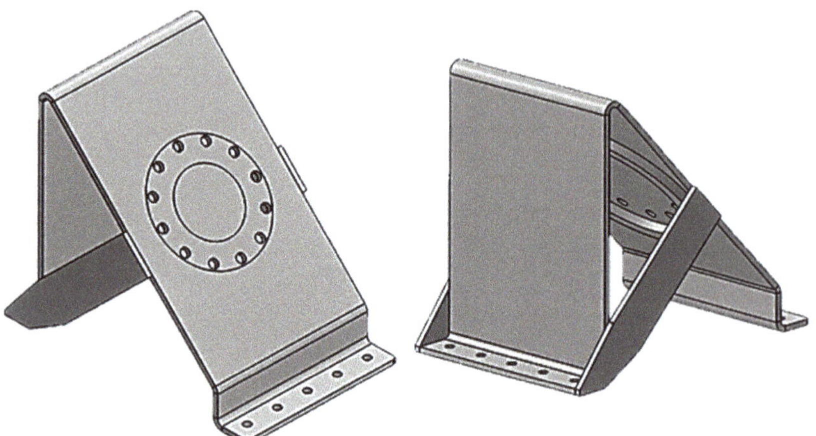

Figure 26. Optimized EOAT backing plate design.

3.2. New Production Process

The exposed design corresponds to one part of the solution for the idealised line, shown in Figure 27. The complete project is a set of mechanical systems, which sums up to an automated line capable of handling 17 winders and able to be expanded. The entire line is formed by various mechanical components, namely, conveyor belts for transporting the final product, robotic arms and rails for their movement, EOAT's, a support structure and a safety system.

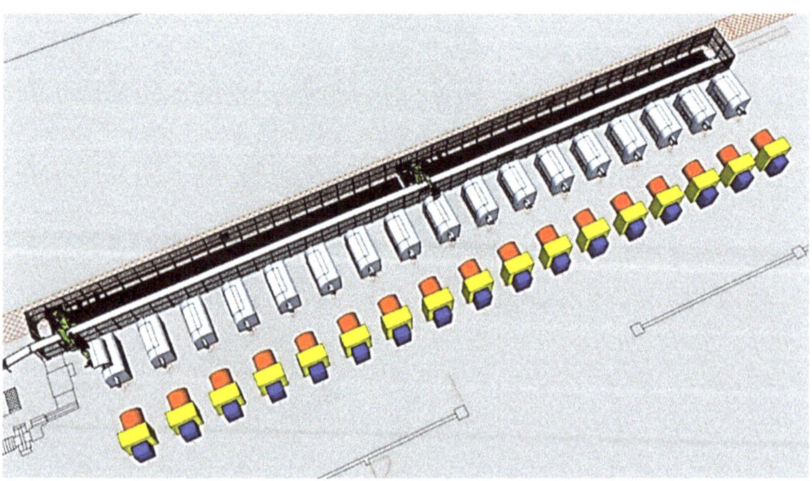

Figure 27. Designed line with 17 winders and their feeders (Image from KUKA SIMPRO).

Before the implementation of this new production process, all the winders were operated manually, with only the human intervention to perform the removal of each coil and replace it with another, in order to begin a new winding of the yarn. This had many problems, such as waste of time, because a single operator was not able to remove all the coils simultaneously from every winding machine, in addition to the time taken to replace it with another. This manual process led to production delays, which caused a huge profitability loss, besides the possibility for mistakes to happen due to the human factors (operator's fatigue, mood variations or lack of attention). Accordingly, an automatic production process solved every single one of these problems: the extraction of a coil and placement of a new one can be made simultaneously in all the 17 winders, eliminating the time loss. In addition, the winders are programmed, so the chance for mistakes to occur in their usual function is reduced to almost zero. This way, this process constitutes a novelty in the yarn rope manufacturing industry, where the Industry 4.0 methods were put to practice in order to overcome the existing obstacles originating from the manual labor.

3.3. Implementation Results

According to data provided by stakeholders of this project, three operators are needed to handle the 17 winders and, in order to meet the production output of the automated solution, three shifts are required. It was empirically assumed that the annual salary of the operators is EUR 14,000.

Another data provided by stakeholders is the time to remove the coil, which, according to them, corresponds to about 60 s for an operator. On the other hand, the robotic arm will be optimised (Equations (1) and (2)) to achieve a time of 40 s per coil removal.

$$Optimisation\ (\%) = 100 - \left(\frac{100 - automatic\ coil\ removal\ time}{manual\ coil\ removing\ time} \right) \tag{1}$$

$$100 - \left(\frac{100 \times 40}{60} \right) = 33\% \tag{2}$$

Taking into account a gain in productivity of about 33%, which is achieved by switching to the automated system, it is also necessary to consider the number of operators required to get the same productivity, shown in Equation (3).

$$\frac{100}{3} = \frac{133,(3)}{x} \rightarrow x \cong 4\ operators \tag{3}$$

Table 2 shows the annual cost data related to this process.

Table 2. Annual cost data with salaries if the automated systems are not implemented.

.	Operators/Line	Cost/Year/Shift	Shifts	Cost/Year
Operator salary	3	EUR 42,000	3	EUR 126,000
Optimization	+1	EUR 14,000	3	EUR 42,000
Total cost/Year				EUR 168,000

Electricity and maintenance costs were disregarded to calculate the return on the investment made in this system. The investment made, considering the line, was 400,000 EUR. The payback period (PBP) is calculated according to Equations (4) and (5):

$$PBP = \frac{New\ equipment\ cost}{Profit\ generated\ per\ year} \tag{4}$$

$$PBP = \frac{400,000}{168,000} = 2.53\ years \tag{5}$$

Assuming empirically that the service life of this line is 10 years, without requiring corrective maintenance, the implementation of this solution brings an expected profit of approximately EUR 1,000,000.

4. Discussion

The industry is experiencing new times of change, through the integration of information and an even greater automation of processes; however, these changes cannot be carried out without the automation of all processes being completed and able to be integrated. This problem had already been addressed by Pinto et al. [33], in a study focused on SMEs. Automation can go through the robotization of processes when several models of the same family of products are manufactured in the same production system and setups are required with high frequency. In spite of this, in many cases, more conventional automation is sufficient to meet the needs at a lower cost, essentially when there is very little variation, or no variation at all, in the product to be manufactured, as referred to by Silva et al. [19] and Araújo et al. [27]. Sometimes, only a mixed solution integrating robotics and conventional automation is able to overcome the real needs of the industry. Indeed, Costa et al. [24] have also mixed a robotic system with conventional automation, generating some novel mechanical solutions able to overcome all requests required by a workstation devoted to the assembly of a semi-product used in widescreen washer engines in the automotive industry. As referred in that work, high-level automated systems have been used around a robot, as well as smart and cheap mechanical solutions, allowing for a high grade of flexibility of the workstation, which became able to assemble a series of different sets of the same family. Castro et al. [14] went further, reusing an out-of-service robot to use it in a welding cell for bus side-body-frames, reusing it and expanding its initial capabilities, through an integrated automatic system that made it possible to extend its range of action, as the bus body-frames have a considerable length. The robot's reprogramming and the integration of its operation with the ability to move along the structure were the key to success in overcoming the need presented by a company that manufactures that type of bus body-frames. Silva et al. [13] also developed a solution based on a robot, due to the need for flexibility and reprogramming, but they used conventional automation integrated with robotics to develop a new system capable of controlling, through artificial vision, cushions and suspension mats used in seats of vehicles, and then proceed to pack them in an organized manner. The solution, although complex, made it possible to guarantee that the quality control became practically 100% reliable and not dependent on possible human errors, complying with the cycle time required by the automatic operation of the existing equipment. Martins et al. [20] and Moreira et al. [21] developed integrated solutions based only on conventional automation, eliminating intermediate stocks of components, reducing internal logistics, and increasing productivity. This is almost always the main motivation of this kind of works, but the development of solutions must be properly supported by an economic study that allows for analyzing the feasibility of its implementation. In the present study, the payback time is around three years, which is at the limit of what is normally acceptable by the industry in general; however, it must still be safeguarded that the reliability of the process is also highly enhanced. This value is much higher than the one found by Magalhães et al. [18], which took just over 8 months, but the complexity of the systems is also quite different. In fact, the complexity of the solution now developed is closer to that of another solution designed by Silva et al. [13], which presented an estimated payback period of 21.5 months. Despite this, in addition to the effort and economic return of the project, it is clear that this solution translates into a production system that can be fully integrated. This is similar to the work presented by Barbosa et al. [34], in which the flexibility of processes and the integration of information allow the systems to approach the concepts usually defined as Smart Manufacturing or Industry 4.0.

Although many studies have as main motivation the reduction in the production cycle time and the cutting of jobs, the main motivation of this work was essentially to ensure a continuous workflow and the integration of a large amount of equipment in the same production line, as this is the reality normally experienced in the textile industry. This contrasts a bit with the reality experienced in the automotive components industry, where competitiveness is in most cases the main motivation, but it is also possible to find works in which the main motivation was quality improvement [22,30], or process integration [13,21,33,34]. There are also cases where automation is modified/improved in order to reduce production stoppages due to breakdowns, as mentioned by Santos et al. [26]. The mechanical solutions developed around the automation used in this work are also noteworthy, being at the level of others developed in similar works [18,23,24,27,28,30], and represent systems that, in addition to solving the problems defined in this work, may constitute extremely interesting solutions to overcome other similar problems in other types of industry. Therefore, the innovative solutions presented by this work can, alone or together, be used to solve other problems. This is one of the main reasons for the presentation of this work, which is intended to enable the integration of these systems, or the development of even more sophisticated ones, which may constitute a starting point for further studies. Video S1 provided as supplementary material, corresponding to the first tests performed with the prototype resulting of this work, will help to understand the operational functioning of the novel robotic manipulator in action.

5. Conclusions

The spinning/textile industry uses coils as systems for storing and transporting yarn between production operations. The winding of these coils, despite being automatic, also included manual operations of starting and ending the winding. Automating this process was the main motivation of this work and, to this end, new solutions were developed to couple to a robotic arm that collects the coils. The developed solution translates in the total automation of the process, allowing it to adapt to a system usually defined as Smart Manufacturing, or Industry 4.0, in which the coils are finished, sent to a common conveyor belt, and the winding of a new coil is started fully automatically, i.e., without human intervention. The developed system made use of several ingenious solutions, which are effectively innovative and, in addition to solving the problem described in this work, can be used to solve other similar problems in other types of industry, which may contain their own specificities. A portfolio of relatively simple and economic solutions is thus created, but interesting enough to be disseminated and included in other solutions aimed at solving other problems. It is worth mentioning that the coils' extraction system was designed to fulfil the necessary functions, always without damaging the yarn used in the process.

The developed equipment is capable of extracting coils of yarn automatically and can be adapted to extract coils of other materials, as well. With the flexibility offered by the robotic arm together with various components in the developed project, this complex operation became feasible. In addition, an equipment capable of cutting and burning the end of the thread was developed, so as not to deprive the product of its quality due to fraying. In order to allow easy cleaning and maintenance, all the equipment has a modular construction. Finally, the cycle time was reduced, since at the end of this project, the execution time was 40 s. The payback time of this solution is about 36 months, which is within the usual acceptable period for this kind of project.

Supplementary Materials: The following supporting information can be downloaded at: https://www.mdpi.com/article/10.3390/machines10100857/s1, Video S1: Prototype working.

Author Contributions: Investigation, R.S. and F.J.G.S.; main research and data collection, R.S.; data curation, R.C. (Rúben Costa), R.C. (Raul Campilho), A.G.P. and L.P.F.; formal analysis, R.C. (Raul Campilho), V.F.C.S., A.G.P. and L.P.F.; writing—original draft, R.C. (Rúben Costa); writing—draft and reviewing, V.F.C.S.; writing—reviewing, F.J.G.S.; conceptualization, F.J.G.S.; supervision, F.J.G.S.; resources, R.C. (Raul Campilho), A.G.P. and L.P.F.; visualization, V.F.C.S. and R.C. (Raul Campilho). All authors have read and agreed to the published version of the manuscript.

Funding: This research received no external funding, and the authors did not receive any funding regarding the development of this work.

Institutional Review Board Statement: Not applicable.

Informed Consent Statement: Not applicable.

Data Availability Statement: No data or materials are available regarding this work.

Acknowledgments: The authors want to thank Jorge Seabra and Carlos Fernandes from CETRIB/INEGI/LAETA due their continuous support.

Conflicts of Interest: The authors declare that they have no known competing financial interests or personal relationships that could have appeared to influence the work reported in this paper. Moreover, the authors declare no conflict of interest.

References

1. Kyosev, Y. *Braiding Technologies for Textiles*; Woodhead Publishing Series in Textiles, No 158; Woodhead Publishing: Sawston, UK, 2015; ISBN 978-0-85709-135-2.
2. Costa, C.; Silva, F.; Campilho, R.; Neves, P.; Godina, R.; Ferreira, S. Influence of textile cord tension in cap ply production. *Procedia Manuf.* **2019**, *38*, 1766–1774. [CrossRef]
3. Fromhold-Eisebith, M.; Marschall, P.; Peters, R.; Thomes, P. Torn between digitized future and context dependent past—How implementing 'Industry 4.0' production technologies could transform the German textile industry. *Technol. Forecast. Soc. Change* **2021**, *166*, 120620. [CrossRef]
4. Majumdar, A.; Garg, H.; Jain, R. Managing the barriers of Industry 4.0 adoption and implementation in textile and clothing industry: Interpretive structural model and triple helix framework. *Comput. Ind.* **2021**, *125*, 103372. [CrossRef]
5. Vaisi, B. A review of optimization models and applications in robotic manufacturing systems: Industry 4.0 and beyond. *Decis. Anal. J.* **2022**, *2*, 100031. [CrossRef]
6. Chemweno, P.; Tor, R.-J. Innovative robotization of manual manufacturing processes. *Procedia CIRP* **2022**, *106*, 96–101. [CrossRef]
7. Säfsten, K.; Winroth, M.; Stahre, J. The content and process of automation strategies. *Int. J. Prod. Econ.* **2007**, *110*, 25–38. [CrossRef]
8. Zheng, C.; Qin, X.; Eynard, B.; Bai, J.; Li, J.; Zhang, Y. SME-oriented flexible design approach for robotic manufacturing systems. *J. Manuf. Syst.* **2019**, *53*, 62–74. [CrossRef]
9. Zheng, C.; Zhang, Y.; Li, J.; Bai, J.; Qin, X.; Eynard, B. Survey on Design Approaches for Robotic Manufacturing Systems in SMEs. *Procedia CIRP* **2019**, *84*, 16–21. [CrossRef]
10. Simões, A.C.; Soares, A.L.; Barros, A.C. Factors influencing the intention of managers to adopt collaborative robots (cobots) in manufacturing organizations. *J. Eng. Technol. Manag.* **2020**, *57*, 101574. [CrossRef]
11. Gultekin, H.; Gürel, S.; Taspinar, R. Bicriteria scheduling of a material handling robot in an -machine cell to minimize the energy consumption of the robot and the cycle time. *Robot Comput.-Integr. Manuf.* **2021**, *72*, 102207. [CrossRef]
12. Gadaleta, M.; Berselli, G.; Pellicciari, M. Energy-optimal layout design of robotic work cells: Potential assessment on an industrial case study. *Robot. Comput.-Integr. Manuf.* **2017**, *47*, 102–111. [CrossRef]
13. Silva, F.J.G.; Swertvaegher, G.; Campilho, R.D.S.G.; Ferreira, L.P.; Sá, J.C. Robotized solution for handling complex automotive parts in inspection and packing. *Procedia Manuf.* **2020**, *51*, 156–163. [CrossRef]
14. Castro, A.F.; Silva, M.F.; Silva, F.J.G. Designing a Robotic Welding Cell for Bus Body Frame Using a Sustainable Way. *Procedia Manuf.* **2017**, *11*, 207–214. [CrossRef]
15. Daniyan, I.; Mpofu, K.; Ramatsetse, B.; Adeodu, A. Design and simulation of a robotic arm for manufacturing operations in the railcar industry. *Procedia Manuf.* **2020**, *51*, 67–72. [CrossRef]
16. Eriksen, R.; Arentoft, M.; Grønbæk, J.; Bay, N.O. Manufacture of functional surfaces through combined application of tool manufacturing processes and Robot Assisted Polishing. *CIRP Ann.* **2012**, *61*, 563–566. [CrossRef]
17. Sujan, V.A.; Dubowsky, S.; Ohkami, Y. Robotic Manipulation of Highly Irregular Shaped Objects: Application to a Robot Crucible Packing System for Semiconductor Manufacture. *J. Manuf. Process.* **2002**, *4*, 1–15. [CrossRef]
18. Magalhães, A.J.A.; Silva, F.J.G.; Campilho, R.D.S.G. A novel concept of bent wires sorting operation between workstations in the production of automotive parts. *J. Braz. Soc. Mech. Sci. Eng.* **2018**, *41*, 25. [CrossRef]
19. Silva, F.J.G.; Soares, M.R.; Ferreira, L.P.; Alves, A.C.; Brito, M.; Campilho, R.D.S.G.; Sousa, V.F.C. A Novel Automated System for the Handling of Car Seat Wires on Plastic Over-Injection Molding Machines. *Machines* **2021**, *9*, 141. [CrossRef]
20. Martins, N.; Silva, F.J.G.; Campilho, R.D.S.G.; Ferreira, L.P. A novel concept of Bowden cables flexible and full-automated manufacturing process improving quality and productivity. *Procedia Manuf.* **2020**, *51*, 438–445. [CrossRef]
21. Moreira, B.M.D.N.; Gouveia, R.M.; Silva, F.J.G.; Campilho, R.D.S.G. A Novel Concept of Production and Assembly Processes Integration. *Procedia Manuf.* **2017**, *11*, 1385–1395. [CrossRef]
22. Araújo, L.M.B.; Silva, F.J.G.; Campilho, R.D.S.G.; Matos, J.A. A novel dynamic holding system for thin metal plate shearing machines. *Robot. Comput.-Integr. Manuf.* **2017**, *44*, 242–252. [CrossRef]
23. Nunes, P.M.S.; Silva, F.J.G. Increasing Flexibility and Productivity in Small Assembly Operations: A Case Study. In *Advances in Sustainable and Competitive Manufacturing Systems*; Azevedo, A., Ed.; Lecture Notes in Mechanical Engineering; Springer: Berlin/Heidelberg, Germany, 2013; pp. 329–340.
24. Costa, R.J.S.; Silva, F.J.G.; Campilho, R.D.S.G. A novel concept of agile assembly machine for sets applied in the automotive industry. *Int. J. Adv. Manuf. Technol.* **2017**, *91*, 4043–4054. [CrossRef]
25. Lindström, V.; Winroth, M. Aligning manufacturing strategy and levels of automation: A case study. *J. Eng. Technol. Manag.* **2010**, *27*, 148–159. [CrossRef]

26. Santos, R.F.L.; Silva, F.J.G.; Gouveia, R.M.; Campilho, R.D.S.G.; Pereira, M.T.; Ferreira, L.P. The Improvement of an APEX Machine involved in the Tire Manufacturing Process. *Procedia Manuf.* **2018**, *17*, 571–578. [CrossRef]
27. Araújo, W.F.S.; Silva, F.J.G.; Campilho, R.D.S.G.; Matos, J.A. Manufacturing cushions and suspension mats for vehicle seats: A novel cell concept. *Int. J. Adv. Manuf. Technol.* **2017**, *90*, 1539–1545. [CrossRef]
28. Vieira, D.; Silva, F.J.G.; Campilho, R.D.S.G.; Sousa, V.F.C.; Ferreira, L.P.; Sá, J.C.; Brito, M. Automating equipment towards Industry 4.0—A new concept for a transfer system of lengthy and low-stiffness products for automobiles. *J. Test. Eval.* **2022**, *50*, 20210721. [CrossRef]
29. Figueiredo, D.; Silva, F.J.G.; Campilho, R.D.S.G.; Silva, A.; Pimentel, C.; Matias, J.C.O. A new concept of automated manufacturing process for wire rope terminals. *Procedia Manuf.* **2020**, *51*, 431–437. [CrossRef]
30. Costa, M.J.R.; Gouveia, R.M.; Silva, F.J.G.; Campilho, R.D.S.G. How to solve quality problems by advanced fully-automated manufacturing systems. *Int. J. Adv. Manuf. Technol.* **2018**, *94*, 3041–3063. [CrossRef]
31. Veiga, N.; Campilho, R.; da Silva, F.; Santos, P.; Lopes, P. Design of automated equipment for the assembly of automotive parts. *Procedia Manuf.* **2019**, *38*, 1316–1323. [CrossRef]
32. Rigger, E.; Shea, K.; Stanković, T. Method for identification and integration of design automation tasks in industrial contexts. *Adv. Eng. Inform.* **2022**, *52*, 101558. [CrossRef]
33. Pinto, B.; Silva, F.; Costa, T.; Campilho, R.; Pereira, M. A Strategic Model to take the First Step towards Industry 4.0 in SMEs. *Procedia Manuf.* **2019**, *38*, 637–645. [CrossRef]
34. Barbosa, M.; Silva, F.; Pimentel, C.; Gouveia, R.M. A Novel Concept of CNC Machining Center Automatic Feeder. *Procedia Manuf.* **2018**, *17*, 952–959. [CrossRef]

MDPI

St. Alban-Anlage 66

4052 Basel

Switzerland

Tel. +41 61 683 77 34

Fax +41 61 302 89 18

www.mdpi.com

Machines Editorial Office

E-mail: machines@mdpi.com

www.mdpi.com/journal/machines